Rough Sets, Fuzzy Sets and
Soft Computing

Rough Sets, Fuzzy Sets and Soft Computing

Editor
S. Bhattacharya Halder

Associate Editor
Subrata Bhowmik

Narosa Publishing House
New Delhi   Chennai   Mumbai   Kolkata

Rough Sets, Fuzzy Sets and Soft Computing
284 pgs.

Editor
S. Bhattacharya Halder

Associate Editor
Subrata Bhowmik

Department of Mathematics
Tripura University (Central University)
Tripura

NAROSA PUBLISHING HOUSE PVT. LTD.

22, Delhi Medical Association Road, Daryaganj, New Delhi 110 002
35-36 Greams Road, Thousand Lights, Chennai 600 006
306 Shiv Centre, Sector 17, Vashi, Navi Mumbai 400 703
2F-2G Shivam Chambers, 53 Syed Amir Ali Avenue, Kolkata 700 019

www.narosa.com

ISBN 978-81-8487-403-7

Published by N.K. Mehra for Narosa Publishing House Pvt Ltd.,
22, Delhi Medical Association Road, Daryaganj, New Delhi 110 002

Printed in India

Preface

This book is the proceeding of the 2nd International conference on Rough Set, Fuzzy Set and Soft Computing. Lots of participants and invitees were present in this conference. Specially Prof. Sankar K. Pal (Padmasree awardee), Prof. Swapan Raha of Viswabharati University, Prof. Anadi Chatterjee, Prof Mihir K. Chakraborty, Prof. Ashok Deshapnde, Prof. B. C. Tripathy, Prof. A. K. Bhunia, Dr. Sujata Ghosh, Prof. Madhumangal Pal, Prof. Helen Saikia. About 90 participants were present in this conference. Some participants from our nearest country Bangladesh also came to join the Seminar and presented their paper. Some papers from abroad also came in this International Conference. More than 66 papers were presented during these three day conference. Pro VC, Prof. B. K. Agarwala inaugurated the programme. The three day programme has been accomplished successfully and the valedictory session was chaired by Hon'ble Vice Chancellor Prof. Arunoday Saha. In the valedictory session all the participants highly appreciated all the technical session including the local hospitality. And the three day 2nd International Conference on Rough set, fuzzy set and soft computing ended in this way.

Editor
S. Bhattacharya Halder

Contents

Soft Computing in Pollution Modeling Nonlinear Natural Environment

Anadi Kumar Chatterjee
Ex-Director, AIEM West Bengal

ABSTRACT

Release of pollutants from different sources in water bodies, land and atmosphere poses serious problems for human society and ecology. The spread of the industrial effluents in rivers from where drinking water is supplied to cities,needs to be carefully monitored. These require the understanding of the mechanism of the diffusion/dispersion processes involved and proper prognostic simulation. The governing equations being nonlinear, exact solutions are not obtainable and the way out is numerical modeling. So long hard computing was utilized efficiently for practical engineering projects. Soft computing methodologies are now attempted to avoid some approximations adopted in hard computing models. The case of Haldia in West Bengal from where effluents are to be discharged in the river Hooghly or in the river Haldi has been discussed in this paper.

1. INTRODUCTION

Haldia in West Bengal nearly hundred km seawards of Kolkata has many industries requiring discharge their industrial effluents in adjacent rivers like the Hooghly or the Haldi. Since the citizens are dependent for their drinking water on these rivers and the questions of survivals of the marine lives and marine eco systems are vital, any bad effect on the water quality of the river will have tremendous damaging impact on the population living on both sides of the river in both upper and lower directions.

Chatterjee (1981, 1984), Chatterjee & Chatterjee (1989), Chatterjee et al (1999) investigated the spread of pollutants in the river Hooghly under different hydrological and meteorological conditions and under storm cyclones. The nonlinear governing equations were solved using finite difference techniques in hard computing environments. In some studies the future scenario in case of new industries coming up were simulated. In such cases imprecision and uncertainties in choosingboundary conditions and selecting important model parameters were unavoidable. Though these were based on realistic physical assumptions, better alternatives were not found.

Use of softcomputing techniques during the last few decades in different engineering problems is now becoming good alternative to allow such uncertainties and imprecision in data and parameter estimations for predictions of future events.

2. THE PROBLEM

Haldia (Fig. 1) is situated on the right bank of the Hooghly estuary at the confluence of rivers Hooghly and the Haldi. During the last two decades several large industries have come up planning their effluents to be released somewhere but not damaging the ecology, human and other living animals and plants. The easiest way is to send them to the sea through the rivers but due to tidal actions the effluents would come back during flood tidal phases though during ebb tidal phases they will go down to the sea. The oscillatory nature of the flow will go on mixing the effluents with the river waters and will make unsuitable for drinking water supply purposes.

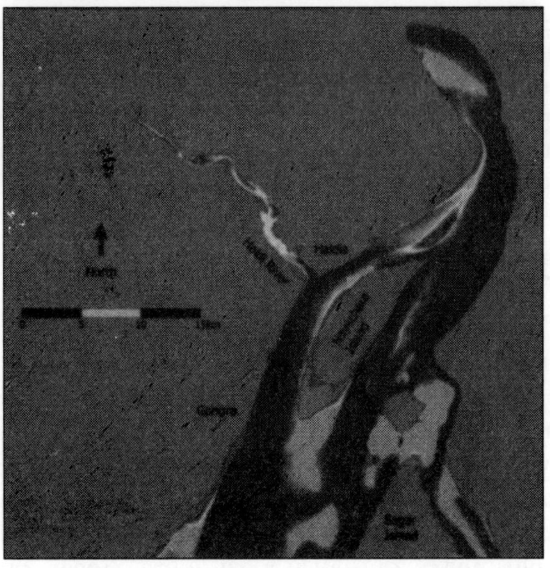

Fig. 1

Since industrialization as well as protection of environment and ecology are both unavoidable, careful analysis and proper assessment has to be made to find the desired solution.

The problem needs 3D modeling of the flow and transport of the pollutants in turbulent environment. The equations describing the flow and transport of effluents along with the turbulent closure schemes , boundary conditions, methods of solution, finite difference grid scheme etc. have been given in (Chatterjee-2011) in some detail.

Many 3D nonlinear numerical modeling techniques with various types of advantages and disadvantages in hard computing environment are available but they require very powerful computers like the Supercomputers which are rare here.

With the advancements of softcomputing methodologies attempts are now made to use them in such cases. Though a full fledged 3D modeling incorporating the advantages of softcomputing techniques is at present a rather very difficult task, attempts have been made first to do it in a 2D environment. Abebe et. al., 2000 indicated the recent applications of Fuzzy set approach in various fields including decision making, control and modeling. Hanss-2002 reported the use of fuzzy transformation method for the simulation and analysis of systems with uncertain parameters. Vikas-2008 used transformation method for the study of environmental risk analysis.

Basic principle of fuzzy modeling is based on Zadeh's 1968 extension principle. The transformation method (Hanss-2002) is followed in the present problem. The transformation method is a special implementation of fuzzy arithmetic based on alpha-cut principle that avoids the well known effect of overestimation which usually arises from use of interval computation for fuzzy arithmetic. The uncertain response reconstructed from a set of deterministic responses, combining the extrema of each interval in every possible way unlike the Fuzzy Alpha Cut technique where a particular level of membership (alpha level) values for uncertain parameters are used for simulation.

The method adopted here describes the use of imprecise data in pollutant transport in unsteady tidal flow simulation using fuzzy set theory. Traditionally, the pollutant transport problems are studied in hardcomputing environments where values of the parameters are defined locally based on measurements where chances of imprecision occur due to human and measurement errors. The present paper is an attempt to use a methodology for utilization of imprecise data (to characterize the overall uncertainty of prediction by fuzzy numbers) and to use fuzzy set theory as a technique to describe vagueness. There are two steps: (1) transforming the governing nonlinear partial differential equations into finite difference algebraic equations and the parameter imprecision to be incorporated in the input as fuzzy numbers resulting the equations to be transformed into equations with fuzzy coefficients at each time step. Since the variables in the equations are fuzzy numbers, the dependent variable is also (2) solving the fuzzy equations to get the membership functions to generate the pollutant concentration as the output at different time steps.

$$\frac{\partial C}{\partial t} + u \frac{\partial C}{\partial x} + v \frac{\partial C}{\partial y} = Dx \frac{\partial^2 C}{\partial x^2} + Dy \frac{\partial^2 C}{\partial y^2} \tag{1}$$

where C is the concentration of the pollutant, Dx and Dy are the two diffusion/ dispersion coefficients in the longitudinal and lateral directions, t is time, x and y are the coordinates in the longitudinal and lateral directions respectively.

The finite difference equation resulting from eqn. (1) is written in the form:

$$Ci, j\,(t + \Delta t) + Ci, j\,(t) + Ui, j\,(t)\,\{Ci + 1, j(t) - Ci - 1, j(t)\}/2.\Delta X + Vi, j(t).$$

$$\{Ci, j + 1(t) - Ci, j -1(t)\}/2.\Delta y = Dx\,\{Ci + 2, j(t) + Ci - 2, j(t) - 2.Ci, j(t)\}$$

$$/4.\Delta x\,\Delta x + Dy.\{Ci, j + 2(t) + Ci, j - 2(t) - 2.Ci, i(t)\}/4.\Delta y.\Delta y \qquad (2)$$

Hanss (2002) proposed two types of transformation methods, one general transformation method and the other the reduced transformation method. They differ in their degree of discretisation of particular interval. The reduced transformation method is used here. In Fig. 2 the membership function used is shown.

The transformation method and the reduced transformation method is briefly as follows:

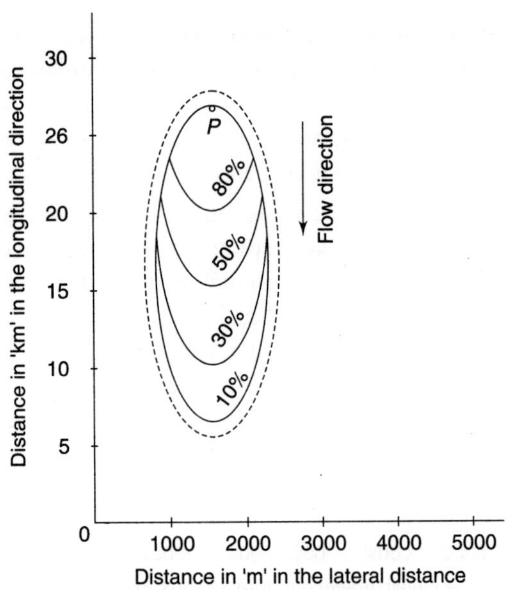

Fig. 2 Membership function

Considering fuzzy numbers Ai, $(i = 1, \cdots n\,)$ the set of input parameters defined on the real line R and supposing xi where $I + 1, 2, \cdot \cdot n$ denote the elements of Ai, if y is the output of the system which depends on n inputs xi by the mapping $y = f(xi)$, the n input parameters are modeled as fuzzy numbers with a membership function $\mu A(x)$ of arbitrary shape. The solution to the fuzzy number. B in y can be obtained by the following steps using transformation method.

1. Using α-sublevel technique, the range of membership $[0, 1]$ is discretised into a finite number of values. So an input parameter Ai can be decomposed into a set of $m + 1$ intervals $Xi(j)$, $j = 0, 1, \ldots m$. The value of the discretisation term, m depends on the degree of accuracy needed in approximation.

2. For each membership level j, the corresponding intervals for A in xi, $I = 1, 2, \ldots n$ is found. These are the supports of the αj cuts of $A1$, $A2$, $\ldots An$ So if $[ai(j), bi(j)]$ is the end points interval of i-th input parameter and for j-th level of membership denoted by $Xi(j)$ then $Ai = [Xi(0), Xi(1), \ldots Xi(m)]$ is set. When αi is equal to Now instead of applying standard interval arithmetic to the interval $Xi(j)$, they are transformed into arrays using a kind of full factorial at each level.

The reduced transformation method is:

The intervals are transformed into arrays $Xi(j)$ of the following forms

$$Xi(j) = \{\alpha i(j), \beta i(j), \alpha i(j), \beta i(j), \alpha i(j), \beta i(j), \ldots \alpha i(j), \beta i(j)\}$$

with

$$\alpha i(j) = \underbrace{\frac{\{ai(j), \ldots, ai(j)\}}{n-1}}_{2 \text{ terms}}, \quad \beta i(j) = \underbrace{\frac{\{bi(j) \ldots bi(j)\}}{n-1}}_{2 \text{ terms}}$$

where $\alpha i(j)$ and $\beta i(j)$ denote the lower and upper bound of the interval at the membership level μj for the ith uncertain parameter. For each interval level, these arrays combine the interval extrema $ai(j)$ and $bi(j)$ in every possible way.

Simulation is carried out by evaluating the expression separately at each of the positions of the arrays using the conventional arithmetic for crisp numbers. Thus if the output B of the system can be expressed in its decomposed and transformed form by the arrays $Bi(j)$, $j = 0, 1, \ldots m$ the kth element $^k bi(j)$ of the array $Bi(j)$ is then given by

$$^k Bi^j = f\{^k X_1^j, {}^k X_2^j, \ldots {}^k X_n^j\}$$

where $^k X_i^j$ denotes the kth element of the array X_i^j

Finally the fuzzy valued result B of the problem can be achieved in its decomposed form

$$B^j = \{a(j), b(j)\}, \quad j = 0, 1, \ldots m.$$

By retransforming the arrays $B_i^{(j)}$ using recursive formula

$$a_k^{(j)} = \min \{b^{(j+1)}, {}^K b^{(j)}\} \quad j = 0, 1, \ldots m-1.$$

$$b_k^{(j)} = \min \{b^{(j+1)}, {}^k b^{(j)}\} \quad j = 0, 1, \ldots m-1.$$

$$a_{kk}^{(m)} = \min \{{}^k b^{(j)}\} = \min \{{}^k b^{(j)}\} = b^{(m)}$$

3. THE MODEL

The Hooghly river from the sea face to Diamond Harbour is considered for the study. Haldia is near the middle. The region is divided into 750 sqm grids. Pollutants are released from the grid point from where effluents are planned to be released in the actual river. The concentration is assumed to be maximum at the point of release and to be zero both at the sea face boundary and the upper boundary as they are considered both far away from the place of release of the effluents. The driving force in the model is the tidal water level variations actually recorded in nature.

A constant amount of effluents was released at the pre selected grid point during the ebb tide period of the tidal cycle i.e. during about six hours to allow the pollutants to move towards the sea. Naturally the concentration profile would show maximum value at the point of release and would also show gradually decreasing trend downward. At the end of the ebb period, when the flood tide starts entering the river from the sea the watermass would try to carry the pollutants back towards the upper direction though not along the same path. Thus with the passage of time i.e. repeat of the ebb and flood tidal cycles the patch of pollutants will gradually disperse and diffuse over bigger and bigger regions in the river. The extent of spreading would depend on the excursion lengths of the ebb and flood tides. It may be noted that the excursion length of the tides will depend on the amplitudes of the tides i.e. it will be more for spring tides with higher amplitudes than neap tides with smaller amplitudes. The pattern of dispersion/diffusion of the pollutants during ebb tide period is shown in Fig. 3. Since the river has a number of islands inside, the to and fro movement of the patch would be influenced by them and with passage of time the pattern of dispersion and convection would gradually cover the river more and more. With the continuous release of the effluents from the industrial units the concentration of these pollutants would go on increasing in the river water excepting the fact that the portion escaping in the sea would not come back as the sea current will carry it out.

At the end of the ebb tidal period of six hours when the flood tide enters the river from the sea, the pollutants will be pushed back by the water and even if at that time no effluents are released the seaward travel of the earlier released pollutants will stop and they will start traveling upwards. The contours of the Fig. 3 pollutant concentrations will get distorted and at the mid flood, the picture

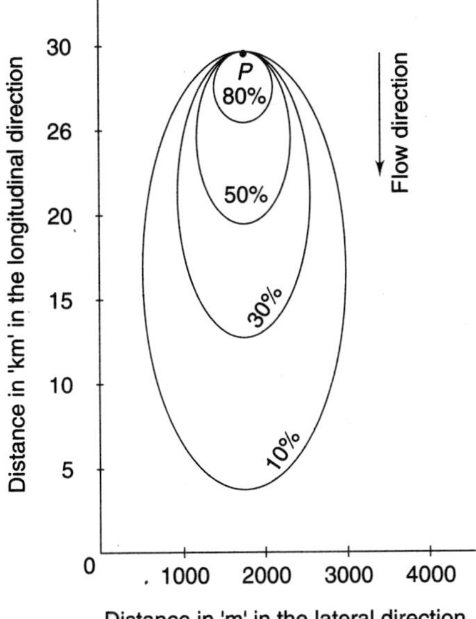

Fig. 3 Spread of pollutants in ebb.

will be as in Fig. 4. However with passage of time i.e. as more and more ebb and flood tidal cycles will pass the pollutants will be more and more dispersed resulting in spread throughout the river with reduced concentration profiles due to the fact that the quantity of pollutants is comparatively much less than the volume of water in the river at that time in that region. Inspite of this fact of quantitatively much smaller volume, the presence of the pollutants in the river water is of serious concern as even small amount of presence of the pollutant may cause havoc in marine eco system and human lives. So the prognostic picture of what will happen due to the release and to control the quantity of release etc. are vital and has to be found from the model results. From the results the amount of pollutants to be released into the river be determined.

The picture is totally different in monsoon months when the river is full of fresh water flow coming down from the upper catchment, thus diluting the pollutant laden water of the river near Haldia. The huge quantity of the monsoon water actually flushes the river and pushes the polluted water to the sea. But this is for about one third of the year and for the rest two third period the pollutant moves up and down with the tides and its being thrown out in the sea is to a smaller extent than that is in monsoon season.

4. CONCLUSION

A model of pollutant transport based on softcomputing technique using Transformation method has been developed. Instead of assuming approximate

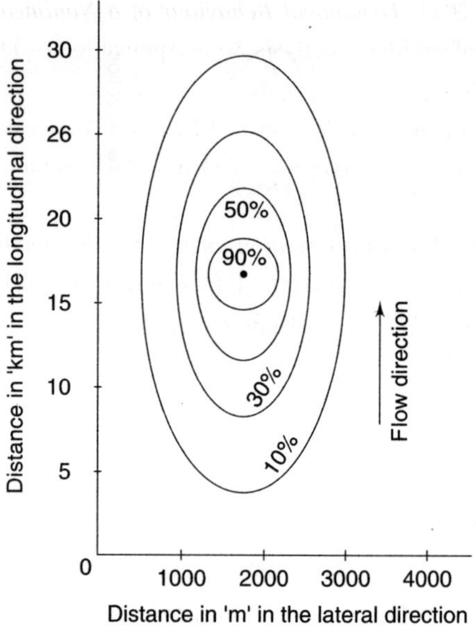

Fig. 4 Spread in flood tide.

important parameters like the Dispersion coefficients and the initial and boundary data adopted in hardcomputing techniques in ad-hoc ways, use of membership functions has been made. The model developed using the soft computing methodology has the scope for further improvements and to develop a 3D prognostic model.

References

1. Abebe, A.J., Guinot V. and Solomatine, D.P. 2000: *Fuzzy alpha-cut vs Monte Carlo techniques in assessing uncertainty in model parameters,* 4[th]IntConfHydroinformat, Iowa, U S A.

2. Chatterjee, A.K. 1981: *A nonlinear Convective Diffusion Model of a Tidal Waterway,* Proc 49th Res & Dev Ses., CBIP, India, Vol-2, pp 333-348.

3. Chatterjee, A.K. 1984: *Mathematical Model of Effluent Distribution in Tidal Rivers,* Jl of RivBeh & Control, India, Vol-XVI, pp 50-56.

4. Chatterjee, A.K. and Chatterjee, C. 1989: *Numerical Model Investigation of the Diffusion/Dispersion Process in the Hooghly Estuary,* ProcIntConf on Hyd & Envi Modelling of Coastal, *Estuarine and River Waters,* Univ of Bradford England, 19-21 Sept.

5. Chatterjee, A.K. et. al., 1999: *Applications of two turbulence closure scheme in the modeling of Tidal Currents and Salinity in the Hooghly Estuary,* Estuarine, Coastal and Shelf Science Vol-48, pp 649-663.

6. Chatterjee, A.K., 2011: *Dynamical Behaviour of a Nonlinear Natural System,* Proc National Conf on Math Analysis & its Applications – Dept of Math, Univ of Tripura, Jan-5&6.

7. Dou, C., Woldt, W., Bogardi, I., Dahab, M., 1997: *Numerical solute transport simulation using fuzzy sets approach,* Journal of Contaminant Hydrology 27, 107.

8. Hanss, A.M., 2002: *The transformation method for the simulation and analysis of systems with uncertain parameters,* Fuzzy Sets Syst. Vol-130, pp 277-289.

9. Vikas Kumar, 2008: *PhD Thesis,* Univ Rovira, Spain.

10. Zadeh, L.A., 1968: *Fuzzy algorithms.* Information and Control 12, pp 94-102.

Rough Sets, Fuzzy Sets and Soft Computing
Editor: S. Bhattacharya Halder

A Rough Set Approach to Predict Coronary Artery Disease from Medical Databases

R.N. Bhaumik

Emeritus Fellow(UGC), Professor of Mathematics(Rtd),
Tripura University, Suryamaninagar, Tripura (W)
E-mail: bhaumik_r_n@yahoo.co.in

ABSTRACT

There is a huge amount of data available within the healthcare system but these are not properly mined and not put to the optimum use. The hospital databases contain patient information that can be used by decision systems to improve medical care, uncover new relationships among data and hidden patterns that can help to predict diseases or suggest more effective treatments. Several mathematical methods were proposed to model data in the form of decision tables. Pawlak's Rough set theory for handling imprecision and uncertainty in data has an advantage over the other techniques, since the rough set philosophy is founded on the assumption that with every object of the universe of discourse we associate some information which may be imprecise, uncertain or incomplete. In this paper we have studied medical data of patients to predict Coronary Artery Disease (CAD) using Rough Set Theory.

1. INTRODUCTION

In order to extract useful information hidden in the sea of data, many methods were proposed. Rough set theory (RST), introduced in 1981 by Z. Pawlak[7], is useful to find relationships in the data. This method emerged as a tool to manage uncertainties, ambiguity and vagueness from incomplete, inexact and noisy information of the real life problem[7, 8].

Databases have also been used to query for heart deceases, as a comprehensive management tool to improve heart decease care and communication among professionals and to provide continuous quality improvement in heart decease care. Coronary Heart Disease(CHD) is the leading cause of death in the world. Heart decease, which is usually called Coronary Artery Disease (CAD), is a broad term that can refer to any condition that affects the heart. In this paper,

Rough sets and ROSETTA software are to be used to the analysis of CAD databases.

2. ROUGH SET

The Rough Set Theory is a way of representing and reasoning imprecision and uncertain information. Such information may be described by a table, the rows correspond to the objects and columns correspond to a set of attributes and each cell gives the value of an object with respect to an attribute. The information system is defined as $S = (U, A, V, I)$ where U is a set of objects(Universe), A is a set of attributes, $V = V = \cup_{a \in A} V_a$, V_a is a set of values of $a (\in A)$ and $I: U \times A \to V$ is an information function such that $I(u, a) \in V_a$ for each $u \in U$, $a \in A$.

Based on the information, it is possible to make some of the objects apart, while others are impossible to distinguish. The latter objects are indiscernible from each other, and form a set. Every $B \in A$ generates a binary relation on U, called indiscernibility relation $IND(B)$ (or simply B), defined by $B = \{(x, y) \in U \times U: a(x) = a(y),), \forall a \in B\}$. Then $[x]_B = \{y \in U: (x, y) \in B\}$ is the equivalent class of U by the indiscernibility relation B. Each set of indiscernible objects forms the building blocks of theory of rough sets.

A rough set is basically an *approximation representation* of a given set in terms of two subsets derived from a crisp partition of the universal set U. The two subsets are called a lower and an upper approximation.

Let $B \subseteq A$ and $X \subseteq U$ in the information system S. We can approximate X using only the information contained in B by constructing the lower and upper approximation of a set X, defined by $B_L X = \cup\{[x]_B : [x]_B \subseteq X, x \in X\}$ and $B_U X = \cup\{[x]_B : [x]_B \cap X \neq \phi, x \in X\}$.

- Thus the lower approximation of a set of attributes is defined when equivalence classes are fully contained in the set we want to approximate.

- The upper approximation of a set of attributes is defined when equivalence classes are at least partially contained in (i.e. overlapwith) the set.

These two approximations can also be used to define the basic concepts of Rough sets. If the lower and upper approximations are equal, then the set is defined crisply.

The set we approximations define three approximation regions:

(i) **Inside(positive) region:** The inside region equals the lower approximation. Cases that are members of the inside region are definite members of the set we want to approximate.

(ii) **Outside(negative) region:** The outside region equals the complement of the upper approximation. The cases that are members of the outside region are definite non-members of the set we want to approximate.

(iii) Boundary region: The boundary region equals the difference between the upper and lower approximations. The cases that are members of the boundary region have a membership status that cannot be ascertained with certainty, at least not on the basis of the attributes that the approximations are built. If this region is non-empty then the set is called **Rough set.**

Remarks: The indiscernibility relation generated in this way is the mathematical basis of rough set theory.

Reduct: An interesting question arises whether there are attributes in the information system which are more important to the knowledge represented in the equivalence class structure than other attributes. If there exists a subset of attributes which can, by itself, fully characterize the knowledge in the database, then this attribute set is called a **reduct**. These to fall reducts in A is denoted $RED(A)$.

The reduct of an information system is *not unique*. Reduct can be defined as the minimal subset of attributes that have the same classification of elements as the universe which is the whole set of attributes. Attributes that are not the elements of reduct are *redundant*.

3. SOME CONCEPTS FOR USING ROSETTA (ROUGH SET TOOLKIT FOR ANALYSIS) OF DATA

All rough set computations were carried out using the ROSETTA software system.

Rules: A decision rule is defined to be a statement of the form "if C then D", where the condition C is a set of elementary conditions connected by "AND", and the decision D is a set of possible outcomes connected by "OR". For example (Fig. 2):

IF BP-Diastolic(80) AND F-VLDL(40) AND F-Chol/HDL(4.60) THEN Decision(1)

IF Lipids(2) AND BP-Systolic(128) AND F-VLDL(27) THEN Decision(0) OR Decision(1)

The decision rules above can be interpreted within the rough set framework. If the then-part of the rule lists more than one possible outcome, that can be interpreted as describing one or more cases that lie in the boundary region of the set approximation. If the then-part of the rule list a single outcome "Yes" (or "No"), that can be interpreted as describing one or more cases that lie in either the inside (or the outside) region of the approximation of the set of patients, when using the attributes listed in the if-part of the rule.

The rule's accuracy and its coverage are derived from support counts, a number of quantities of interest. Let d denote one of the possible outcome values listed in the then-part of the rule. Then

$$\text{Accuracy} = \frac{\text{number of cases that match both } C \text{ and } d}{\text{number of cases that match } C}$$

$$\text{Coverage} = \frac{\text{number of cases that match both } C \text{ and } d}{\text{number of cases that match } d}$$

Rules from ROSETTA

The rules are sorted by RHS Coverage (i.e., RHS is Right Hand Side or THEN part of the rule, LHS is Left Hand Side or IF-part of the rule). For each rule, the following statistics are given:

LHS support: Number of objects in the training set matching the IF-part.

RHS support: Number of objects in the training set matching the IF-part *and* the

THEN-part.

(LHS and RHS support is the same unless the THEN-part contains several decisions).

RHS Accuracy: RHS support divided by LHS support.

(Accuracy is 1.0 unless the THEN-part contains several decisions).

LHS Coverage: LHS support divided by the number of objects in the training set.

RHS Coverage: RHS Support divided by the number of objects in the decision class

listed in the THEN part of the rule.

[*RHS Stability:*Not applicable for the Johnson algorithm (always 1.0)].

*LHS Length:*Number of attributes in the IF-part of the rule.

RHS Length: Number of decisions in the THEN-part of the rule.

The rules can now be used to classify the objects in the test set.

Theoverallrule-basedmodelingconsistsofthe four basicstepslistedbelow.

Discretization: The rough set approach is a logically founded approach based on indiscernibility. This means that we do not need a notion of "distance" between attribute values. Discretization is a step that is not specific to the rough set approach, most rule or tree induction algorithms currently require it. An example of such an algorithm is based on Boolean reasoning and rough Sets.

Exploring: In exploring which discretization method works best on the training set for the Johnson algorithm is examined.

Ruleinduction: Compute if-then rules from the discretized decision table.

Rule application: Apply the rules to classify new cases.

Model evaluation: The classificatory performance of the rules are evaluated. In medicine, measures derived from **Receiver Operating Characteristic(ROC)** curves are common for binary outcomes.

Creating Reducts: In creating reducts, the subset of attributes facilitate rule generation with minimal subsets. First, we choose the *Johnson reducer algorithm* and options selected from this algorithm were discernibility = object related (universe = all objects).

Classification: A classifier k, when applied to an $x \in U$ in an information system S, assigns a classification $d'_k (x)$ to x where $d'_k : U \to V_d$ [$V_d = \{0, 1\}$ and $d(x)$ is the actual classification of x].

The next step is applying a classification method to the test set and we choose the batch classifier with the standard/tuned voting method (RSES, i.e., Rough set Explorer System). Then "confusion matrix" produced by the program is described below (1 = Heart Disease, 0 = No Heart Disease).

Confusion Matrix

The confusion matrix contains information about actual and predicted classifications done by a classification system. Each column of the matrix represents the instances in a predicted class, while each row represents the instances in an actual class. In artifical intelligence, a confusion matrix is a specific table that allows visualization of the performance of an algorithm.

The following table shows the confusion matrix for a two class (0 and 1) classifier.

Table 1

		Predicted d'	
		Negative 0	Positive 1
D_0	Negative	TN	FP
Actual 1	Positive	FN	TP

Here

TN(True Negative) = no. of correct predictions that an instance is negative.

FP(False Positive) = incorrect predictions that an instance is positive.

FN(False Negative) = incorrect predictions that an instance is negative.

TP(True positive) = no. of correct predictions that an instance is positive.

The entries in the confusion matrix have the following meaning in the context of our study. Several standard terms have been defined for this matrix:

Specificity is defined as : TN/(TN + FP)

Sensitivity is defined as : TP/(TP + FN)

Positive predictive value : TP/(TP + FP)

Negative predictive value : TN/(TN + FN)

Overall accuracy = (TN + TP)/(TN + FP + FN + TP)

It is customary to randomly divide the set U of objects into two disjoint subjects – a training set and a test set. The training set is used to construct k, while the test set is used to assess its performance.

4. APPLICATION OF ROSETTA SOFTWARE FOR CAD PATIENTS

The Rosetta is a toolkit developed by Alexander Øhrn[5] used for data analysis using Rough Sets Theory. The Rosetta toolkit is confirmed by a computational kernel and a GUI (Graphical User Interference). First, we download ROSETTA software for data mining and knowledge discovery based on rough set theory from www.idi.ntnu.no/~aleks/rosetta/(Øhrn and Komorowski 1997).

The heart disease data set of 253 patients of Tripura is used in this study. The dataset has 27 attributes, 17 of which are relevant as shown in Table 2. The data is available in Microsoft Excel spread sheet elsewhere in this paper.

Now the rough set approach, used in this general scheme, is typically able to make use of the data stored in a relational database directly. Still, some preprocessing of the data usually should be done. In a rough set analysis the preprocessing typically consists of two steps. The first step is to deal with missing values. In the present analysis the completion algorithm was simply to remove all objects containing missing values.

Table 2 Description of the attributes in the heart disease dataset

No.	Attributes Name	Values
1	Chief Complaints	1 Specific
		2 Non specific
2	Anti-Lipids	1 Yes
		2 No
3	FHO-CAD	1 Positive
		2 Negative

Table Contd...

Table Contd...

4	BP-Systolic (mm Hg)		Integer (Continuous)
5	BP-Diastolic (mm Hg)		Integer (Continuous)
6	ECG	1	Abnormal
		2	Normal
7	Echo-EF	1	>55
		2	45-54
		3	30-44
		4	< 30
8	TMT	1	Abnormal Positive
		2	Normal Negative
9	Fasting Blood Sugar (mg/dl)	1	< 100
		2	101-150
		3	151-200
		4	201-250
		5	251-300
		6	351-400
10	Creatinine (mg/dl)		Integer (Continuous)
11	F-Cholesterol (mg/dl)		Integer (Continuous)
12	Triglyceride (mg/dl)		Integer (Continuous)
13	LDL Cholesterol (mg/dl)		Integer (Continuous)
14	F-HDL (mg/dl)		Integer (Continuous)
15	F-VLDL (mg/dl)		Integer (Continuous)
16	F-Chol/HDL (ratio)		Integer (Continuous)
17	Decision Class	0	No
		1	Yes

Now, we randomly divided the 253 cases into a training set ($n = 177$), and a test set ($n = 76$). If a data table is used both for training and testing, it is very important that the test set is not used to decide upon the discretization intervals.

Doing this would brake with the principle of not letting the test set be involved in any steps of the generation of a classifier. Because of this, the preprocessing step of discretization is done after the splitting of the table. The test set must be discretized according to the discretization found on the training set.

After splitting the objects, training table is computed reducts by **"Genetic Algorithm"** with **"object-related"**reduct. After reducts, rules are generated from the reducts and decision tables. After these steps of computation and rule generation, **"weak reducts"** and **"rules"** are filtered. After experiment with genetic algorithms we see that the experiment generate **2408 rules** with **246 reducts** as shown in Figure 2. Then we discretize the numerical attributes. As the

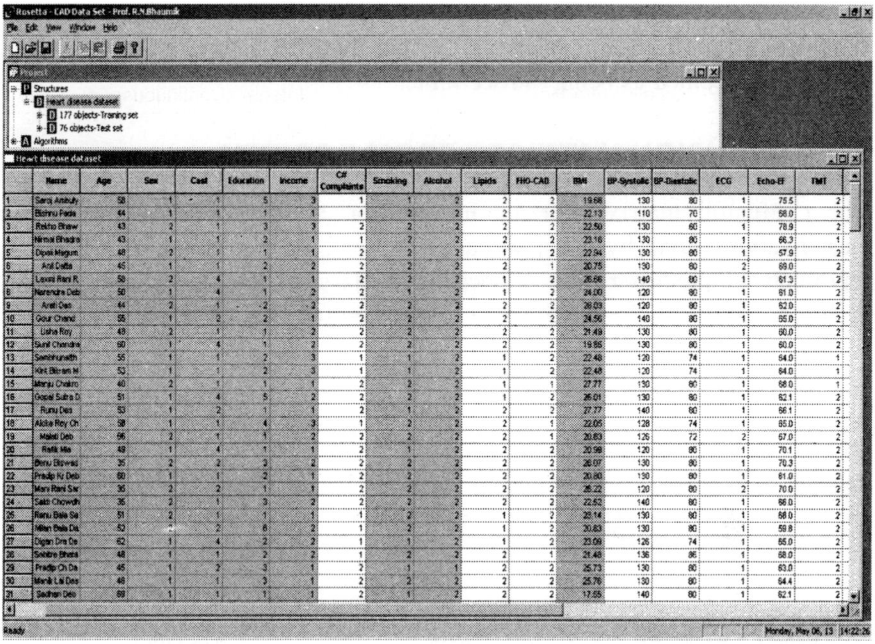

Fig. 1 Image of ROSETTA software of heart disease data sets.

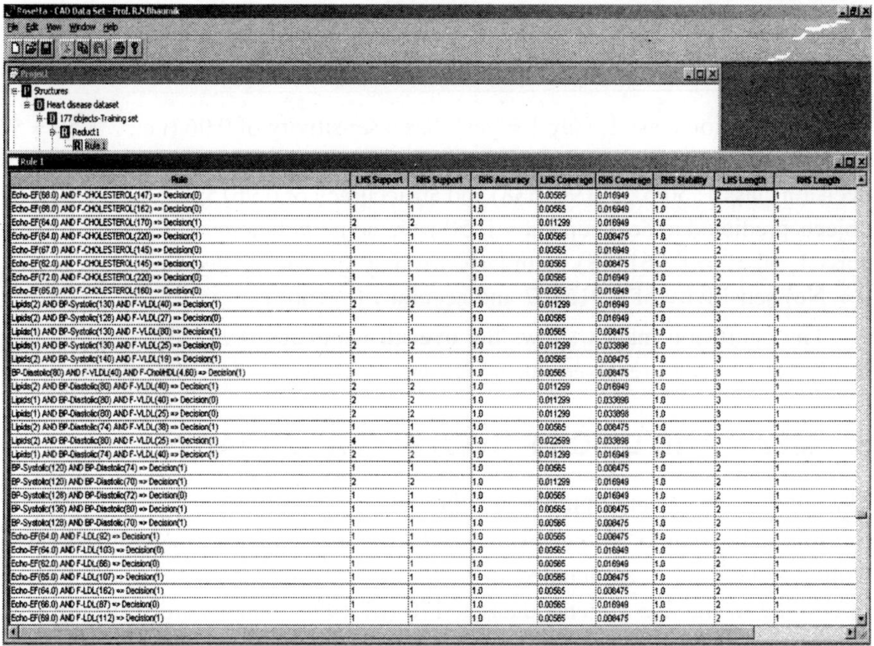

Fig. 2 Image of ROSETTA software on training data sets.

rough set theory is based on the concept of indiscernibility, numerical attributes should be discretized into intervals, so that numbers falling within the same interval are deemed as being indiscernible.

5. CONFUSION MATRICES FROM ROSETTA

The confusion matrix contains information about *actual* and *predicted* classifications done by a classification system. Each column of the matrix represents the predicted value of decision by classifier for a tested object, while each row represents the actual value of decision for a tested object.

The confusion matrix shows the overall accuracy as well as sensitivity and accuracy for each class. Now the 76 objects in the test set are classified according to the genetic algorithms method of reduct generation.

Table 3

		Predicted		
		0	1	
	0	22	2	0.91667
Actual	12	50	0.961538	
	0.916667	0.961538	0.947368	
		Class	1	
		Area	0.995994	
ROC		Std. error	0.006367	Thr (0, 1)0.764
		Thr. Acc.	0.128	

The decision class 1, (say 1 = yes), has a sensitivity of 0.96 (i.e., 2 + 50 = 52 objects actually belonging to the decision class 1, 50 was correctly classified as decision class 1: 50/52 = 0.961538) and an accuracy of 0.961538 (i.e., 2 + 50 = 52 objects predicted to decision class 1, 50 were actually belonging to their decision class 1, 50/52 = 0.961538). [similarly we may get for the decision class 0 (say, 0 = No) sensitivity of 0.916667 and accuracy 0.91667].

The confusion matrix shows the **overall accuracy** 0.947368 which is best result that has been received during Rosetta analysis for decision tables.

6. MEAN AND DISCRIPTIVES FOR ACCURACY

The software has many options and choices.

From the Figure 3, the mean is 67.19% and 47.04% standard deviation with correlation = 1 as detailed in the Statistics printout in the above Figure. According to the 253 patients dataset, 170 (67.19%) patients are suffering with Heart disease and 83 (32.806%) are not suffering with Heart disease.

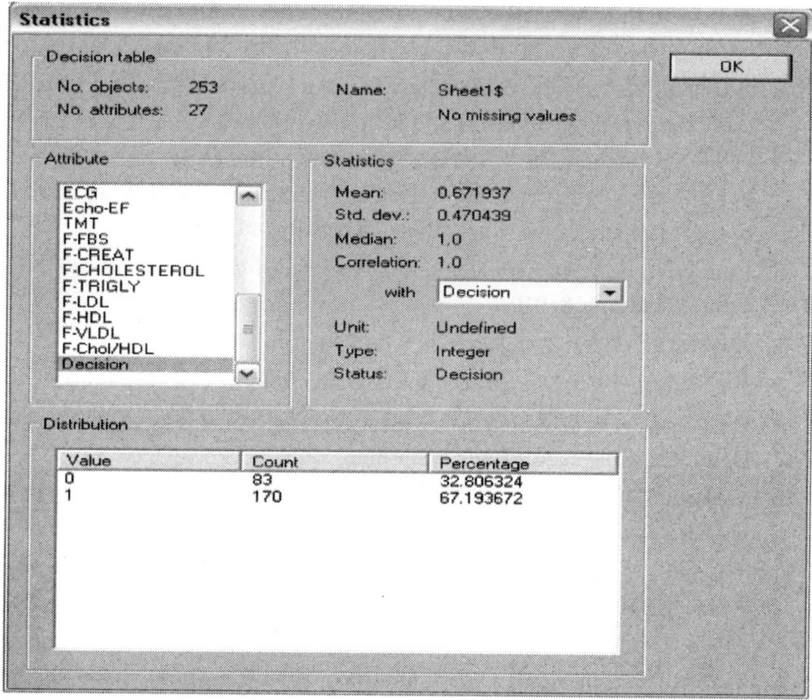

Fig. 3

7. CONCLUSIONS

In this paper, we have provided a brief overview of rough sets and their use in CAD data. Although rough sets represent a relatively recent approach, a number of effective applications have demonstrated their potential and it is only to be expected that research will continue to improve upon and extend these techniques.

To the analysis of CAD databases, some generalizations of rough sets and the Rosetta software both are useful. The accuracy of predicting heart disease status are 94.7% on the initial random sample. So we can say that rough set theory is the best method for predicting data set based on ROSETTA software.

Reference

1. Bhaumik, R.N., Gangwal, C., (2012): *An Inconsistency Checking Method for Relational Database: A Rough Set Approach*; The J. of fuzzy Maths; **20(1)**, 63-70.

2. Gangwal, C., Bhaumik, R.N., (2012): *Intuitionistic Fuzzy Rough Relation in some Medical Applications,* Int. J. of Adv. Research in Comp. Engg. and Tec.; **1(6)**, 28-32.

3. Hassanien, A.E., Abraham, A. et. al., (2008): *Overview of rough-hybrid approaches in image processing*. In IEEE Conference on Fuzzy Systems; 2135-2142.

4. Komorowski, J., Øhrn, A., (1999): *Modeling prognostic power of cardiactests using rough sets*, ArtifIntellMed; **15**(2), 167-191.

5. Ohrn, A., Komorowski, J., ROSETTA, (1997): *A Rough Set Toolkit for Analysis of Data*; Proc. of the Third Int. Joint Conf. on Inform. Sciences. 3., USA, Dept. of Electrical and Comp. Engg, Duke University; 403-407

6. Paterson, G. I., (1995): *A rough sets approach to patient classification in medical records*, Medinfo **8**, 910

7. Pawlak, Z., (1981): *Roughsets*, ICS PAS Reports (431).

8. Pawlak, Z., (1982): *Rough sets*; Int. J. Comp. Inform. Sci., **11**, 341-356.

9. Pawlak, Z., (1991): *Rough Sets*; Theoretical Aspects of Reasoning About Data; Dordrecht; Kluwer Academic Publishing.

10. Setiawan, N.A., et. al., (2008): *A comparative study of imputation methods to predict missing attribute values in CAD data set*; Proc. 21 Biomed.; 266-269.

11. Tsumoto, S., (1998): *Automated induction of medical expert system rules Fromclinical databases based Rough set theory*; Information Sciences; **112**, 67-84.

12. Tsumoto, S., Tanaka, H., (1994): *Induction of medical expert system rules based on Roughsetsandresamplingmethods*; ProcAnnuSympComputApplMedCare; 1066-1070.

[* The author is thankful to the University Grant Commission, New Delhi for selecting as Emeritus Fellow and the grant during this study. The author is also thankful to Dr. S. Punia, Agartala Medical College for his help in collecting the medical data.]

Relation among Fuzzy Set, Rough Set, Soft Set and their Generalizations

Anjan Mukherjee

Department of Mathematics, Tripura University
E-mail: anjan2002_m@yahoo.co.in

1. INTRODUCTION

Many fields deal daily with the uncertain data that may not be successfully modeled by the classical mathematics. There are some mathematical tools for dealing with uncertainties, few of them are fuzzy set theory, developed by Zadeh (1965)[18], Rough set thoery introduced by Pwlak (1982)[15] and soft set theory, introduced by Molodtsov (1999)[11], that are related to our work. A fuzzy set allows a membership value other than 0 and 1. A rough set uses there membership functions, a reference set and its lower and upper approximation in an approximation space. There are extensive studies on the relationships between rough sets and fuzzy sets. Many proposals have been made for the combination of rough set and fuzzy set. The result of these studies lead to the introduction of the notion of rough fuzzy sets and fuzzy rough sets. Dubois and Prade[4] investigated the problem of combining fuzzy sets with rough sets. In general a rough fuzzy set is the approximation of a fuzzy set in a crisp approximation space, where as a fuzzy rough set is the approximation of a crisp set or a fuzzy set in a fuzzy approximation space. Base on a Pawlak's approximation space, the approximation of a soft set was proposed to obtain a hybrid model called rough soft sets (Feng et. al., 2009)[5]. Alternatively a soft set instead of an equivalence relation was used to generalize the universe. This lead to a deviation of Pawlak approximation space called a soft approximation space in which soft rough approximation and soft rough sets were introduced (Feng et. al., 2009). Further they considered approximation of a fuzzy set in a soft approximation space and initiated a concept called soft rough fuzzy sets which was the extension of Dubois and Prade's rough fuzzy sets. The aim of this talk is to study relation among fuzzy set, rough set and soft set also to introduce Intuitionistic fuzzy soft rough sets (IFSRs) and establish some results on them.

We now give some ready references for further discussion:

Definition 1.1 [18]   Let U be a universe. A fuzzy set X over U is defined by a function μ_X representing a mapping $\mu_X: U \to [0, 1]$. Here μ_X is called the membership function of X and the value $\mu_X(u)$ is called the grade of membership of $u \in U$. The value represents the degree of u belonging to the fuzzy set X. Thus a fuzzy set X over U can be represented as follows:

$$X = \{(\mu_X(u)/u): u \in U\} \text{ or} \{(U, \mu_X(u)): u \in U\}$$

or $\{(\mu_X(u), u): u \in U\}$ or $\{(U/\mu_X(u)): u \in U\}$

where $\mu_X(u) \in [0, 1]$.

Definition 1.2 [15]   Let R be an equivalence relation on the universal set U. Then the pair (U, R) is called a Pawlak approximation space. An equivalence class of R containing x will be denoted by $[x]_R$. Now for $X \subseteq U$, the lower and upper approximation of X with respect to (U, R) are denoted by respectively R_*X and R^*X and are defined by

$$R_*X = \{x \in U: [x]_R \subseteq X\} \text{ or } \cup\{[x]_R: [x]_R \subseteq X\}, x \in U$$

$$R^*X = \{x \in U: [x]_R \cap X \neq \phi\} \text{ or } \cup\{[x]_R: [x]_R \cap X \neq \phi\}, x \in U$$

Now if $R_*X = R^*X$, then X is called definable; otherwise X is called a *rough set*.

In 1999, Molodtsov [11], has introduced the concept of soft set in the following way:

Definition 1.3 [11]   Let U be an initial universal set, and E be the set of parameters. Let $P(U)$ denotes the power set of U and $A \subseteq E$, then the pair $\xi = (F, A)$ is called a soft set over U is a parameterized family of subsets of the universe U. For $e \in A$, $F(e)$ may be considered as a sets of e – approximate elements of the soft set (F, A) , where $F: A \rightarrow P(U)$.

2.   RELATION BETWEEN FUZZY SET AND SOFT SET

Theorem 2.1[2]   Every fuzzy set may be considered a soft set.

In order to better understand the relationship, let us consider the following example:

Example 2.2:   Suppose that there are six alternatives in the universe of houses $U = \{h_1, h_2, h_3, h_4, h_5, h_6\}$ and we consider the single parameter "quality of the houses" to be a linguistic variable. For this variable we define the set of linguistic terms T(quality)={best, good, fair, poor}. Each linguistic term is associated with its own fuzzy set. Let ws consider two of them as follows:

$$F_{[best]} = \{(h_1, 0.25), (h_2, 0.6), (h_5, 0.9), (h_6, 1.0)\} \text{ and}$$

$$F_{[poor]} = \{(h_1, 0.9), (h_2, 0.4), (h_3, 1.0), (h_4, 1.0), (h_5, 0.25)\}..$$

Now the α-level sets of $F_{[poor]}$ are

$$F_{[poor]}(0.25) = \{h_1, h_2, h_3, h_4, h_5\}$$

$$F_{[poor]}(0.4) = \{h_1, h_2, h_3, h_4\}.$$

$$F_{[poor]}(0.9) = \{h_1, h_3, h_4\}.$$

$$F_{[poor]}(1.0) = \{h_3, h_4\}.$$

The values $A = \{0.25, 0.4, 0.9, 1.0\} \subset [0.1]$ can be treated as a set of parameters, such that the mapping $F_{[poor]}: A \to P(U).$ gives approximate value set $F_{[poor]}(\alpha)$ for $\alpha \in A$. Thus we can write the equivalent soft set

$$(F_{[poor]}, [0,1]) = \{0.25, \{h_1, h_2, h_3, h_4, h_5\}), (0.4, \{h_1, h_2, h_3, h_4\}),$$

$$(0.9, \{h_1, h_3, h_4\}), (1.0, \{h_3, h_4\})\}.$$

3. RELATION BETWEEN ROUGH SET AND SOFT SET

Theorem 3.1 [2]:   Every Rough set may be considered a soft set.

In order to better understand the relationship, let us consider the following example:

Example 3.2:   Suppose that a subset of five houses $X = \{h_1, h_2, h_3, h_4, h_5\}$ in the universe of $U = \{h_1, h_2, h_3, h_4, h_5, h_6\}$ are under consideration. We construct the information table:

House	h_1	h_2	h_3	h_4	h_5	h_6
Quality	Fair	Fair	Best	Good	Best	Fair
Price	Cheap	Cheap	Middle	Expensive	Middle	Cheap
Place	City	City	Village	City	Village	City

The rows of the table are leveled by attributes, and the table entries are the attribute values for each home.

Each column of the table can thus be seen as summarizing the available information on a specific home. The table evaluates all six houses in terms of three attributes, 'quality'. 'price', 'place'. These three attributes are characterized by the value sets

{best, Good, fair, poor}, {expensive, middle, cheap},{village, city} respectively.

Now, the equivalence classes are

$$[h_1]_R = [h_2]_R = [h_6]_R - \{h_1, h_2, h_6\}.$$

$$[h_3]_R = [h_5]_R = \{h_3, h_5\}.$$

$$[h_4]_R = \{h_4\}.$$

Thus $\qquad R_*(X) = \{h_1, h_4, h_5\}$

and $\qquad\qquad R^*(X) = \{h_1, h_2, h_3, h_4, h_5, h_6\}.$

$$R(X) = \{\{h_1, h_4, h_5\}, \{h_1, h_2, h_3, h_4, h_5, h_6\}\}.$$

Thus every rough set $R(X)$ of X may also be considered a soft set with the representation

$$(F, E) = \{(p_1(x), R_*(X)), (p_2(x), R^*(X))\}.$$

Before going to next section we need the following background:

Nanda and Majumdar (1993)[14] introduced the notion of fuzzy rough sets. In 1998, Chakrabarty et al.'s [3] approached intuitionistic fuzzy rough sets (IF rough set)), they constructed an IF rough set (A, B) of the rough set (P, Q), where A and B are both IF sets in X such that $A \subseteq B$ i.e., $\mu_A \leq \mu_B$ and $\nu_A \geq \nu_B$. From this point of view the lower approximation A and the upper approximation B are both IF sets. Jena and Ghosh (2002) [4] reintroduced the same notion. Samanta and Mondal (2001) [17] also introduced this notion but they called it a rough IF set. They also defined the concept of IF rough set According to them an IF rough set is a couple (A, B) such that A and B are both fuzzy rough sets (in the sense of Nanda and Majumdar [6]) and A is included in the complement of B According to Samanta and Mondal (2001) [17] an intuitionistic fuzzy rough set (A, B) is a generalization of an IF set in which membership and non-membership functions are no longer fuzzy sets but fuzzy rough sets A and B. On the other hand, for Chakrabarty et al. [3], an intuitionistic fuzzy rough set (A, B) is a generalization of a fuzzy rough set in which upper and lower approximation are no longer fuzzy sets but IF sets A and B. Rizvi et. al., (2002) [16] described their proposal as "Rough intuitionistic fuzzy set" in which the lower and upper approximations themselves are not Intuitionistic fuzzy sets in X but intuitionistic fuzzy sets in the class of equivalence classes.

4. A NEW HYBRID MODEL (INTUITIONISTIC FUZZY SOFT ROUGH SETS)

In this section we shall consider lower and upper intuitionistic fuzzy soft rough approximation of intuitionistic fuzzy sets in an intuitionistic fuzzy soft approximation space, and obtain a new hybrid model called intuitionistic fuzzy soft rough sets which can be seen as an extension of Dubois and Prade's rough fuzzy sets and also an extension of Feng et. al's soft rough fuzzy sets.

F. Feng et. al., [5] introduced the soft approximation space and soft rough set in the following way:

Let $\xi = (F, A)$ be a soft set over U. Then the pair $P = \{U, \xi\}$ is called a soft approximation space based on P. The two operators

$$aPr_P(x)\!\downarrow = \{h \in U : \exists\, e \in A\ h \in F(e) \subseteq X\},$$

and $$aPr_P(x)\!\uparrow = \{h \in U : \exists\, e \in A\ h \in F(e);\ F(e) \cap X \neq \Phi\},$$

assigning to every subset $X \subseteq U$. The two sets $aPr_P(x)\!\downarrow$ and $aPr_P(x)\!\uparrow$ are called the lower and upper soft rough approximation of X in P respectively. More over POS$_P(x) = aPr_P(x)\!\downarrow$; Neg$_P(x) = U - aPr_P(x)\!\uparrow$; Bnd$_P(x) = aPr_P(x)\!\uparrow - aPr_P(x)\!\downarrow$, are called the soft positive, soft negative and soft boundary region of X respectively. If $aPr_P(x)\!\downarrow = aPr_P(x)\!\uparrow$, X is said to be soft definable otherwise X is called a soft rough set.

Definition 4.1 An intuitionistic fuzzy soft set (IFsoft set) $\xi = (F, A)$ over the universe U is called, a full IFsoft set if $\cup\, F(e) = U,\ e \in A$

Definition 4.2 A full IFsoft set $\xi = (F, A)$ over U called a covering IFsoft set if $F(e) \neq \Phi,\ \forall\, e \in A$.

Definition 4.3 Let $\xi = (F, A)$ be a full IFsoft set over U and $S = (U, \xi)$ be a IF soft approximation space. For an IFset $\alpha \in P(U)$ ($P(U)$ is the family of all IFsets in U), the lower and upper intuitionistic fuzzy soft rough approximation of α with respect to S are defined by IFSap$_S(\alpha)\!\downarrow$ and IFSap$_S(\alpha)\!\uparrow$ respectively, which are IFsets in U given by

$$\text{IFSap}_S(\alpha)\!\downarrow = \wedge \{\alpha(y) : \exists\, e \in A\ \{x, y\} \subseteq F(e)\},$$

and

$$\text{IFSap}_S(\alpha)\!\uparrow = \vee \{\alpha(y) : \exists\, e \in A\ \{x, y\} \subseteq F(e)\} \text{ for all } x \in U.$$

The operators IFSap$_S\!\downarrow$ and IFSap$_S\!\uparrow$ are called the lower and upper IF softrough approximation operators on IF sets. If IFSap$_S\!\downarrow = $ IFSap$_S\!\uparrow$, α is said to be intuitionistic fuzzy soft definable, otherwise α is called an IF soft rough set. Here $\alpha(y) = \{(y, \mu_A(y), \nu_A(y) : y \in X)\}$. Moreover,

$$\text{Pos}_S(x) = \text{IFSap}_S(x)\!\downarrow,\ \text{Neg}_S(x) = U - \text{IFSap}_S(x)\!\uparrow,$$

$$\text{B nd}_S(x) = \text{IFSap}_S(x)\!\uparrow - \text{IFSap}_S(x)\!\downarrow,$$

are called the IF soft positive, IF soft negative and IF soft boundary region of X respectively.

By definition we immediately have that $X \subseteq U$ is an IF soft definable set if

$$\text{B nd}_S(x) = \Phi.$$

Also it is clear that IFSap$_S(x)\!\downarrow \subseteq X$ and IFSap$_S(x)\!\downarrow \subseteq$ IFSap$_S(x)\!\uparrow$ for all $X \subseteq U$.

Theorem 4.4: Let $\xi = (F, A)$ be an IF soft set over U, $S = \{U, \xi\}$ be an IF soft approximation space, and $\alpha, \beta \in P(U)$, then we have the following:

(a) IFSap $_S(\alpha)\downarrow \subseteq \alpha \subseteq$ IFSap $_S(\alpha)\uparrow$.

(b) IFSap $_S(\Phi)\downarrow =$ IFSap $_S(\Phi)\uparrow = \Phi$.

(c) IFSap $_S(U)\downarrow =$ IFSap $_S(U)\uparrow = U$.

(d) (IFSap $_S(\alpha)\uparrow)^C =$ (IFSap $_S(\alpha^C))\downarrow$, if $\alpha = \{(x, \mu_A(x), \nu_A(x) : x \in U)\}$then $\alpha^C = \{(x, \nu_A(x), \mu_A(x) : x \in U)\}$.

(e) (IFSap $_S(\alpha)\downarrow)^C =$ (IFSap $_S(\alpha^C))\uparrow$.

(f) (IFSap $_S(\alpha \cap \beta)\downarrow) =$ IFSap $_S(\alpha)\downarrow \cap$ IFSap $_S(\beta)\downarrow$.

(g) (IFSap $_S(\alpha \cup \beta)\downarrow) \supseteq$ IFSap $_S(\alpha)\downarrow \cup$ IFSap $_S(\beta)\downarrow$.

(h) (IFSap $_S(\alpha \cup \beta)\uparrow) =$ IFSap $_S(\alpha)\uparrow \cap$ IFSap $_S(\beta)\downarrow$.

(i) (IFSap $_S(\alpha \cap \beta)\uparrow) \subseteq$ IFSap $_S(\alpha)\uparrow \cap$ IFSap $_S(\beta)\uparrow$.

(j) $\alpha \subseteq \beta \Rightarrow$ IFSap $_S(\alpha)\downarrow \subseteq$ IFSap $_S(\beta)\downarrow$.

(k) $\alpha \subseteq \beta \Rightarrow$ IFSap $_S(\alpha)\uparrow \subseteq$ IFSap $_S(\beta)\uparrow$.

Example 4.5: Suppose $U = \{h_1, h_2, h_3, h_4, h_5\}$ is the universe , consisting of five houses and set of parameters is given by $E = \{e_1, e_2, e_3, e_4, e_5\}$ where e_i ($i = 1, 2, 3, 4, 5$) stands for beautiful, modern, cheep, in green surrounding and in good repair respectively. Let us consider an IF soft set (F, E) which describes the attractiveness of houses that Mr. Z is considering for purchase. In this case to define the IF soft set means to point out beautiful houses, modern houses and so on. Consider the mapping F given by "houses (.)" where (.) is to be filled in by one parameter $e_i \in E$ for instance $F(e_1)$ means houses (beautiful) and its functional values is the set consisting of all the beautiful houses in U. Let $F(e_1) = h_5$, $F(e_2) = \{h_1, h_4\}$, $F(e_3) = \{h_1, h_2, h_3\}$ and $F(e_4) = \{h_3, h_5\}$. Let $\xi = (F, E)$ be the IF soft set over U. Let $S = \{U, \xi\}$ be an IF soft approximation space. Then for Intuitionistic fuzzy set $\alpha = \{h_1/(0.8, 0.2), h_2/(0.5, 0.3), h_3/(0.7, 0.3), h_4/(0.2, 0.7), h_5/(0.3, 0.6)\}$, then by definition we calculate IFSap $_S(\alpha)\downarrow$ and IFSap $_S(\alpha)\uparrow$ as follows:

IFSap $_S(\alpha)\downarrow = \{h_1/(0.2, 0.7), h_2/(0.5, 0.3), h_3/(0.3, 0.6), h_4/(0.2, 0.7), h_5/(0.3, 0.6)\}$, and IFSap $_S(\alpha)\uparrow = \{h_1/(0.8, 0.2), h_2/(0.8, 0.2), h_3/(0.8, 0.2), h_4/(0.8, 0.2), h_5/(0.7, 0.3)\}$. Similarly for the IF set, $\beta = \{h_1/(0.1, 0.8), h_2/(0.3, 0.6), h_3/(0.6, 0.4), h_4/(0.8, 0.2), h_5/(0.5, 0.5)\}$, we calculate IFSap $_S(\beta)\downarrow$ and IFSap $_S(\beta)\uparrow$ as follows: IFSap $_S(\beta)\downarrow = \{h_1/(0.1, 0.8), h_2/(0.1, 0.8), h_3/(0.1, 0.8), h_4/(0.1, 0.8), h_5/(0.5, 0.5)\}$, and IFSap $_S(\beta)\uparrow = \{h_1/(0.8, 0.2), h_2/(0.6, 0.4), h_3/(0.6, 0.4), h_4/(0.8, 0.2), h_5/(0.6, 0.4)\}$. We first observe that

IFSap $_S(\alpha)\downarrow \subseteq$ IFSap $_S(\alpha)\uparrow$ and IFSap $_S(\beta)\downarrow \subseteq$ IFSap $_S(\beta)\uparrow$.

Now, (IFSap $_S(\alpha)\downarrow)^C = \{h_1/(0.7, 0.2), h_2/(0.3, 0.5), h_3/(0.6, 0.3), h_4/(0.7, 0.2), h_5/(0.6, 0.3)\}$, also $\alpha^C = \{h_1/(0.2, 0.8), h_2/(0.3, 0.5), h_3/(0.3, 0.7), h_4/(0.7,$

0.2), $h_5/(0.6, 0.3)$}, so IFSap $_S(\alpha^C)\!\uparrow = \{h_1/(0.7, 0.2), h_2/(0.3, 0.5), h_3/(0.6, 0.3), h_4/(0.7, 0.2), h_5/(0.6, 0.3)\}$,

Thus, $(\text{IFSap }_S(\alpha)\!\downarrow)^C = \text{IFSap }_S(\alpha^C)\!\uparrow$.

Now, $\alpha \cap \beta = \{h_1/(0.1, 0.8), h_2/(0.3, 0.6), h_3/(0.6, 0.4), h_4/(0.2, 0.7), h_5/(0.3, 0.6)\}$, we calculate IFSap $_S(\alpha \cap \beta)\!\downarrow$ as follows:

IFSap $_S(\alpha \cap \beta)\!\downarrow = \{h_1/(0.1, 0.8), h_2/(0.1, 0.8), h_3/(0.1, 0.8), h_4/(0.1, 0.8), h_5/(0.3, 0.6)\}$, also IFSap $_S(\alpha)\!\downarrow \cap$ IFSap $_S(\beta)\!\downarrow = \{h_1/(0.1, 0.8), h_2/(0.1, 0.8), h_3/(0.1, 0.8), h_4/(0.1, 0.8), h_5/(0.3, 0.6)\}$.

Thus, IFSap $_S(\alpha \cap \beta)\!\downarrow = $ IFSap $_S(\alpha)\!\downarrow \cap$ IFSap $_S(\beta)\!\downarrow$.

Now, we calculate IFSap $_S(\alpha \cap \beta)\!\uparrow$ as follows:

IFSap $_S(\alpha \cap \beta)\!\uparrow = \{h_1/(0.6, 0.4), h_2/(0.6, 0.4), h_3/(0.6, 0.4), h_4/(0.2, 0.7), h_5/(0.6, 0.4)\}$. Also we obtain IFSap $_S(\alpha)\!\uparrow \cap$ IFSap $_S(\beta)\!\uparrow = \{h_1/(0.8, 0.2), h_2/(0.6, 0.4), h_3/(0.6, 0.4), h_4/(0.8, 0.2), h_5/(0.6, 0.4)\}$.

Thus, IFSap $_S(\alpha \cap \beta)\!\uparrow \subseteq$ IFSap $_S(\alpha)\!\uparrow \cap$ IFSap $_S(\beta)\!\uparrow$.

Now, $\alpha \cup \beta = \{h_1/(0.8, 0.2), h_2/(0.5, 0.3), h_3/(0.7, 0.3), h_4/(0.8, 0.2), h_5/(0.5, 0.5)\}$. We obtain, IFSap $_S(\alpha \cup \beta)\!\downarrow = \{h_1/(0.5, 0.3), h_2/(0.5, 0.3), h_3/(0.5, 0.5), h_4/(0.8, 0.2), h_5/(0.5, 0.5)\}$. Also we calculate IFSap $_S(\alpha)\!\downarrow \cup$ IFSap $_S(\beta)\!\downarrow = \{h_1/(0.2, 0.7), h_2/(0.5, 0.3), h_3/(0.3, 0.6), h_4/(0.2, 0.7), h_5/(0.5, 0.5)\}$.

Thus, IFSap $_S(\alpha \cup \beta)\!\downarrow \supseteq$ IFSap $_S(\alpha)\!\downarrow \cup$ IFSap $_S(\beta)\!\downarrow$.

Now, IFSap $_S(\alpha \cup \beta)\!\uparrow = \{h_1/(0.8, 0.2), h_2/(0.8, 0.2), h_3/(0.8, 0.2), h_4/(0.8, 0.2), h_5/(0.7, 0.3)\}$. Also, IFSap $_S(\alpha)\!\uparrow \cup$ IFSap $_S(\beta)\!\downarrow = \{h_1/(0.8, 0.2), h_2/(0.8, 0.2), h_3/(0.8, 0.2), h_4/(0.8, 0.2), h_5/(0.7, 0.3)\}$.

Thus, IFSap $_S(\alpha \cup \beta)\!\uparrow = $ IFSap $_S(\alpha)\!\uparrow \cup$ IFSap $_S(\beta)\!\uparrow$.

Now, $\alpha \cup \beta = \{h_1/(0.8, 0.2), h_2/(0.5, 0.3), h_3/(0.7, 0.3), h_4/(0.8, 0.2), h_5/(0.5, 0.5)\}$. We obtain, IFSap $_S(\alpha \cup \beta)\!\downarrow = \{h_1/(0.5, 0.3), h_2/(0.5, 0.3), h_3/(0.5, 0.5), h_4/(0.8, 0.2), h_5/(0.5, 0.5)\}$. Also we calculate IFSap $_S(\alpha)\!\downarrow \cup$ IFSap $_S(\beta)\!\downarrow = \{h_1/(0.2, 0.7), h_2/(0.5, 0.3), h_3/(0.3, 0.6), h_4/(0.2, 0.7), h_5/(0.5, 0.5)\}$.

Thus, IFSap $_S(\alpha \cup \beta)\!\downarrow \supseteq$ IFSap $_S(\alpha)\!\downarrow \cup$ IFSap $_S(\beta)\!\downarrow$.

Now, IFSap $_S(\alpha \cup \beta)\!\uparrow = \{h_1/(0.8, 0.2), h_2/(0.8, 0.2), h_3/(0.8, 0.2), h_4/(0.8, 0.2), h_5/(0.7, 0.3)\}$. Also, IFSap $_S(\alpha)\!\uparrow \cup$ IFSap $_S(\beta)\!\uparrow = \{h_1/(0.8, 0.2), h_2/(0.8, 0.2), h_3/(0.8, 0.2), h_4/(0.8, 0.2), h_5/(0.7, 0.3)\}$.

Thus, IFSap $_S(\alpha \cup \beta)\!\uparrow = $ IFSap $_S(\alpha)\!\uparrow \cup$ IFSap $_S(\beta)\!\uparrow$.

Remark: In this talk we first defined intuitionistic fuzzu soft rough sets (IFSRsets). We provided an example that demonstrated that this method can be successfully worked [Intuitionistic fuzzy soft rough sets, The Journal of Fuzzy Mathematics, 20(2), (2012), 423-432]. It can be applied to problems of many fields that contain uncertainty. However the approach should be more comprehensive in the future to solve the related problems. Now if $\alpha = \{(x, \mu_\alpha(x), \nu_\alpha(x) : x \in X)\}$ be an IF set then taking $\nu_\alpha(x) = 1 - \mu_\alpha(x)$, it is clear that IF soft rough sets are soft rough fuzzy sets due to Feng et. al. Thus IF soft rough sets are the extension of soft rough fuzzy sets. Further Feng et.al showed that soft rough fuzzy sets are the extension of rough fuzzy sets due to Dubois and Prade. Thus our work is the extension of both the previous work of Dubois, prade and Feng et.al.

References

1. Atanassov, K., 1986: Intuitionistic *Fuzzy sets*, Fuzzy Sets and Systems 20 87-96.

2. Aktas, H., Cagman, N., 2007: *Soft sets and Soft groups*, Information Sciences, 1(77), 2726-2735.

3. Chakrabarty, K., Gedeon T., and Koczy, I., 1998: *Intuitionistic fuzzy rough sets*, in Proceeding of the Fourth Joint Conference of Information Sciences, Durham, NC, JCTS, 211-214.

4. Dubois, D., and Prade, H., 1990: *Rough fuzzy sets and fuzzy rough sets*, Int.J.Gen. Syst. V-17, 191-209.

5. Feng, F., Li, Changxing, Davvaz, B., and Ali, M.I., 2009: *Soft sets combined with fuzzy sets and rough sets:a tentative approach*, published on line (Springer) (Soft comput DOI 10.1007/s00500-009-0465-6).

6. Jena, S.P., and Ghosh, S.K., 2002: *Intuitionistic fuzzy rough sets*, Notes on Intuitionistic Fuzzy Sets8(1) 1-18.

7. Maji, P.K., Biswas, R., and Roy, A.R., 2001: *Fuzzy soft sets*, Journal of Fuzzy Mathematics, 9(3), 589-602.

8. Maji, P.K., Biswas, R., and Roy, A.R., 2001: *Intuitionistic fuzzy soft sets*, Journal of Fuzzy Mathematics, 9(3), 677-691.

9. Maji, P.K., Biswas, R., and Roy, A.R., 2003: *Soft set theory*, Computers and Mathematics with Applications, 45, 555-562.

10. Maji, P.K., Biswas, R., and Roy, A.R., 2002: *An application of soft sets in a decition making problem*, Computers and Mathematics with application, 44, 1077-1083.

11. Molodtsov, D.A., 1999: *Soft set theory-first results*, Computers and Mathematics with applications, 37, 19-31.

12. Mukherjee, A., and Chakraborty, S.B., 2007: *Relation on intuitionistic fuzzy soft sets & their applications*, International Conference on Modeling and Stimulation CIT, Coimbator, (671-677).

13. Mukherjee, A., and Chakraborty, S.B., 2008: *On intuitionistic fuzzy soft relations*, Bulletin of Kerala Mathematics Association, 5 (1), 35-42.

14. Nanda S., and Majumdar, S., 1992: *Fuzzy rough sets*, Fuzzy Sets and Systems 45, 157-160.

15. Pawlak, Z., 1982: *Rough sets*, Int. J. Infor and Comp. Sc. 11, 341-356.

16. Rizvi, S., Nagvi H.J., and Nadeem, D., 2002: *Rough intuitionistic fuzzy sets*, in Proceeding of the 6th Joint Conference on information Sciences, Durham, NC, JCTS, 101-104.

17. Samanta, S.K., and Mondal, T.K., 2001: *Intuitioistic fuzzy rough sets and rough intuitionistic fuzzy sets*, Journal of Fuzzy Mathematics 9(3), 561-582.

18. Zadeh, L.A., 1965: *Fuzzy Sets*, Information and Control, 8, 338-353.

Rough Sets, Fuzzy Sets and Soft Computing
Editor: S. Bhattacharya Halder
Copyright © 2015, Narosa Publishing House, New Delhi

Medical Diagnosis based on Evidence Theory

Tazid Ali

Department of Mathematics, Dibrugarh University
Dibrugarh, Assam
E-mail: tazidali@yahoo.com

ABSTRACT

In the field of medical diagnosis uncertainty and imprecision are inherent concepts. Uncertainty characterizes a relation between symptoms and diseases, while imprecision is associated with the symptom representation. A single disease may manifest itself quite differently in different patients and at different disease stages. As Dempster-Shafer Theory (DST) of Evidence is a powerful and flexible mathematical tool for handling uncertainty, imprecision and incomplete information, it can be exploited in medical diagnosis. By representing the uncertainty and the imprecision of the symptoms via the notion of evidence, belief can be committed to a single disease. In this paper we have considered an extended version of DST by treating focal elements as both crisp and fuzzy sets.

We have illustrated the use of the model with a case study of diagnosis of a patient for different forms of anaemia.

Keywords: Uncertainty, Dempster-Shafer Theory, Medical Diagnosis

1. INTRODUCTION

A diagnosis can be regarded as an attempt at classification of an individual's condition into separate and distinct categories that allow medical decisions about treatment and prognosis to be made. It is the art of determining a person's pathological status from available set of symptoms. It is the art of information processing. With the increased value of information available to physicians from new medical technologies, the process of classifying different sets of symptoms under a single name and determining appropriate therapeutic actions becomes increasingly difficult. A single disease may manifest itself quite differently in different patients and at different disease stages. Furthermore a single symptom may be indicative of several different diseases, and the presence of several diseases in single patients may disrupt the expiated symptom pattern of any one of them.

The best and most useful descriptions of disease entities often used linguistic terms that are irreducibly vague. Although medical knowledge concerning the symptom disease relationship constitutes one source of imprecision and uncertainty in the diagnostic process, the knowledge concerning the state of the patient constitutes another. The physician generally gathers knowledge about the patients from the past history, physical examination, laboratory test results and other investigative procedures such as X-rays and ultrasonography. The knowledge provided by each of these sources carries with it varying degrees of uncertainty. The past history offered by the patient may be subjective, exaggerated, underestimated or incomplete. Mistake may be made in the physical examination, and the symptoms may be overlooked. The measurements provided by laboratory tests are often of limited precision, and the exact borderline between normal and pathological is often unclear. X-rays and other similar procedures require correct interpretation of the results. Thus, the state and symptoms of patients can be known by the physician with only a limited degree of precision. In the face of the uncertainty concerning the observed symptoms of the patients as well as the uncertainty concerning the relation of the symptoms to a disease entity, it is nevertheless crucial that the physician determine the diagnostic label that will entail the appropriate therapeutic line of treatment.

Different theories like fuzzy set theory [4], probability theory are being used to take care of the uncertainty associated diagnosis process.

2. EVIDENCE THEORY BASED MODEL FOR MEDICAL DIAGNOSIS

Diagnosis support systems operate on rules with fuzzy premises, which represent imprecise symptoms. In this section we discuss a model of diagnosis in which disease risk changes monotonously with gradually appearing symptoms. The question of whether the symptoms match patient's state precisely enough to be included in inference is taken into consideration. The disease uncertainty is represented in the framework of the Dempster-Shafer theory of evidence. As each disease is associated with a number of symptoms, each diagnosis is based on multiple symptoms. The symptoms associated to a particular disease make a set of focal elements for the diagnosis.

Focal elements defined in the theory are represented both by crisp as fuzzy set. The belief of the theory is used for evaluation of diagnostic hypotheses. The diagnosis with the greatest belief is considered as the final conclusion. A focal element may refer to single medical parameter (single focal element) or it may be related to several parameters (complex element). Medical diagnostic support system may be described as:

If X is $\overline{X}$, and Y is $\overline{Y}$, ..., and Z is $\overline{Z}$, then Diagnosis is D (1)

where X, Y, ..., Z, are medical parameters (laboratory tests for example), $\overline{X}$, $\overline{Y}$, ..., $\overline{Z}$ are linguistic variables (symptoms) that describe values of the medical parameters (e.g., high, low) and D is the diagnostic inference or conclusion. Such a rule gives the knowledge base of the diagnosis. Belief value is associated to the credibility of the diagnosis for an examined patient. The credibility of the conclusion is assessed by the *Bel* and *Pl* which gives an interval. For simplicity sake we will confine our self with *Bel* measure only. The BPA value can result from statistical investigation or can be stated according to a diagnostic index. It may also be provided by an expert (physician).

The truth or falsity of a focal element is crucial in *Bel* calculation. If the focal element is expressed by a crisp notion then truth will be assigned to the two-element set $\{0, 1\}$. However if the focal element is expressed by a fuzzy concept then truth will be interpreted in the $[0, 1]$ interval. Each focal element is defined using one (in case of single focal element) or several (for complex focal element) membership or characteristic functions. To each focal element a_i we associate a number $\eta_i \in [0, 1]$ which indicates the degree of presence of the symptom (corresponding to the focal element) in the patient. It is made on the basis of observation on the patient. It indicates the level of certainty that a particular symptom occurs in the patient. For single focal element the certainty level is defined as

$$\eta_i = (x^*) \text{ or } \eta_i = \xi_i\,(x^*)$$

where μ is the membership function representing the focal element (symptom) and x^* is the observed value of that symptom, ξ is the characteristic function of symptom's presence. $\xi_i\,(x^*)$ is 1 if the symptom is present and 0 otherwise.

For complex fuzzy focal element, say, $a_i = \{a_i^1, a_i^2, ..., a_i^k\}$

$$\eta_i = \min\,\{\mu_{i_1}\,(x^*_1), \mu_{i_2}\,(x^*_2), ..., \mu_{i_k}\,(x^*_k)\},$$

where μ_{i_1}, μ_{i_2}, ..., μ_{i_k} are the membership functions (characteristic function) of the of a_i^1, a_i^2, ..., a_i^k respectively and x^*_1, x^*_2, ..., x^*_k are the corresponding observed values. The rule (1) defined for a particular disease makes a knowledge base of the diagnosis. The set of symptoms associated to a particular disease make a set of focal elements for the diagnosis.

The BPA of the focal elements will be provided either by experts based on medical knowledge or will be obtained based on data. The BPA of a focal element indicates the relative importance of that symptom in the diagnosis of the disease under consideration. If a_1, a_2, ..., a_k, are the focal elements and $m(a_1)$, $m(a_2)$, ..., $m(a_k)$ are the BPAs, then

$$\sum_{i=1}^{k} m(a_i) = 1$$

If the truth of each focal element is one, i.e., each symptoms shows full membership in the focal element, then the Belief is the sum of the BPAs and that is 1, and we conclude that the patient under consideration is suffering from that particular disease. However in reality, we observe that presence of symptoms is hardly total, it is partial. To take account of this imprecision we perform the dot product between the vectors $(m(a_1), m(a_2), ..., m(a_k))$ and $(\eta_1, \eta_2, ..., \eta_k)$ to obtain new weight, say, $(w_1, w_2, ..., w_k)$ of the foal elements based on observations. Then belief that a person is suffering from a disease is given by

$$\text{Bel}(D) = \sum_{i=1}^{k} w_i$$

When belief values of all possible diagnose is calculated, the final diagnosis can be determined as a result of their comparison.

3. AN APPLICATION

Suppose we want to diagnose an ill patient for five possible form of anaemia, viz. Aplastic Anaemia, Mylopthisic Anaemia, Chronic Anaemia, Megaloblastic Anaemia, Iron deficiency Anaemia. These diseases are functions of six symptoms or parameters viz. Haemoglobin(HGLN), Hematocrit(HMCT), Mean Corpuscular Volume(MCV), Mean corpuscular haemoglobin concentration(MCHC), Reticulate count(RCC), White blood cells(WBC). For each disease there is a rule [Table 1] that determines the presence of that disease in a person.

Table 1

	HGLN	HMCT	MCV	MCHC	RCC	WBC	Diagnosis
Rule 1	Very Low	Very Low	Normal	Normal	Low	Very Low	Aplastic Anaemia (D_1)
Rule 2	Low	Low	Normal	Normal	Low	Normal	Myelophthisic Anaemia (D_2)
Rule 3	Very Low	Very Low	Low	Normal	Low	Normal	Chronic Anaemia (D_3)
Rule 4	Very Low	Very Low	High	Normal	Low	Normal	Megaloblastic Anaemia (D_4)
Rule 5	Very Low	Very Low	Low	Low	Low	Normal	Iron Deficiency Anaemia (D_5)

Each symptom is fuzzy in nature. The membership functions of the symptoms are expressed in Table 2 to 7. {Data taken from [2]}

Table 2 Haemoglobin (HGLN) gram/100 L or gram%

	4	8	12	14	16	20
High	0.0	0.1	0.2	0.4	0.6	1.0
Normal	0.0	0.3	0.6	1.0	0.6	0.3
Low	1.0	0.8	0.8	0.6	0.4	0.1
Very High	0.0	0.0	0.0	0.2	0.3	0.6
Very Low	1.0	0.9	0.6	0.0	0.0	0.0

Table 3 Hematocrit (HMCT) % of RBC in blood sample

	20	25	30	35	40	45
High	0.0	0.0	0.1	0.2	0.4	1.0
Normal	0.0	0.2	0.6	1.0	1.0	0.6
Low	1.0	1.0	0.4	0.3	0.2	0.0
Very High	0.0	0.0	0.0	0.0	0.2	0.6
Very Low	1.0	1.0	0.2	0.1	0.0	0.0

Table 4 Mean corpuscular volume (MCV) Cubic microns (red blood cell volume)

	50	65	80	95	110	125
High	0.0	0.0	0.1	0.2	0.4	1.0
Normal	0.0	0.3	0.8	1.0	0.7	0.3
Low	1.0	0.9	0.4	0.2	0.1	0.0
Very High	0.0	0.0	0.0	0.3	0.6	1.0
Very Low	1.0	1.0	0.2	0.0	0.0	0.0

Table 5 Mean Corpuscular Haemoglobin concentration (MCHC) gHb/dl of RBC or % Hb/cell

	20	25	30	35	40	45
High	0.0	0.1	0.2	0.4	0.8	1.0
Normal	0.0	0.3	0.8	1.0	0.7	0.2
Low	1.0	0.9	0.4	0.2	0.1	0.0
Very High	0.0	0.0	0.1	0.2	0.6	1.0
Very Low	1.0	0.8	0.2	0.0	0.0	0.0

Table 6 Reticulocyte count (RCC) % of erythrocytes or 1000 cells/ltr.

	0.2	0.5	1.0	1.5	5	10
High	0.0	0.0	0.2	0.4	0.8	1.0
Normal	0.0	0.8	1.0	0.8	0.4	0.2
Low	1.0	0.7	0.4	0.2	0.1	0.0
Very High	0.0	0.0	0.0	0.2	0.6	1.0
Very Low	1.0	0.4	0.2	0.0	0.0	0.0

Table 7 White blood cells (WBC) Nos./cu.mm

	500	2K	4K	6K	8K	15K
High	0.0	0.1	0.2	0.3	0.4	1.0
Normal	0.0	0.2	0.6	0.9	1.0	0.4
Low	1.0	0.9	0.8	0.7	0.6	0.0
Very High	0.0	0.0	0.0	0.1	0.2	0.4
Very Low	1.0	1.0	1.0	0.9	0.7	0.6

For the disease Aplastic anaemia, the set of focal elements is $\{a_1$ = very low HGLN, a_2 = very low HMCT, a_3 = normal MCV, a_4 = normal MCHC, a_5 = low RCC, a_6 = very low WBC$\}$. Here there are no complex focal elements. Suppose medical tests indicate the following measurement of the six parameters in the person concerned.

HGLN	HMCT	MCV	MCHC	RCC	WBC
12%	25%	50	30	0.5%	4K

Then we have $\eta_1 = 0.6$, $\eta_2 = 1$, $\eta_3 = 0$, $\eta_4 = 0.8$, $\eta_5 = 0.7$, $\eta_6 = 1$.

Suppose for convenience we assume each symptom is equally important in the diagnosis of the disease, i.e., each focal element is assigned a BPA of 1/6. Then after calculation we obtain the Belief value for disease D_1 as

$$\text{Bel } (D_1) = 0.6831$$

Similarly we calculate belief for other form of anaemia and obtain the values as:

$$\text{Bel } (D_2) = 0.6498$$

$$\text{Bel } (D_3) = 0.7831$$

$$\text{Bel } (D_4) = 0.3499$$

$$\text{Bel } (D_5) = 0.4498$$

Since the Belief value for the disease D_3 is maximum, we conclude that the person is suffering from Chronic Anaemia.

4. CONCLUSION

In this paper we have discussed a model for medical diagnosis which is based on DST. Any disease is associated with a set of symptoms, which gives a rule for the diagnosis. The set of symptoms forms the set of focal elements in the frame work of DST. The membership functions of the focal elements will have to be constructed depending upon medical knowledge and in consultation with medical expert. Based on the evidence of the presence of symptoms in the patient, weight is calculated for each focal element. Belief for a particular disease is then obtained by adding all the weights of the focal elements. Then diagnostic inference is done based on the maximum *Bel* value. Sometimes BPA of the focal elements (symptoms) may be assigned by more than one expert (physician). Then the two bodies of evidence can be combined [1, 3] to give a new body of evidence. Finally the diagnosis can be made as discussed in the previous section. Using our proposed method a numerical example is solved. The fuzzy sets corresponding to the focal elements are taken as discrete. It can easily be converted to continuous functions.

References

1. Dutta, P. and Ali, T., 2011: *Fuzzy Focal Elements in Dempster-Shafer Theory of Evidence: Case Study in Risk Assessment*; Int. J. Comp. Appl.; **34**(1), 46-53.

2. Kumar, K.S. and Unnikrishnan, A., 1992: *On an Efficient Selection of Rules to be fixed in fuzzy rule based system*; NDPL, RDDO; Cochin, India.

3. Straszecka, E., 2008: *Combining Basic Probability Assignments for Fuzzy Focal Elements*; Proceedings of the 9th International Conference on Artificial Intelligent and Soft Computing; 341-350.

4. Yuan, B. and Klir, G.J., 2003: *Fuzzy Sets and Fuzzy Systems: Theory and Applications*; Prentice Hall of India Pvt. Ltd.

Rough Sets, Fuzzy Sets and Soft Computing
Editor: S. Bhattacharya Halder
Copyright © 2015, Narosa Publishing House, New Delhi

Fusion of Wavelet Coefficients for Classification of Human Face Images Using Kernel Independent Component Analysis and Different SVM Kernels

Priya Saha[1], Mrinal Kanti Bhowmik[1], Debotosh Bhattacharjee[2], Barin Kumar De[3], Mita Nasipuri[2]

[1]*Department of Computer Science and Engineering*
Tripura University (A Central University)
Suryamaninagar, Tripura
[2]*Department of Computer Science and Engineering*
Jadavpur University, Kolkata
[3]*Department of Physics*
Tripura University (A Central University)
Suryamaninagar, Tripura

ABSTRACT

This paper integrates a non-linear feature extraction method namely Kernel Independent Component Analysis (KICA) and a Support Vector Machine (SVM) as a classifier for human face classification. In this proposed approach, wavelet based image fusion is used to fuse the face images in optical and infrared spectrum. Forward wavelet transform up to level five has been done before fusion of the coefficients. Inverse Daubechies wavelet transform of those coefficients generates fused face images. Since wavelet transform keeps high and low frequency sub-band components separately, reconstructed fused images, therefore, have the information in both low and high frequencies. Features of the fused images are extracted using KICA. Finally, we have used SVM to classify extracted features of face using three different kernels: linear, polynomial and RBF. Detailed experimentation is carried out using IRIS Thermal/Visual Face Database containing different subjects under different pose, illumination and expression. Experimental results show that the performance of the approach presented here gives success rate of 96.25%, 64.14%, and 96.06% on average for the three kernels respectively.

Keywords: Fusion; Wavelet transform; Kernel Independent Component Analysis (KICA); Support Vector Machine (SVM).

1. INTRODUCTION

Face recognition has been extensively spread within the research and development communities in the last decade. This technology has several commercial and law enforcement application. These applications range static matching of photographs such as passports, credit cards, photo ID, driver's licenses (Chellappa et al., 1995). Most of the face recognition techniques have been evolved in order to prevail over two main challenges: illumination and pose variation. These two problems can cause severe performance degradation in face recognition system (webpage). Visual image based face recognition are used the most at present, but in different lighting condition visual images loses their quality. To tackle these problems, thermal images are gaining much interest in these days because of their illumination invariant properties. Thermal images also have some limitations like facial hair, glasses, or cosmetics. Therefore, to extract the advantages from both types of images, fusion is the most constructive notion as it combines multiple information sources together and produces a more informative representation of the data (Maruthi and Sankarasubramanian, 2005).

Hanif et al. (Hanif and Ali, 2006) have discussed data fusion of thermal and visual images to overcome the drawbacks present in individual thermal and visual images. Bebis et al. (Bebis et al., 2006) discussed infrared (IR) imagery offers a promising alternative to visible imagery, due to its relative insensitivity to variations in face appearance caused by illumination changes. Thermal IR images have several shortcomings including that it is opaque to glass. Nikolov et al. (Nikolov et al., 2001) have presented some recent results on the use of wavelet algorithms for image fusion. A detailed overview on wavelet transform for image fusion using image decomposition and reconstruction has been proposed by M. K. Bhowmik et al. in (Bhowmik et al., 2010a). An image fusion technique based on the weighted average of Daubechies wavelet transform (db2) has been presented. The reason behind the use of wavelet decomposition and reconstruction is to separate the high frequency and low frequency information of an image and generate fused image of high and low frequency sub-bands separately in wavelet transform. M. K. Bhowmik et al. (Bhowmik et al., 2010b) have used the process of fusion of visual and thermal images using different wavelet transformations. Coefficients of discrete wavelet transforms from both optical and thermal images were computed separately and combined. They also used inverse discrete wavelet transformation in order to obtain fused face image. M. K. Bhowmik et al. (Bhowmik et al., 2010c) have investigated Quotient based Fusion of thermal and visual images, which were passed separately through level-1 and level-2 multi-resolution analyses. This approach is based on a definition of illumination invariant signature image that enables an analytic generation of the image space with varying illumination. Principal Component Analysis (PCA) was used for dimension reduction of quotient fused images and then, those images are classified using a multilayer perceptron (MLP). In another paper of M. K. Bhowmik et al. (Bhowmik et al., 2011), an image fusion technique based on the weighted average of Daubechies wavelet transform have presented, and a comparative study has conducted for dimensionality reduction based on Principal Component Analysis (PCA) and Independent Component Analysis (ICA). Different researchers have examined the performance of KICA on face images which has shown better recognition result than the other linear methods. J. H. Cao et al. (Cao et al., 2010) investigated face recognition method using KICA and Kernel-based improved PSVM that results better performance over native ICA method. Y. Huang et al. (Huang et al., 2010) proposed a new gabor based Kernel Independent Component analysis (GKICA) which got higher recognition rate than ICA and KICA.

The contribution of the paper is to provide systematic KICA based method to classify the fused images of visual and infrared images for efficient face recognition. To cope up with the complex non-linear variations due to illumination changes, and facial expressions, that influence the recognition performance nonlinear feature extraction method, KICA has introduced in this paper. The paper is organized as follows: system overview is given in section II; experimental results with discussions on them as well as with a comparative study are given in section III; and the conclusions are drawn in section IV.

2. SYSTEM OVERVIEW

In this work, thermal and visual images are used to decompose with level five and from these images; the fused image is generated using db4. In the first step, decomposition of both the infrared and visual images up to level five has been done using wavelet decomposition. Then fused image is generated from both the decomposed images. In this process, Kernel Independent Component Analysis (KICA) is performed for feature extraction, and then Support Vector Machine (SVM) classifier is used to classify input images.

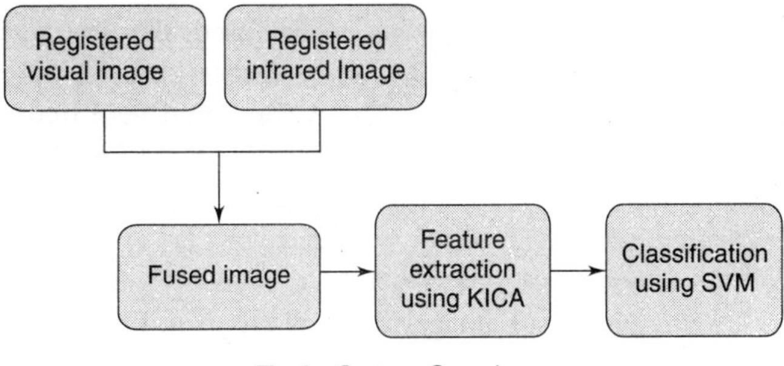

Fig 1 System Overview

Generating Registered Images

Image Registration allows geometrical transformation which aligns points in one view of an object with corresponding points in another view of that object or another object. In this paper, we have used registered visible and registered thermal IR images before they are fused. First, we have taken the frontal image as the base image and another unregistered image for registration in case of both thermal and visual images (separately). In the second step, pair of control points was selected from both the base and unregistered images. Next task is to choose the transformation type, infer its parameters, and transform the unregistered image to generate the registered image. Affine transformation is used in this work as it has six degrees of freedom and is equivalent to the combined effect of translation, rotation, isotropic scaling, and shear (non-uniform scaling in some direction). This image registration process was applied for the visual and thermal images separately as well as for each expression and illumination type of each person i.e. out of the 11 images of different rotations; we have considered the frontal image as the base image and registered the other 10 images with respect to the frontal image.

Image Decomposition and Reconstruction

The Daubechies (db4) is used for decomposition and reconstruction of the images in case of generation of fused imaging technique. Daubechies wavelets, a wavelet used to convolve image data, can be orthogonal, having scaling functions with same number of coefficients as the wavelet functions, or biorthogonal with different number of coefficients. The main idea behind the fusion algorithm: (a) the two images are to be processed and re-sampled to the one with the same size; and (b) they are respectively decomposed into the sub-images using forward wavelet transform, which have the same resolution at the same levels and different resolution among different levels; and (c) information fusion is performed based on the high-frequency sub-images of decomposed images; and finally the resultant image is obtained using inverse wavelet transform (Shu-long, 2004).

Let $A(x, y)$ and $B(x, y)$ be the images to be fused, the decomposed low-frequency sub-images of $A(x, y)$ and $B(x, y)$ be respectively $lA_j (x, y)$ and $lB_j (x, y)$ (j is the parameter of resolution). The decomposed high-frequency sub-images of $A(x, y)$ and $B(x, y)$ be respectively $hA^k_j (x, y)$ and $hB^k_j (x, y)$ (j is the parameter of resolution and $j = 1, 2, ..., J$. For every j, $k = 1, 2, 3$). Then the fused high-frequency sub-images $F^k_j (x, y)$ are: if $hA^k_j (x, y) > hB^k_j (x, y)$ then $F^k_j (x, y) = hA^k_j (x, y)$ and if $hA^k_j (x, y) < hB^k_j (x, y)$ then $F^k_j (x, y) = hB^k_j (x, y)$; and the fused low-frequency sub images $F_j (x, y)$ are as follows:

$$F_j (x, y) = k_1 \cdot lA_j (x, y) + k_2 \cdot lB_j (x, y) \qquad (1)$$

In Eqn. 1, k_1 and k_2 are given parameters, if the image B is fused into A, then $k_1 > k_2$ and vice-versa (Shu-long, 2004).

Fusion of low frequency sub-images are computed as weighted sum of corresponding coefficients, found from both the images (thermal and its corresponding visual) and for high-frequency sub-images, larger coefficient is considered as final output. After generating the decomposed fused image, the inverse Daubechies Wavelet is applied to generate synthesized fused image. When the decomposition scheme is being repeated more, the approximation image more concentrates in the low frequency energy. Finally, the reconstructed image is used as an input to feature extracted algorithm.

Kernel Independent Component Analysis

Kernel Independent Component Analysis (KICA) is a method based on kernel function, which allows data space to be mapped into high dimensional feature space, and also an improved feature extraction method from existing ICA. The KICA algorithm developed by F.R. Bach and M.I. Jordan was primarily used for separating randomly mixed auditory signals (Martiriggiano et al., 2005). However, this method has been begun to use in bidimensional images. The concept of kernel in ICA has been introduced in face images due to the failure of linear methods of feature extraction. Linear methods like Principal Component Analysis (PCA), Linear Discriminant Analysis (LDA), and Independent Component Analysis (ICA), which are three powerful tools largely used for data reduction, and feature extraction in the appearance-based approaches (Turk and Pentland, 1991; Belhumeur et al., 1997) are inadequate to describe complex non-linear variations due to illumination changes, facial expressions and aging. Non-linear extension of linear methods has been initiated to eliminate these problems. The steps of KICA are presented here as follows:

1. Centering of Fused Image, 2. Whitening and Initializing of De-mixing matrix, 3. Memory Initialization for storing de-mixing matrix and Hilbert Space Independent Component (HSIC) at each iteration where maximum iteration is 20, 4. Define the width of Gaussian kernel, sigma as 1, 5. Memory utilization by Incomplete cholesky, and default Convergence threshold value is 1 for cholesky precision, 6. Independence measures based on RKHS covarience operator, 7. Computation of Euclidean Gradient, 8. Compute Approximate Hessian, 9. Check for convergence and find the final De-mixing matrix.

Support Vector Machine (SVM)

Support Vector Machine is originally a model for binary classification that performs classification tasks by constructing optimal separating hyperplanes in multidimensional space (webpage1). Although, the roots of the SVM approach were about to classify linearly separable data, but recently it is widely used in nonlinear data classification (Sivanandam et al., 2006). In this paper, we have used multiclass SVM to carry out recognition on face images, and Quadratic Programming (QP) optimization method to train the SVM.

Linear Classification

Consider a given training set $\{x_k, y_k\}_{k=1}^{N}$ with input data $x_k \in \Re^n$ and output data $y_k \in \Re$ with class labels $y_k \in \{+1, -1\}$, where x_k is the input and y_k is the output vector and finally the linear classifier can be shown as, $y(x) = \text{sign}[w^T x + b]$, where w is weight of the classifier and b is the threshold of classifier.

Nonlinear Classification

In the extension from linear SVM classifier to nonlinear SVM classifier, we can formally replace x by $\varphi(x)$ and apply the kernel trick (continuous function $k(x, z) = \varphi(x)^T \varphi(z)$). The primal problem in non-linear SVM cannot be solved in the same way as in linear SVM because the unknown w can be infinite dimensional (Suykans at al., 2002).

Examples of kernels that have been widely used for different classification tasks are polynomial and Gaussian.

Polynomial Kernel: The function for the polynomial kernel is given by the expression, $k(x_i, x_j) = (1 + x_i^T x_j)^p$, where p is the polynomial degree.

Gaussian Kernel: The expression for gaussian kernel is,

$$k(x_i, x_j) = \exp(-\|x_i - x_j\|^2 / 2\sigma^2), \text{ where } \sigma \text{ is the radius.}$$

Linear Kernel: The expression for linear kernel is, $k(x_i, x_j) = (1 + x_i^T x_j)$.

3. EXPERIMENT RESULTS AND DISCUSSIONS

The benchmark database OTCBVS (Object Tracking and Classification Beyond Visible Spectrum) contains both thermal and visual face images under the same situation like pose, illumination etc. The whole experiment consists of expression, illumination, and full datasets with 660 images in 20 classes, 748 images with 17 classes and 1626 images with 28 classes respectively. For the sake of experiment, the registered visual and thermal images are then cropped into 50×50 dimensions. The cropped visual and thermal images are fused using wavelet transformation, which is described previously. After that, all the fused images are taken for feature extraction using KICA. These images are taken for training and testing purposes from different classes of different datasets. From 660 face images, support vector machine multiclass classifier considers 20 persons as 20 classes taking 33 images from each class. The experiment has done using 'k-fold' cross validation where expression, illumination and full dataset are tested using '3-fold', and '4-fold' respectively. Number of folds is taken as per the total number of images in each class in each dataset. In K-fold cross-validation, K-1 folds are used for training and the last fold is used for evaluation. This process is repeated K times, leaving one different fold for evaluation each time. Linear kernel, Polynomial kernel with degree 3, Gaussian kernel with scaling factor 1 and penalty factor c with value 1 are used in our SVM kernel functions. The accuracy of a classification process defined as the portion of true positives, and true negatives in the population of all instances, classification accuracy A = (TP+TN)/(TP+TN+ FP+FN). TP=True Positives, TN=True Negatives, FP=False positives and FN=False Negatives. All these values have been taken from the confusion matrix acquired from the classifier performance. "True positives" are active compounds which have correctly been classified as active, and "false positives" are inactive compounds which have wrongly been classified as Active (Janecek et al., 2008). The maximum classification accuracies of three SVM kernels are listed below. It has been observed from the conducted experiment that linear kernel in both datasets, expression and illumination, have performed better. Linear kernel is performed 0.15% and 0.40% better than other two kernels in expression and illumination dataset respectively. Linear kernel and polynomial kernel both generate 99.08% correct rate in Full datasets. In Table 2, a comparative study has been shown based on different techniques used by different researchers.

Table 1 Experiment Results

Dataset	Recognition Rates of three SVM Kernels		
Expression	Linear	Polynomial	Gaussian RBF
	95.15%	95.00%	95.00%
Illumination	Linear	Polynomial	Gaussian RBF
	94.52%	94.12%	94.2%
Full Datasets	Linear	Polynomial	Gaussian RBF
	99.08%	4.80%	99.08%

Table 2 Comparison between different methods of Face Recognition

Methods		Classification accuracy
Present Methods	Linear	96.25%
	Polynomial	64.14%
	RBF	96.06%
Simple Spatial Fusion (Hanif and ali, 2006)		91.00%
Fusion of Thermal and Visual (Singh et al., 2004)		90.00%
Gabor based KICA using Polynomial kernel (Huang et al., 2010)		ORL-95% YALE-87.78%
Fusion of Visual and LWIR + PCA (Socolinsky and Selinger, 2004)		87.87%

4. CONCLUSION

This paper formulates Daubechies wavelet (db4) for fusion of optical and infrared image, KICA for feature extraction method and SVM for classification purpose in human face classification. We have used IRIS thermal/visible face database. The visual and thermal face images have their own advantages and disadvantages. Therefore, we have tried to combine the advantages from visual and thermal image. The experimental results show that linear kernel gives better classification accuracy than polynomial and RBF kernels. We can conclude that our approach has achieved efficient classification performance for face recognition purpose. We will further investigate different wavelet fusion methods and other classifiers for face recognition.

Acknowledgement

The research has been supported by the grant from DIT, MCIT, Govt. of India, Vide No. 12(2)/2011-ESD, dated 29/03/2011.

References

1. Bebis, G., Gyaourova, A., Singh, S. and Pavlidis, I., 2006: *Face Recognition by Fusing Thermal Infrared and Visible Imagery*, Image and Vision Computing, **24**, 727-742.

2. Belhumeur, P.N., Hespanha, J.P. and Kriegman, D.J., 1997: *Eigenfaces vs. Fisherfaces: recognition using class specific linear projection* IEEE Trans. Pattern Analysis and Machine Intelligence, **19**, 711–720.

3. Bhowmik, M. K., Bhattacharjee, D., Basu, D. K., Nasipuri, M., 2011: *Independent Component Analysis (ICA) of fused Wavelet Coefficients of thermal and visual images for human face recognition*, Proc. SPIE Defense, Security, and Sensing.

4. Bhowmik, M.K., Bhattacharjee, D., Nasipuri, M., Basu, D.K. and Kundu, M., 2010a: *Fusion of Daubechies Wavelet Coefficients for Human Face Recognition*, Int. J. Recent Trends in Engineering, **3**, 137-143.

5. Bhowmik, M.K., Bhattacharjee, D., Nasipuri, M., Basu, D.K. and Kundu, M., 2010b: *Fusion of Wavelet Coefficients from Visual and Thermal Face Images for Human Face Recognition – A Comparative Study*, Int. J. Image Processing, **4**, 12-23.

6. Bhowmik, M.K., Bhattacharjee, D., Nasipuri, M., Basu, D.K. and Kundu, M., 2010c: *Quotient Based Multiresolution Image Fusion of Thermal and Visual Images Using Daubechies Wavelet Transform for Human Face Recognition*, Int. J. Computer Science Issue, **7**, 18-27.

7. Cao, J.H., Ran, Y.Z. and Xu, Z.J. 2010: *A New Kernel Function Based Face Recognition Algorithm* Proc. Int. Conf. Computer, Mechatronics, Control and Electronic Engineering, 226 – 230.

8. Chellappa, R., Wilson, C.L. and Sirohey, S., 1995: *Human and Machine Recognition of Faces: A Survey*, Proc. IEEE, **83**, 705-740.

9. Hanif, M. and Ali, U., 2006: *Optimized visual and thermal image fusion for efficient face recognition,* Proc. IEEE Conf. on Information Fusion, 1-6.

10. Huang, Y., Li, M., Lin, C. and Tian, L., 2010: *Gabor-based Kernel Independent Component Analysis for Face Recognition* Proc. Int. Conf. Intelligent Information Hiding and Multimedia Signal Processing, 376-379.

11. Janecek, A.G.K., Gansterer, W.N., Demel, M.A. and Ecker, G.F., 2008: *On the Relationship Between Feature Selection and Classification Accuracy* Proc. JMLR Workshop and Conference Proceedings, **4**, 90-105.

12. Martiriggiano, T., Leo, M., D''Orazio, T. and Distante, A., 2005: Face *Recognition by Kernel Independent Component Analysis* In Innovations in Applied Artificial Intelligence, Ali, M. and Esposito (eds), Vol. 3533, pp. 55-58, Springer Berlin Heidelberg.

13. Maruthi, R. and Sankarasubramanian, K., 2007: *Multi focus image fusion based on the information level in the region of the images*, J. theoretical and applied information technology, **3**, 80-85.

14. Nikolov, S., Hill, P., Bull, D. and Canagarajah, N., 2001: *Wavelet for image fusion* In Wavelets in Signal and Image Analysis: from Theory to Practice, Petrosian, A.A. and Meyer, F.G. (eds), Vol. 19, Springer.

15. Shu-long, Z., 2004: *Image Fusion using Wavelet Transform*, Symp. Geospatial Theory, Processing and Applications, 5-9.

16. Singh, S., Gyaourva, A., Bebis, G. and Pavlidis, I., 2004: *Infrared and Visible Image Fusion for Face Recognition* Proc. *SPIE*, **5404**, 585-596.

17. Sivanandam, S.N., Sumathi, S. and Deepa, S.N., 2006: *Introduction to Neural Networks using Matlab 6.0* Tata McGraw-Hill Pub. Co.: New Delhi.

18. Socolinsky, D.A. and Selinger, A., 2004: *Thermal Face Recognition in an Operational Scenario* Proc. IEEE Computer Society Conf. Computer Vision and Pattern Recognition, **2**, 1012-1019.

19. Suykans, J.A.K., Gestel, T.V., Brabanter, J.D., Moor, B.D. and Vandewalle, J., 2002 *Least Squares Support Vector Machines* World Scientific Pub. Co.: Singapore.

20. Turk, M.A. and Pentland, A., 1991: *Eigenfaces for recognition* J. Cognitive Neuroscience, **3**, 71–86.

21. Webpage: http://www.fingertec.com/download/tips/whitepaper-02.pdf

22. Webpage1: http://www.statsoft.com/textbook/support-vector-machines.

A Fuzzy Approach Based on Borda Count to Select the Consensual Alternative for Customer Relationship Management

Th. Biren Singh[1] and Ksh. Mangijaobi Devi[2*]
[1]*Oriental College, Imphal*
[2]*Department of Mathematics, Manipur University*
Canchipur, Manipur
E-mail: mangijaobiksh@gmail.com[*]

ABSTRACT

The selection of products or services or facility locations is one of the important decision making issues for logistic managers. Customers often commonly take relevant attributes in account for selection of products or services or facility locations. Further, attribute assessment of a product or service or facility location is often presented by a sequence of linguistic data sequences and the linguistic data sequences constitute a linguistic data matrix .To access the linguistic data matrix of customers' assessment on a product or service or facility location a proper decision making method is essential and proposed in this paper. In the decision making method the entries of the data matrix are represented by LR- fuzzy numbers and a fuzzy Borda-type group decision making method is developed and the selection can be made through fuzzy preference relations. The procedure representing the selection through the preference relations of the customers will be the foundation of developing customer relationship management (CRM).

Keywords: Linguistic data sequence, LR-fuzzy numbers, Customer relationship management, Borda count.

1. INTRODUCTION

In real world, for the selection of an indicated facility location, customers often have to assess related attributes of the facility location. The related attributes are so many and their important degrees (Wang, 2010), such as Very low (VL), Low (L), Between low and medium (BLM), Medium (M), Between medium and high (BMH), High (H), Very high (VH) are varied for different customers. These linguistic terms can be transformed into LR-fuzzy numbers as shown in

Table 3.1. Further, attributes ratings will stand for customers' preferences that are presented with linguistic labels (Delgado et al., 1993; Garcia-Lapresta et al., 2001) such as, prefers in definite degree(PDD), prefers in high degree(HD), prefers in low degree(LD), indifference(ID), as given in Table 3.2. Thus, a customer's assessment on several attributes can be shown with linguistic data matrices. To discover essential decision, linguistic data matrices of customers' preferences have to be processed.

In the information procedure, one of the most important techniques is group decision making technique. For a given site location, group decision making methods can process linguistic data matrices of customers' assessments to obtain most preferred decisions. The optimal decision represents preferences of different customers in the site location. Managers will develop related customer relationship management (CRM) and provide necessary assists for different customers to their preferences. Thus, CRM can be constructed on a group decision making method.

There were a variety of methods in group decision making (see e.g., Bellman and Zadeh, 1970; Bezdek et al., 1979; Biswas, 1994; Dummett, 1998). Among the methods, in this paper we begin by focusing our attention on the Borda count. Orginally, the Borda count method was proposed by Jean Charles de Borda in 1970 even though McLean and Urken (McLean and Urken, 1995a, b) indicated that Nicolaus Cusanus (XVth C) foresaid the idea of the Borda count.

The Borda count rule has been re-considered many times, inspiring other group decision making procedures. The generalizations of the Borda count by using fuzzy preferences can be found in (Garcia-Lapresta et al., 2001; Marchant, 2000). Most of the previous Borda count methods were utilized in crisp values, no matter how the scale of values to score the alternatives belong to large sample or small sample, long term or short term. In (Zadeh, 1975) we find that individuals tend to express their preferences in linguistic manner rather than in precise numerical values. Considering this view, in this paper we also extend the Borda count by utilising linguistic labels for the customers to compare the alternatives. The linguistic labels assigned by the individual customers are represented with the triangular fuzzy numbers of LR-type. The profile of the individual preferences is constructed by using the concepts of signed distances of the triangular fuzzy numbers of LR-type from the zero triangular fuzzy number of LR-type.

The paper is organised as follows. In Section 2, we consider the classic Borda count. The basic notations, definitions: triangular fuzzy number of LR-type, linguistic variable, fuzzy preference relations are given in Section 3. In Section 4, a Borda-type procedure based on the linguistic labels for selection of facility location is proposed. An example is solved to illustrate the efficiency of the proposed method for CRM, in Section 5.

2. THE BORDA COUNT

The Borda count was designed assuming the customers' (agents') preferences to be linear ordering. The marks assigned to the alternatives ought to be arranged, from the best to the worst, in the sequential manner. However, this may appear in agents' preferences, and then, the previous scheme could not be used, although the original idea can be extended in several ways to be applied in this situation (Gärdenfors, 1973).

Our proposal is one of these possibilities that generalized the original Borda count in a natural way, assigning to each alternative the number of alternatives worse than that to be scored. This can be formalized as follows.

Suppose $A_1, A_2, ..., A_m$ are m agents and $x_1, x_2, ..., x_n$ are n alternatives. Let p^1, $p^2, ..., p^m$ be the preference relations (asymmetric binary relations) of m agents over n alternatives $x_1, x_2, ..., x_n$. Each agent, k gives a mark to each alternative, according to the number of alternatives worse than it: $r^k(x_i) = \#\{x_j \mid x_i\, p^k x_j\}$ is the score given by agent k to the alternative x_i, (Garcia- Lapresta et al., 2001). The same result is obtained by considering, for each agent k, its preference matrix:

$$
\begin{pmatrix}
r^k_{11} & r^k_{12} & \cdots & r^k_{1n} \\
r^k_{21} & r^k_{22} & \cdots & r^k_{2n} \\
\vdots & \vdots & \cdots & \vdots \\
r^k_{n1} & r^k_{n2} & \cdots & r^k_{nn}
\end{pmatrix}
\tag{2.1}
$$

where
$$
r^k_{ij} = \begin{cases} 1, & \text{if } x_i\, p^k x_j \\ 0, & \text{otherwise} \end{cases}
$$

In this way, the agent k gives the alternative x_i the following mark:

$$
r^k(x_i) = \sum_{x_i p^k x_j} r^k_{ij}
\tag{2.2}
$$

We noted that the possible score range is contained in the set of values $\{0, 1, ..., n - 1\}$. It is also noted that some of the upper values might not be reached if there is indifference (absence of preference) among distinct alternatives. However, if the orderings were linear, the above-mentioned set would exactly be ranged.

With these individual counts a collective one is obtained:

$$
r(x_i) = \sum_{k=1}^{m} r^k(x_i)
\tag{2.3}
$$

So, the collective preference relation p^B is defined by:

$$x_i p^B x_j \Rightarrow r(x_i) > r(x_j),$$

which is negatively transitive. Thus, the highest scored alternatives (maximal for p^B) will be chosen.

Nevertheless, it is reasonable for the Borda count to require the fulfilment of the following property of monotonicity: when two alternatives are compared by an agent, the highest scored must be the preferred one.

Definition 2.1: The count r^k is monotonic if and only if $x_i p^B x_j \Rightarrow r^k(x_i) > r^k(x_j)$ for all pairs of alternatives $x_i, x_j \in X$.

Proposition 2.1: If the preference relation p^k is transitive, then the count r^k is monotonic.

Proof: Suppose $x_i p^k x_j$. Being p^k is transitive, if $x_j p^k x_p$, then $x_i p^k x_p$ and consequently, $\{x_p \mid x_j p^k x_p\} \subset \{x_p \mid x_i p^k x_p\}$. The inclusion is strict, because x_j belongs to the second set not to the first one. Finally, $r^k(x_i) = \#\{x_p \mid x_i p^k x_p\}$ $= \#\{x_p \mid x_i p^k x_p\}$.

3. LR- FUZZY NUMBERS, LINGUISTIC VARIABLES AND PREFERENCE RELATIONS

In the following, we briefly review some basic definitions of fuzzy sets theory from (Garcia-Lapresta et al., 2001; Tanino, 1984; Wang, 2010; Zadeh, 1965; Zimmermann, 1991). The basic definitions and notations below will be used throughout the paper unless otherwise stated.

Definition 3.1: Let U be a universal set. A fuzzy set $\tilde{A}$ of U is defined with a membership function $\mu_{\tilde{A}}(x) \rightarrow [0, 1]$, where $\mu_{\tilde{A}}(x)$, $\forall x \in U$ indicates the degree of membership of x in $\tilde{A}$.

Definition 3.2: A fuzzy set $\tilde{A}$ is said to be normal iff $\underset{x \in U}{\text{Sup}} \, \mu_{\tilde{A}}(x) = 1$ where U is the universal set.

Definition 3.3: A fuzzy set $\tilde{A}$ of the universal set U is said to be convex if $\mu_{\tilde{A}}(\lambda x + (1 - \lambda)) y \geq (\mu_{\tilde{A}}(x) \wedge \mu_{\tilde{A}}(y))$, $\forall x, y \in U, \lambda \in [0, 1]$, where $\wedge$ denotes the minimum operator.

Definition 3.4: A fuzzy set $\tilde{A}$ is a fuzzy number if it is both convex and normal on U.

Definition 3.5: For a given triplet of real numbers (m, p, q) such that $p > 0$, $q > 0$, the LR- fuzzy number $\tilde{M}$ of the universal set E (of real numbers) associated with (m, p, q) is defined by its membership function

$$\mu_{\tilde{M}} : E \to [0, 1] : \mu_{\tilde{M}}(x) = \begin{cases} L\left(\dfrac{m - x}{p}\right), & \text{for } x \leq m \\ R\left(\dfrac{x - m}{q}\right), & \text{for } x \geq m \end{cases},$$

where m is called the core value of $\tilde{M}$, it is denoted as $\mathrm{cor}(\tilde{M})$, p and q are called the left and right spreads respectively. Symbolically $\tilde{M}$ is denoted by $\tilde{M} = (m, p, q)_{LR}$.

Definition 3.6: An LR- fuzzy number, $\tilde{M} = (m, p, q)$ is called a triangular fuzzy number of LR-Type if reference functions $L(x)$ (and $R(x)$) are linear functions and the membership function is defined as:

$$\mu_{\tilde{M}}(x) = \begin{cases} L(x) = \dfrac{x - (m - p)}{p}, & \text{for } m - p \leq x \leq m \\ 1, & \text{for } x = m \\ R(x) = \dfrac{m + q - x}{q}, & \text{for } m \leq x \leq m + q \\ 0, & \text{otherwise} \end{cases}$$

It is denoted by $\tilde{M} = (m, p, q)_{LR}$. It is noted that when $p = 0$, $q = 0$ then $\tilde{M}$ becomes a crisp number m, i.e., $\tilde{M} = (m, 0, 0)_{LR} = m$. When $p = 0$ and $q = 0$ then the reference functions L(and R) are respectively 0.

Let F_S be a family of triangular fuzzy numbers of LR- type on E, then following (Dubois and Prade, 1980; Zimmermann, 1991), we define the following definitions and results.

Definition 3.7: For each $\alpha \in [0, 1]$, the α-level fuzzy interval of $\tilde{M} \in F_S$, is defined as $\tilde{M}(\alpha) = \left\{ x \mid \mu_{\tilde{M}} \geq \alpha \right\} = [\underline{M}(\alpha), \overline{M}(\alpha)]$, which is unique and $\underline{M}(\alpha) = m - p + p\alpha$ and $\overline{M}(\alpha) = m + q - q\alpha$ are respectively left-point, right point of fuzzy interval $\tilde{M}(\alpha) : \tilde{M} = \bigcup_{0 \leq \alpha \leq 1} \alpha \tilde{M}(\alpha)$. The integrations of $\underline{M}(\alpha)$ and $\overline{M}(\alpha)$ on the common range $[0, 1]$ exist.

Definition 3.8: The signed distance of $\tilde{M}$ from zero triangular fuzzy number of LR-type, $\tilde{O} = (0, 0, 0)_{LR}$, is defined as

$$d(\tilde{M}, \tilde{O}) = \frac{1}{2} \int_0^1 [\underline{M}(\alpha) + \overline{M}(\alpha)] \, d\alpha.$$

Definition 3.9: (Dubois and Prade, 1980). Suppose $\tilde{M}, \tilde{N} \in F_S$ with $\tilde{M} = (m, p, q)_{LR}$, $\tilde{N} = (n, p', q')_{LR}$. Then we define

(i) $\tilde{M} \oplus \tilde{N} = (m, p, q)_{LR} \oplus (n, p', q')_{LR} = (m + n, p + p', q + q')_{LR}$;

(ii) $-(m, p, q)_{LR} = (-m, p, q)_{LR}$;

(iii) $(m, p, q)_{LR} \ominus (n, p', q')_{LR} = (m - n, p + p', q + q')_{LR}$

Definition 3.10: For $\tilde{M} = (m, p, q)_{LR}$ and $\tilde{N} = (n, p', q')_{LR}$ belong to F_S, the following ordering is defined:

$$\tilde{M} \prec \tilde{N} \Rightarrow \begin{cases} d(\tilde{M}, \tilde{O}) < d(\tilde{N}, \tilde{O}) \\ \text{or} \\ d(\tilde{M}, \tilde{O}) = d(\tilde{N}, \tilde{O}) \text{ and } m + q < n + q' \\ \text{or} \\ d(\tilde{M}, \tilde{O}) = d(\tilde{N}, \tilde{O}), \\ m + p = n + q' \text{ and } m - p < n + q' \end{cases}$$

If $p = q$ and $p' = q'$ then

$$d(\tilde{M}, \tilde{O}) < d(\tilde{N}, \tilde{O}) \Rightarrow \begin{cases} m < n \\ \text{or} \\ m = n \text{ and } m + q < n + q' \end{cases}$$

Definition 3.11: Let $\tilde{R}$ be a fuzzy binary relation on F_S, where F_S is a set of triangular fuzzy numbers of LR-type. The following conditions may hold for $\tilde{R}$. Denoting $\mu_{\tilde{R}}(\tilde{x}, \tilde{y})$, $\forall \tilde{x}, \tilde{y} \in F_S$, we have

(i) $\tilde{R}$ is reflexive if $r(\tilde{x}, \tilde{x}) = 1 \ \forall \ \tilde{x} \in F_S$

(ii) $\tilde{R}$ is symmetric if $r(\tilde{x}, \tilde{y}) = r(\tilde{y}, \tilde{x}) \ \forall \ \tilde{x}, \tilde{y} \in F_S$

(iii) $\tilde{R}$ is transitive if $r(\tilde{x}, \tilde{z}) \geq V_{\tilde{y}} \in F_S \ (r(\tilde{x}, \tilde{y}) \wedge r(\tilde{y}, \tilde{z})) \ \forall \tilde{x}, \tilde{y}, \tilde{z} \in F_S$

where V and Λ denote the maximum operator and minimum operator respectively. If $\tilde{R}$ is reflexive, symmetric and transitive, $\tilde{R}$ is an equivalence relation on F_S.

Definition 3.12: An $n \times n$ fuzzy relation matrix $\tilde{A} = [r_a(\tilde{x}_i, \tilde{x}_j)]_{n \times n}$ is a matrix where $0 \leq r_a(\tilde{x}_i, \tilde{x}_j) \leq 1$ indicates the relation between two fuzzy numbers $\tilde{x}_i$ and $\tilde{x}_j$ and $1 \leq i \leq n$, $1 \leq j \leq n$. It is reflexive if $r_a(\tilde{x}_i, \tilde{x}_i) = 1$ for $0 \leq i \leq n$. Matrix $\tilde{A} = [r_a(\tilde{x}_i, \tilde{x}_j)]_{n \times n}$ is symmetric if $r_a(\tilde{x}_i, \tilde{x}_j) = r_a(\tilde{x}_j, \tilde{x}_i)$. The matrix $\tilde{A}$ is called max-min transitive if $r_a(\tilde{x}_i, \tilde{x}_k) \geq \min (r_a(\tilde{x}_i, \tilde{x}_j), r_a(\tilde{x}_j, \tilde{x}_k))$

Definition 3.13: A linguistic variable is a variable whose values are linguistic terms (Zadeh, 1977). Let $X = \{x_1, x_2, ..., x_n\}$ be a set of n alternatives.

Suppose that there are m customers who have to show their preferences over each pair of alternatives $x_i, x_j \in X$, $1 \leq i \leq n$, $1 \leq j \leq n$.

Let $L = \{l_0, l_1, ..., l_s\}$ be a set of linguistic variables (labels), ranked by a linear order: $l_0 < l_1 <, ..., l_s$. We consider that individuals use these labels for declaring their preferences over the pairs of X.

Suppose that there is an intermediate label which represents indifference and the rest of labels are around it in a symmetrical way, so that the number of labels, $s + 1$ will be odd and, consequently, $l_{s/2}$ is the central label.

Definition 3.14: A fuzzy binary relation $\tilde{R}$ defined by $\mu_{\tilde{R}} : X \times X \to L$ is a linguistic preference relation over X based on L, with $\mu_{\tilde{R}}(\tilde{x}, \tilde{y}) = r_{ij}$ satisfying $r_{ij} = l_h \to r_{ji} = l_{s-h}$, for every $i, j \in \{1, 2, ..., n\}$ and for every $h \in \{0, 1, ..., s\}$.

In which follows, r_{ij} will denote the label of preference of x_i over x_j, according to the following scheme:

1. $r_{ij} = l_s$, if x_i is definitely preferred to x_j,
2. $l_{s/2} < r_{ij} < l_s$, if x_i is slightly preferred to x_j,
3. $r_{ij} = l_{s/2}$, if x_i is indifferent to x_j, i.e., if x_i is not preferred to x_j and x_j is not preferred to x_i.
4. $l_0 < r_{ij} < l_{s/2}$, if x_j is slightly preferred to x_i,
5. $r_{ij} = l_0$, if x_j is definitely preferred to x_i.

In the above scheme, we can use [0, 1] instead of L(Bezdek et al.,1979, Kacprzyk, 1986).

Definition 3.15: For each customer $k \in \{1, 2, ..., m\}$, the linguistic preference relation over X based on L, $\tilde{R}^k$, induces an ordinary relation over X defined by:

$x_i > x_j \Leftrightarrow r_{ij}^k > l_{s/2}$, for all $i, j \in \{1, 2, ..., n\}$. Thus, $\tilde{R}^k$ is an $n \times n$ matrix whose entries belong to L.

Note that the relation $\underset{k}{>}$ is asymmetric over X and the indifference relation associated with $\underset{k}{>}$ is reflexive and symmetric, i.e., $r_{ij}^k = l_{s/2}$, and $r_{ij}^k = l_{s/2} \to r_{ji}^k = l_{s/2}$ for every $i, j \in \{1, 2, ..., n\}$.

Definition 3.16: A linguistic preference relation $\tilde{R}^k$ over X based on L is weak max-max transitive (Tanino, 1984) if and only if it is defined by:

$$\left(x_i \underset{k}{>} x_j \text{ and } x_j \underset{k}{>} x_p \right) \to r_{ip}^k \geq \max\{r_{ij}^k, r_{jp}^k\} \text{ for every } i, j, p \in \{1, 2, ..., \}.$$

We denote the set of weak max-max transitive linguistic preference relations over X based on L, by $\tilde{T}_L(X)$. Following (Garcia-Lapresta et al., 2001; Tanino, 1984), in the paper we use the linguistic terms and labels, as given in Table 3.1 and Table 3.2 respectively.

Table 3.1 The linguistic terms and corresponding to triangular fuzzy numbers of LR-type

Linguistic terms employed for expressing importance	Triangular fuzzy number of LR-type
Very low (VL)	(0.0, 0.0, 0.2)
Low (L)	(0.2, 0.2, 0.2)
Between low and medium (BLM)	(0.4, 0.2, 0.1)
Medium (M)	(0.5, 0.1, 0.1)
Between medium and high (BMH)	(0.6, 0.1, 0.2)
High (H)	(0.8, 0.2, 0.2)
Very high (HV)	(1.0, 0.2, 0.0)

Table 3.2 The linguistic labels and corresponding to LR-fuzzy numbers

Agent	Label		Triangular fuzzy number of LR-type
$r_{ij}^k = l_6$	x_i prefers to x_j in definite degree		(1.0, 0.0, 0.0)
$r_{ij}^k = l_5$	x_i prefers to x_j in high degree		(0.9, 0.1, 0.1)
$r_{ij}^k = l_4$	x_i prefers to x_j in low degree		(0.7, 0.2, 0.2)
$r_{ij}^k = l_3$	x_i prefers to x_j indifferently		(0.5, 0.2, 0.2)
$r_{ij}^k = l_2$	x_i prefers to x_j in low degree		(0.3, 0.2, 0.2)
$r_{ij}^k = l_1$	x_i prefers to x_j in high degree		(0.1, 0.1, 0.1)
$r_{ij}^k = l_0$	x_i prefers to x_j in definite degree		(0.0, 0.0, 0.0)

The graphical representation of linguistic labels in triangular fuzzy numbers of LR – type of Table 3.2 is given in Fig. 3.1.

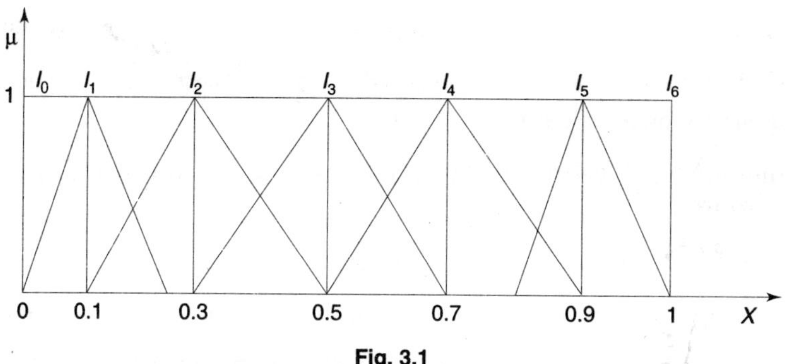

Fig. 3.1

4. A BORDA COUNT BASED ON LINGUISTIC LABELS

We shall define a new count for each customer (agent) to evaluate each alternative according to intensities of preferences in pair wise comparisons where the intensities of preferences are derived from the linguistic terms(labels) assigned to the alternatives by the agent

In the classic Borda count, the values assigned by an individual count is the number of alternatives considered worse than it for an ordinary preference relation p^k. Now, analogously, this role will be played by the ordinary relation $\underset{k}{>}$ induced by the linguistic preference relation $\tilde{R}^k$ over X based on L.

Let n different alternatives x_1, x_2, ..., x_n be assessed by m customers, C_1, C_2, ..., C_m in terms of linguistic terms. It is assumed that the number of alternatives is equal to the number of available linguistic terms.

In what follows, we shall suppose that each linguistic term is related to a triangular fuzzy number of LR-type (see Table 3.1) and then calculate the signed-distances of the fuzzy numbers from the Zero triangular fuzzy number of LR-type. From the matrix of signed-distances, we construct the individual preference relation $\tilde{R}^k$ based on L and then applied the Borda count. The process can be established as shown in the following.

Suppose that D is an $m \times n$ data matrix corresponding to m customers and n alternatives, where each entry of D is a linguistic term assigned to an alternative by a customer.

By changing linguistic terms of D into triangular fuzzy numbers of LR-type (Table 3.1) and then calculating the signed-distances, we construct the matrix,

$$\overline{D} = (d_{ij})_{m \times n},$$

where d_{ij} is the signed-distance corresponding to a linguistic term in D.

From $\overline{D}$, we are to evaluate the individual fuzzy preference, $\tilde{R}^k$ based on L as discussed in the following.

Suppose $C_k = (d_{k1}, d_{k2}, ..., d_{kn})$ is the kth row of $\overline{D}$, then individual preference relation, $\tilde{R}^k : C_k \times C_k \to L$ is constructed as $\tilde{R}^k = (l_s)_{n \times n}$, where l_s' are calculated by using the following algorithm.

Algorithm 4.1:   We calculate $\tilde{R}^k (d_{kj}, d_{ku})$, $j = 1, 2, ..., n$; $u = 1, 2, ..., n$, as shown below:

1.   Set $\tilde{R}^k (d_{kj}, d_{ku}) = l_{s/2}$, for $d_{kj} - d_{ku} = 0$.

2.   For $d_{kj} < d_{ku}$, calculate $|d_{kj} - d_{ku}| = h$ (say) and $(cor\,(l_{s/2}) - h) = h'$ (say). If $\mu_{l_{s'}} (h') = \max_{s'} \left(\mu_{l_{s'}}(h') \right)$, $s' = 0, 1, .., s/2 - 1$ then set $\tilde{R}^k (d_{kj}, d_{ku}) = l_{s'}$.

3.   For $d_{kj} > d_{ku}$, calculate $|d_{kj} - d_{ku}| = h$ (say) and $cor\,(l_{s/2}) + h = h'$ (say). If $\mu_{l_{s'}}(h') = \max_{s'} \left(\mu_{l_{s'}}(h') \right)$, $s'' = s/2 + 1, ..., s$ then set $\tilde{R}^k (d_{kj}, d_{ku}) = l_{s'}$.

In this way we can calculate k, ($k = 1, 2, ..., m$) individual preference relations, $\tilde{R}^k$. Note that the notation r_{ij}^k, l_s and triangular fuzzy number of LR-type can be interchangeable according to the context as given in Table 3.2.

Next, the introduced individual count will add up the preference intensities among a fixed alternative and those considered worse than it according to $\underset{k}{>}$. Taking into account individual counts, agent k gives the alternative x_i the value

$$r_k(x_i) = \begin{cases} \underset{x_i \underset{k}{>} x_j}{\Sigma}\ r_{ij}^k, & \text{if } \{j \in \{1, 2, ..., n\} \mid r_{ij}^k > l_{s/2}\} \neq \Phi \\ l_0, & \text{otherwise} \end{cases}$$

With these individual counts a collective one is obtained:

$$r(x_i) = \sum_{k=1}^{m} r^k(x_i) \tag{4.1}$$

The collective preference relation is defined by

$$x_i P^{LB} x_j \rightarrow r(x_i) > r(x_j),$$

which is negatively transitive. In this way the highest scored alternatives (maximal for) will be chosen.

We have to point out that the classic Borda count gives individual scores belonging to the set $\{0, 1, ..., n - 1\}$ or, belonging to the interval $[0, 1]$, while the fuzzy Borda count takes its values in the set of triangular fuzzy numbers of LR- type belonging to F_S. In this linguistic framework, we also consider a monotonicity property. As in section 2, when two alternatives are compared by an agent, the highest assessment must be given to the preferred alternative according to.

Definition 4.1: (Blin, 1974). The count r^k is monotonic if and only if $x_i \underset{k}{>} x_j \rightarrow r^k(x_i) > r^k(x_j)$, for all pairs of alternatives $x_i, x_j \in X$.

Definition 4.2: Let $\tilde{R}^k$ be a linguistic preference relation over X based on L. Then $\tilde{R}^k$ is said to be linear if and only if it is satisfied $x_i \neq x_j \Rightarrow r_{ij}^k \neq l_{s/2}$ for all $x_i, x_j \in X$.

Proposition 4.1. If $\tilde{R}^k \in \tilde{T}_L(X)$ is linear, then for all, $x_i, x_j, x_l \in X$ it is satisfied

(i) $r_{il}^k > r_{ij}^k > l_{s/2} \Rightarrow r^k(x_i) - r^k(x_l) > r^k(x_i) - r^k(x_j)$

(ii) $r_{il}^k > r_{jl}^k > l_{s/2} \Rightarrow r^k(x_i) - r^k(x_l) > r_k(x_j) - r_k(x_l)$

Proof:

(i) Here, we will see $r_{jl}^k > l_{s/2}$.

If $r_{jl}^k < l_{s/2}$ i.e. $r_{jl}^k > l_{s/2}$, then we would have $r_{jl}^k \geq \max\left\{ r_{il}^k, r_{lj}^k \right\}$, hence $r_{ij}^k \neq r_{il}^k$

a contradiction to the hypothesis. On the other hand, if $r_{jl}^k = l_{s/2}$, then

$x_j = x_l$, in contradiction with the hypothesis. We have $r^k(x_j) > r^k(x_l)$ and

consequently, $r^k(x_i) - r^k(x_l) > r^k(x_i) - r^k(x_j)$.

(ii) Analogously, the result can be proved.

Theorem 4.1: **Suppose $\widetilde{R}^k$ is a linguistic preference relation derived from the signed-distance matrix, $\overline{D}$. Then $\widetilde{R}^k$ is linear iff each linguistic labels of the kth row of the data matrix, D are all different from each other.**

Proof: $\widetilde{R}^k$ is linear, implies $l_{s/2}$ is unique element in each row of $\widetilde{R}^k$, implies the kth row of $\overline{D}$ is linear, that is, the entries of the kth row of $\overline{D}$ are all distinct. This implies that the linguistic labels of the kth row of D are all distinct.

Conversely, suppose the linguistic labels of the kth row of D are all distinct. $\Rightarrow$ the kth row of $\overline{D}$ has distinct elements. Thus, there is only one $l_{s/2}$ in each row of $\widetilde{R}^k$.

In the following section we investigate an empirical group decision making problem for customer relationship management based on classical Borda count and the proposed fuzzy model.

5. EMPERICAL EXAMPLE

A manager of a business establishment desires to select a suitable location for a new investment. The selection process is done based on the evaluation of the alternatives by a group of fifteen customer related agents (customers), C_1, C_2, ..., C_{15}. After preliminary screening, seven alternatives x_1, x_2, x_3, x_4, x_5, x_6, x_7 remain for further evaluation. In the evaluation process, the manager allows the customers to use seven linguistic terms:

(1) very low (VL),

(2) low (L),

(3) between low and medium (BLM),

(4) medium (M),

(5) between medium and high (BMH),

(6) high (H),

(7) very high (VH),

in such a way that no agent is allowed to assign equal linguistic labels among the alternatives.

The ratings of the seven alternatives according to the individual assessment of the customers are given in the form of data matrix, D:

Alternatives/ Customers	x_1	x_2	x_3	x_4	x_5	x_6	x_7
C_1	VL	L	BLM	M	BMH	H	VH
C_2	VL	BLM	L	BMH	VH	M	H
C_3	VL	M	L	BLM	H	VH	BMH
C_4	VL	M	BMH	L	BLM	VH	H
C_5	VL	L	BLM	H	VH	M	BMH
C_6	L	BLM	VL	BHM	M	H	VH
C_7	L	VL	BLM	H	M	VH	BHM
C_8	L	M	VL	BLM	BMH	H	VH
C_9	L	VL	M	H	BLM	VH	BMH
C_{10}	L	BLM	VL	H	VH	M	BMH
C_{11}	BLM	VL	L	VH	BMH	M	H
C_{12}	BLM	L	VL	H	VH	M	BMH
C_{13}	BLM	VL	M	L	H	VH	BMH
C_{14}	BLM	M	VL	L	VH	H	BMH
C_{15}	BLM	VL	L	H	VH	M	BMH

$$(5.1)$$

Replacing the linguistic terms of data matrix (5.1) by the appropriate triangular fuzzy numbers of LR-type from Table 3.2 we calculate the signed distance of each fuzzy number of D (e.g., by Definition 3.8, d (BLM) = d(0.4, 0.2, 0.1) = 1/4 (4 0.4 + 0.1 – 0.2) = 0.375). Then the matrix of signed-distances, $\overline{D}$ is constructed as given below:

$$
\begin{bmatrix}
0.05 & 0.2 & 0.375 & 0.5 & 0.625 & 0.8 & 0.95 \\
0.05 & 0.375 & 0.2 & 0.625 & 0.95 & 0.5 & 0.8 \\
0.05 & 0.5 & 0.2 & 0.375 & 0.8 & 0.95 & 0.625 \\
0.05 & 0.5 & 0.625 & 0.2 & 0.375 & 0.95 & 0.8 \\
0.05 & 0.2 & 0.375 & 0.8 & 0.95 & 0.5 & 0.625 \\
0.2 & 0.375 & 0.05 & 0.625 & 0.5 & 0.8 & 0.95 \\
0.2 & 0.05 & 0.375 & 0.8 & 0.5 & 0.95 & 0.625 \\
0.2 & 0.5 & 0.005 & 0.375 & 0.625 & 0.8 & 0.95 \\
0.2 & 0.05 & 0.5 & 0.8 & 0.375 & 0.95 & 0.625 \\
0.2 & 0.375 & 0.05 & 0.8 & 0.95 & 0.5 & 0.625 \\
0.375 & 0.05 & 0.2 & 0.95 & 0.625 & 0.5 & 0.8 \\
0.375 & 0.2 & 0.058 & 0.8 & 0.95 & 0.5 & 0.625 \\
0.375 & 0.05 & 0.5 & 0.2 & 0.8 & 0.95 & 0.625 \\
0.375 & 0.5 & 0.05 & 0.2 & 0.95 & 0.8 & 0.625 \\
0.375 & 0.05 & 0.2 & 0.8 & 0.95 & 0.5 & 0.625
\end{bmatrix}
\quad (5.2)
$$

Using the algorithm 4.1 we calculate the individual preference relation $\tilde{R}^k$ ($k = 1, 2, ..., 15$) corresponding to each row of the matrix $\overline{D}$, (e.g. in $\tilde{R}^1$, $\tilde{R}^1$ $(0.05, 0.05) = l_3$, $\tilde{R}^1$ $(0.05, 0.2) = l_2$, for $|0.05 - 0.2| = 0.15$, $0.5 - 0.15 = 0.35$, $\mu_{l_s}(0.35) = \max (\mu_{l_{s'}}(0.35))$, $(s' = 0, 1, 2) \Rightarrow s' = 2$; $\tilde{R}^1$ $(0.625, 0.375) = l_4$, for $|0.625 - 0.375| = 0.25$, $0.5 + 0.25 = 0.75$ $\mu_{l_s}(0.75) = \max\limits_{s'} (\mu_{l_{s'}}(0.75))$, $(s'' = 4, 5, 6) \Rightarrow s'' = 4$). Thus, we obtained 15 fuzzy preference relations:

$$\tilde{R}^1 = \begin{vmatrix} l_3 & l_2 & l_1 & l_1 & l_0 & l_0 & l_0 \\ l_4 & l_3 & l_2 & l_2 & l_1 & l_0 & l_0 \\ l_5 & l_4 & l_3 & l_2 & l_2 & l_1 & l_0 \\ l_5 & l_4 & l_4 & l_3 & l_2 & l_2 & l_1 \\ l_6 & l_5 & l_4 & l_4 & l_3 & l_2 & l_1 \\ l_6 & l_6 & l_5 & l_4 & l_4 & l_3 & l_2 \\ l_6 & l_6 & l_6 & l_5 & l_5 & l_4 & l_3 \end{vmatrix}, \; ..., \; \tilde{R}^{15} = \begin{vmatrix} l_3 & l_5 & l_4 & l_1 & l_0 & l_2 & l_2 \\ l_1 & l_3 & l_2 & l_0 & l_0 & l_1 & l_0 \\ l_2 & l_4 & l_3 & l_0 & l_0 & l_2 & l_1 \\ l_5 & l_6 & l_6 & l_3 & l_2 & l_4 & l_4 \\ l_6 & l_6 & l_6 & l_4 & l_3 & l_5 & l_5 \\ l_4 & l_5 & l_4 & l_2 & l_1 & l_3 & l_2 \\ l_4 & l_6 & l_5 & l_2 & l_1 & l_4 & l_3 \end{vmatrix},$$

From the above individual preference matrices, using Eqn (4.1) we can obtain the following triangular fuzzy numbers of LR-type as follows:

$$r(x_1) = (12.4, 2.6, 2.6),$$

$$r(x_2) = (16.1, 3.5, 3.5),$$

$$r(x_3) = (14.8, 3.0, 3.0),$$

$$r(x_4) = (45.5, 6.1, 6.1),$$

$$r(x_5) = (59.3, 6.5, 6.5),$$

$$r(x_6) = (58.3, 7.1, 7.1),$$

$$r(x_7) = (59.1, 7.2, 7.2).$$

From above, we obtain $r(x_5) > r(x_7) > r(x_6) > r(x_4) > r(x_2) > r(x_3) > r(x_1)$.

Thus, the ranking of the alternatives is obtained as

$$x_5 > x_7 > x_6 > x_4 > x_2 > x_3 > x_1.$$

6. CONCLUSION

We have considered a Borda-type decision procedure based on linguistic labels for customer relationship management. In our approach the linguistic preference modelling of the alternatives is derived from the linguistic labels through the signed-distances of the triangular fuzzy numbers of LR-type representing the linguistic labels of the alternatives. The linguistic preferences, so obtained, allow linearity as well as weak max-max transitivity. We assumed that the alternatives were all different and no agent assigned equal linguistic labels to the alternatives. An example has been shown. Nevertheless, other linguistic labels can be used, depending to the nature and characteristics of the alternatives and the agents, in each concrete decision problem.

The advantage of the proposed approach is its flexibility of graduation allowing the agents to express their preferences more faithfully than in the classic Borda count, through different linguistic labels. The probability of ties appearance in total scores of the Borda count based on linguistic labels is lowered than in the classic Borda count.

References

1. Bellman R.E, Zadeh, L.A., 1970: *Decision making in a fuzzy environment;* Management Science; **17(4)**, 141-1645.

2. Bezdek J.C., Spillman B., Spillman R., 1979: *Fuzzy relation spaces for group decision theory : An application*; Fuzzy Sets and Systems; **4**, 5-14.

3. Biswas T., 1994: *Efficiency and consistency in group decisions*; Public Choice; **80**, 23-34.

4. Blin, J.M., 1974: *Fuzzy relations in group decision theory*; J. of Cybernetics; **4(2)**, 12-22.

5. Delgado, M., Verdegay, J.L., Vila, M.A., 1993: *Linguistic decision making models;* International Journal of Intelligent Systems; **7**, 479-492.

6. Dubois, D., Prade, H., 1980: *Fuzzy Sets and Systems: Theory and Applications;* New York, London, Toronto; 55-58.

7. Dummett M., 1998: *The Borda count and agenda manipulation;* Social Choice and Welfare; **15**, 289-296.

8. Garcia-Lapresta J.L., Martinez-Panero, M., Lazzari, L.L., 2001: *A Group Decision Making Method Using Fuzzy Triangular Numbers;* Fuzzy Sets in Management; Economics and Marketing; 35-50.

9. Gärdenfors, P., 1973: *Positionalist voting functions*; Theory and Decision; **4**, 1-24.

10. Kacprzyk J., 1986, *Group decision making with a fuzzy linguistic majority*; Fuzzy Sets and Systems; **11**, 105-118.

11. Marchant T., 2000, *Does the Borda rule provide more than a ranking?;* Social Choice and Welfare; **17**, 381-391.

12. McLean, I. and Urken, A.B., (eds); 1995 a: *Classics of Social Choice*; Ann Arbor – The University of Michigan Press.

13. McLean, I. and Urken, A.B., 1995 b: *Independence of irrelevant alternatives before Arrow*; Mathematical Social Sciences; **30**, 107-126.

14. Tanino, T., 1984: *Fuzzy preference orderings in group decision making*; Fuzzy Sets and systems; **12**, 117-131.

15. Wang, Y.J., 2010: *A Clustering method based on fuzzy equivalence relation for customer relationship management*; Expert Systems with Applications; 6421-6428.

16. Zadeh L.A., 1965: *Fuzzy sets*; Information and Control; **8(3)**, 338-353.

17. Zadeh L.A., 1975: *The concept of a linguistic variable and its application to approximate reasoning*; Information Sciences. Part I: **8**, 199-249. Part II: **8**, 301-357. Part III: **9**, 43-80.

18. Zadeh, L.A., 1977: *Linguistic characterization of preference relations as a G basis for choice in social systems*; Erkenntnis; **11(3)**, 383-410.

19. Zimmermann, H.J., 1991: *Fuzzy SetsTheory And Its Applications*; Kluwer Academic Publish, New York.

Rough Sets, Fuzzy Sets and Soft Computing
Editor: S. Bhattacharya Halder

Uncertainty Quantification in Radiological Risk Assessment based on P-Box Approach: A Case Study in Inhalation of I-131

Tazid Ali[*] and Hrishikesh Boruah[#]

*Department of Mathematics, Dibrugarh University, Dibrugarh, Assam
E-mail: tazidali@yahoo.com[*]; h.boruah07@gmail.com[#]*

ABSTRACT

Radiological Risk Assessment comprehensively explains methods used for estimating risk to people exposed to radioactive materials released to the environment by nuclear facilities. Radiological Risk Assessment attempts to quantify the risks to human health. Probabilistic risk assessment studies use probability distribution for one or more variables of the risk equation in order to quantitatively characterize the uncertainty. However, there are situations where one cannot specify parameter values for input distribution or/and precise probability shape is/are not known. In such situation, an advanced technique called the probability box (p-box) can be applied. In this paper we have made a case study of inhalation risk due to I-131 using p-box approach. We have further compared the results with Double Monte Carlo (DMC) approach for uncertainty propagation.

Keywords: Double Monte-Carlo, p-box, Radiological Risk Assessment, Uncertainty.

1. INTRODUCTION

Uncertainty analysis is a systematic study in which "a neighborhood of alternative assumptions is selected and the corresponding interval of inferences is identified" [3]. Uncertainty plays a critical role in the analysis for a wide and diverse set in various fields. Ideals and concepts of uncertainty have long been associated with gambling and games. The Greek in the 4th century BC were the first recorded civilization to have considered uncertainty. There are two kinds of uncertainty. One kind arises as variability (or aleatory uncertainty) resulting from inherent variability, natural stochasticity, environmental or structural variation across space or through time, manufacturing or genetic heterogeneity among components or individuals, and variety of other sources of randomness. It is also called randomness, stochastic uncertainty, objective uncertainty, dissonance, or irreducible uncertainty. The standard representation of variability is the probability distribution function. Another kind is uncertainty (or epistemic uncertainty). This is defined as uncertainty which arises from incompleteness of knowledge about the world. Sources of epistemic uncertainty include measurement uncertainty, small sample size, detection limits and data censoring, ignorance about the details of the physical mechanisms and processes involved and other imperfection in scientific understanding.

These two kinds of uncertainty can propagate through various mathematical expressions with different calculation method. Probability Bounds Analysis (PBA)[2, 6] is related to one of these methods. It is a combination of Probability theory and Interval Analysis. Probability theory is used to propagate aleatory uncertainty (or variability) and Interval Analysis is used to propagate epistemic uncertainty (or uncertainty). Probabilistic approaches characterize the uncertainty in the parameter by a probability distribution. There uncertainty consists purely of variability. We don't know which particular value will come up in a random selection, but we know how likely each value is. In Interval approach uncertainty is represented with an interval having lower bounds and upper bounds. Probability bounds analysis allows one to obtain fully rigorous results even when the empirical information is very poor. The idea of bounding probability has a very long tradition in probability theory. The uncertainty is characterized by a probability distribution, or by a p-box in the case of the probability bounds approach. P-box arises from probability distribution when there is uncertainty about the values of the defining parameters (mean, standard deviation etc.).

Radioactive materials can flow accidently or intentionally. Accidental releases occur many times at commercial nuclear power plants and nuclear waste disposal sites. When radioactive materials are released in to the environment, radio nuclides will be moved into the body by inhalation and ingestion, which cause internal exposure.

In the case study, we take an inhalation risk model and evaluate the inhalation risk due to dispersion of Iodine-131 from pressurized Iodine-131 tank. The variables of the risk model are described in terms of probability distribution and the parameters of these variables are in some intervals. The case involves pressurized Iodine storage, consisting of a pressurized tank and a pipe line. The objective is to obtain probability bounds for the uncertainty propagation. We first evaluate the mass discharge rate of Iodine-131, taking fixed atmospheric pressure and from fixed pressurized Iodine tank. After that, we evaluate the concentration as well as radiological risk for the radionuclide Iodine-131 for different Stability class of weather category, viz., B, D, E, and F. From the results of the radiological risk we may conclude that radiological risk is high for the stability class B and low for the stability class F.

Stability class of weather category (A, B, C, D, E, F) is temperature dependent and it is defined with respect to the temperature gradient i.e. rate of change of temperature with respect to the displacement. The stability class A is basically very unstable and F is extremely stable. Weather category D is neutral (calm condition). F appears at night time whereas A appears in the day time. C is slightly unstable, E is moderately stable.

2. PROBABILITY BOX (P-BOX)

A probability box or "p-box "consists of a pair of such functions that are used to circumscribe an imprecisely known distribution function F. Suppose $\overline{F}$ and $\underline{F}$ are nondecreasing functions from the real line R into [0, 1] and $\underline{F}(x) \le \overline{F}(x)$ for all $x \in R$. Let $[\overline{F}, \underline{F}]$ denote the set of all nondecreasing functions F from real into [0, 1] such that $\underline{F}(x) \le F(x) \le \overline{F}(x)$. When the function $\overline{F}$ and $\underline{F}$ circumscribe an imprecisely known probability distribution, we call $[\overline{F}, \underline{F}]$, specified by the pair of functions, a "Probability box" or "p-box" for that distribution. This means that, if $[\overline{F}, \underline{F}]$ is a p-box for a random variable X whose distribution F is unknown except that it is within the p-box, then $\underline{F}(x)$ is a lower bound on $F(x)$ which is the (imprecisely known) probability that the random variable X is smaller than x. Likewise, $\overline{F}(x)$ is an upper bound on the same probability. In the probability box $[\overline{F}, \underline{F}]$, the gap between $\overline{F}$ and $\underline{F}$ reflects the incomplete nature of the knowledge.

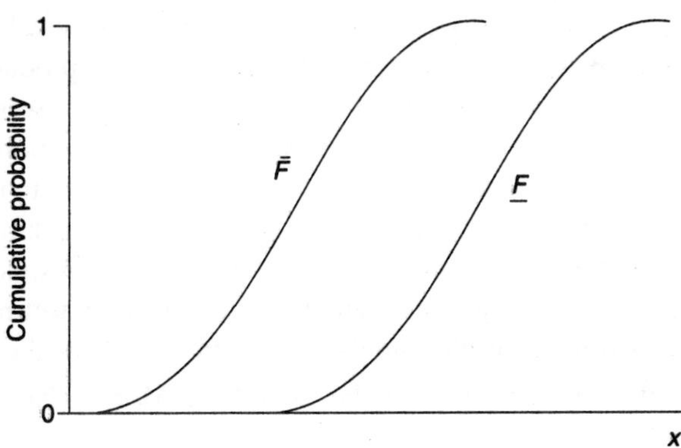

Fig. 1 A p-box with upper and lower bounds

The idea of p-boxes is that the output p-box will contain all possible output distributions that could result from the input distributions, assuming the distributions of the random quantities actually lie in their respective p-boxes. We can use a p-box to circumscribe the uncertainty about a probability distribution in the same way that an ordinary interval is used to circumscribe uncertainty about a scalar number. If the probability distribution is completely specified, these bounds are coincident. A *p*-box represents both incertitude and variability in the same time.

Probability boxes can be used wherever a probability distribution can be used. *p*-boxes can be freely combined with scalars, interval and probability distribution in mathematical expression. Advantages of *p*-box are that there is a method for combining *p*-boxes for different random quantities without assuming anything about the dependence between random quantities and they are useful tools for risk and sensitivity analysis.

3. DOUBLE MONTE CARLO METHOD (DMC)

Monte Carlo simulation (MCS) is one of several techniques currently employed to carry out risk assessments. In the 1940s MCS, originated as a random sampling technique for solving difficult deterministic equations. An overview of the history of Monte Carlo Simulation can be found in [5]. Since then Monte Carlo methods have continued to evolve and due to advances in computing they can now be used in many applications. Two-dimensional Monte Carlo simulation (2D MCS) is used in a wide range of applications including human health risk assessment, avian risk assessment, environmental flood risk assessment, and microbial risk assessment.

Two-dimensional Monte Carlo simulation (2D MCS) or Two-phase Monte Carlo simulation or DMC is an extension of Monte Carlo simulation. In 2D MCS there are two loops (as opposed to just one in MCS) allowing variability and uncertainty to be modeled separately. The variability is modeled in the inner loop and the uncertainty in the outer loop. 2D MCS can produce bounds on the output of a particular model at any credible level and it takes into account parameter uncertainty for each random quantity in the model. Some advantages of Monte Carlo methods are include the availability of software to implement the procedure and that it can be used as a sensitivity analysis by making adjustments to the model and then comparing the results from each adjustment to see the effect of changes. Model uncertainty can also be included by setting up different models and comparing or enveloping the results from each of them.

The procedure for carrying out the two-phase Monte Carlo is explained in [4].

4. A CASE STUDY OF INHALATION RISK

In this section we discuss a case study of risk due to inhalation of I-131. The framework of the problem is adopted from [1].

Risk Model: Risk equation for Inhalation is

$$R_{inh} = C \times BR \times CF \times RF \tag{1}$$

where R_{inh} = Risk due to Inhalation (per year)

C = Air concentration (Bq-s/m^3)

BR = Breathing rate (m^3/year)

CF = Conversion factor (sv/Bq)

RF = Risk factor (per Sv)

The case involves pressurized Iodine storage, consisting of a pressurized tank and a pipe line. The assessment will consider the release of pressurized Iodine to the surrounding, and the assessment end-point is concentration of Iodine at a geographical location (x, y) [x is horizontal and y is vertical line of the tank]. Two different physical models [1] are used, i.e. a discharge model for estimating the mass of discharge and a dispersion model for estimating the concentration of Iodine at a grid point.

The average concentration of Iodine-131 at grid point $(x, y, 0)$, following a discharge is calculated using equation (2) and (3) below. In the grid point x is the downwind distance from the discharge to the geographical point of interest; and y is the cross-wind distance from the centre of the gas plume. Here x is set to 300 meters and y is set to 0 meter. Here we consider function of stability class B, D, E and F. Breathing rate is taken as a triangular distribution (8760 m^3/yr, 10512 m^3/yr, 13140 m^3/yr). Risk factor and conversion factor is taken as 2.43E-12 per Sv and 2.20E-08(sv/Bq) respectively.(as per FGR13)

The average concentration of Iodine-131 at grid point (x, y) following a discharge is given by the following dispersion model. [Here we consider U for uniform distribution, N for normal distribution, and T for triangular distribution.]

$$C = \frac{Q}{\pi \, \sigma_y \sigma_z \, W[N([\mu_1, \mu_2], [\sigma_1, \sigma_2])]} \tag{2}$$

where C = Concentration of Iodine-131

Q = Mass discharge /release rate [Kg/s]

σ_1 and σ_2 = Dispersion coefficient (function of stability class B, D, E and F)[m]

W = Wind speed [m/s]

From the above equation it is seen that C, the ground level ($z = 0$) concentration of a pollutant at a point (x, y) downwind from a source is proportional to the rate of emission (Q). Further, it is inversely proportional to the wind speed (W) and to the parameters $\sigma_y(m)$ and $\sigma_z(m)$.

The equation used for mass discharge rate is given by

$$Q = C_d \, [U(a, b)]A \, [T(l, m, u)] \sqrt{\frac{2(p_0 - p_a)}{v_f}} \tag{3}$$

where C_d = Discharge coefficient

A = Area of the hole [m^2]

p_0 = Pressure in the tank [N/m^2] = 5×10^5

p_a = Atmospheric pressure [N/m^2] = 1×10^5

v_f = Specific volume of liquid [m/kg] = 1/617 = 0.00162075

Wind speed discharge coefficient and hole area are considered as uncertain variables. In Table 1, the distributions used for uncertainty inputs are given. Computer code for triangular distribution in Matlab software. Inhalation risk have been obtained and shown in figure below for different stability classes.

Table 1 Distributions of uncertain inputs

Parameter	Probability distribution
Wind speed, *W*, (stability class *B*)	Normal ([3.9, 4.1], [0.3, 0.4])
Wind speed, *W*, (stability class *D*)	Normal ([4.5, 5.5], [1.0, 1.3])
Wind speed, *W*, (stability class *E*)	Normal ([3.9, 4.1], [0.3, 0.4])
Wind speed, *W*, (stability class *F*)	Uniform (1, 2)
Hole area, *A*	Triangular (0.0012, .0018, .0025)
Discharge coefficient, C_d	Uniform (0.7, 0.9)

Briggs' formula is used to determine σ_y and σ_z.

Table 2 shows the Briggs' formula for different stability classes and Table 3 shows the results for x = 300 meters for *B*, *D*, *E* and *F* classes.

Table 2 $\sigma_y(m)$ and $\sigma_z(m)$ for different stability classes

Pasquill Type	$\sigma_y(m)$	$\sigma_z(m)$
A	$\dfrac{0.22x}{\sqrt{1 + 0.0001x}}$	$0.20x$
B	$\dfrac{0.16x}{\sqrt{1 + 0.001x}}$	$0.12x$
C	$\dfrac{0.11x}{\sqrt{1 + 0.0001x}}$	$\dfrac{0.080x}{\sqrt{1 + 0.0002x}}$
D	$\dfrac{0.08x}{\sqrt{1 + 0.0001x}}$	$\dfrac{0.060x}{\sqrt{1 + 0.0015x}}$
E	$\dfrac{0.06x}{\sqrt{1 + 0.0001x}}$	$\dfrac{0.030x}{1 + 0.0003x}$
F	$\dfrac{0.04x}{\sqrt{1 + 0.0001x}}$	$\dfrac{0.016x}{1 + 0.0003x}$

Table 3 $\sigma_y(m)$ and $\sigma_z(m)$ for different stability classes for $x = 300$ meters.

Pasquill Type	$\sigma_y(m)$	$\sigma_z(m)$
B	47.29581	36
D	23.64791	14.94819
E	17.73593	8.620437
F	11.82395	4.597566

5. RESULTS AND DISCUSSION

Inhalation risk obtained for different stability classes are shown Figs. 2 to 5. Table 1 gives the information for uncertain inputs of the risk model. Table 3 gives the value of $\sigma_y(m)$ and $\sigma_z(m)$ for different stability classes (i.e. for B, D, E and F) for $x = 300$ meters. Taking the parameters of wind speed for stability class B, D, E and F (in Table 1) as uniform distribution, we evaluate the inhalation risk using Double Monte Carlo method. Here, we have developed a computer code for triangular distribution in Matlab software. Figure 2(a) shows the inhalation risk for stability class B using p-box method and the average wide of the risk is **2.0818e + 015.** In Fig. 2(b) inhalation risk for stability class B is shown by using Double Monte Carlo method and the average wide is **4.5284e + 007.** Similarly Fig. 3(a), 4(a) and 5(a) gives the inhalation risk for stability class D, E and F using p-box method and the average wide of the risk are **1.5219e + 015, 1.8694e + 014** and **2.0067e + 010** respectively. Also Fig. 3(b), 4(b) and 5(b) gives the inhalation risk for stability class D, E and F using Double Monte Carlo method and the average wide of the risk are **5.9657e + 008, 5.0519e + 008** and **2.0019e + 010** respectively. From the above results we may conclude that the average wide of different stability class from B to F using p-box method is decreasing. No such trend is seen when calculation is performed using Double Monte Carlo method. However the average width which are obtained using Double Monte Carlo method is smaller, compare to the wide of p-box in each stability class. So we can conclude that in p-box approach uncertainty is more as compared to Double Monte Carlo method.

Acknowledgement

The authors would like to express their gratitude to the funding Agency Board of Research in Nuclear Sciences, Department of Atomic Energy, Govt. of India, for providing financial support under a Research Project.

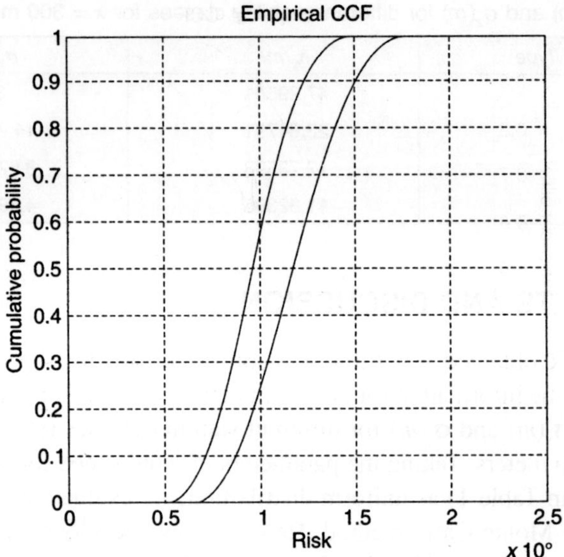

Fig. 2(a) Inhalation risk for stability class *B* (using *p*-box method)

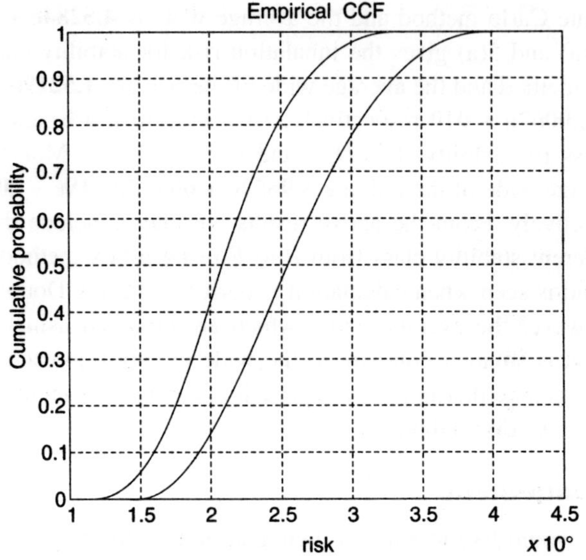

Fig. 2(b) Inhalation risk for stability class *B* (using DMC method)

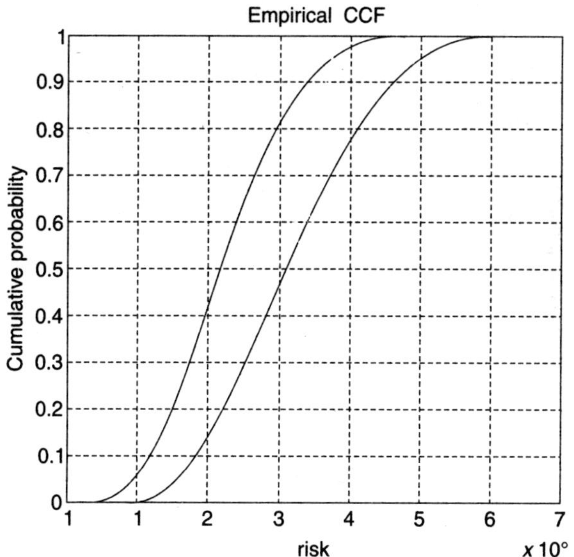

Fig. 3(a) Inhalation risk for stability class *D* (using *p*-box method)

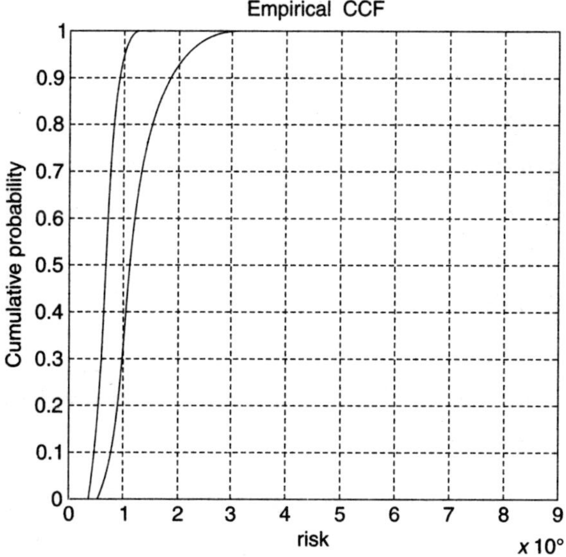

Fig. 3(b) Inhalation risk for stability class *D* (using DMC method)

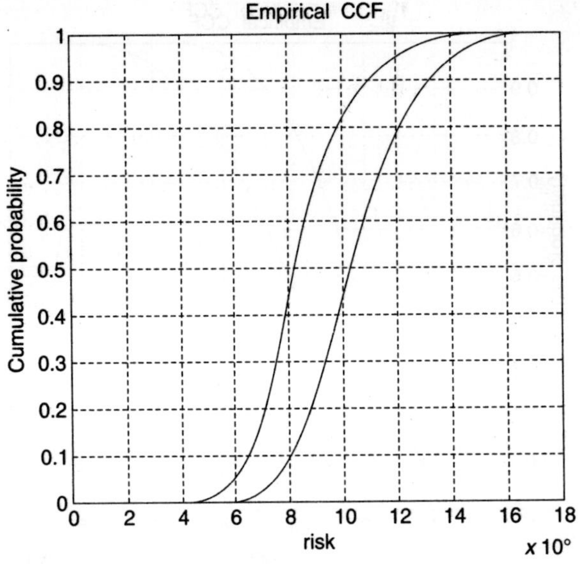

Fig. 4(a)   Inhalation risk for stability class *E* (using *p*-box method)

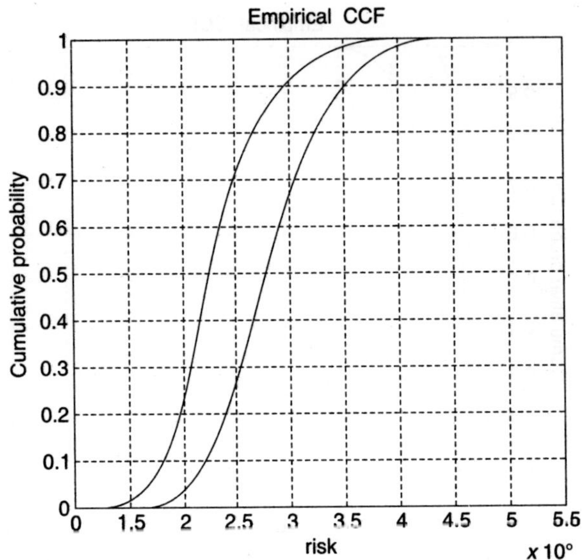

Fig. 4(b)   Inhalation risk for stability class *E* (using DMC method)

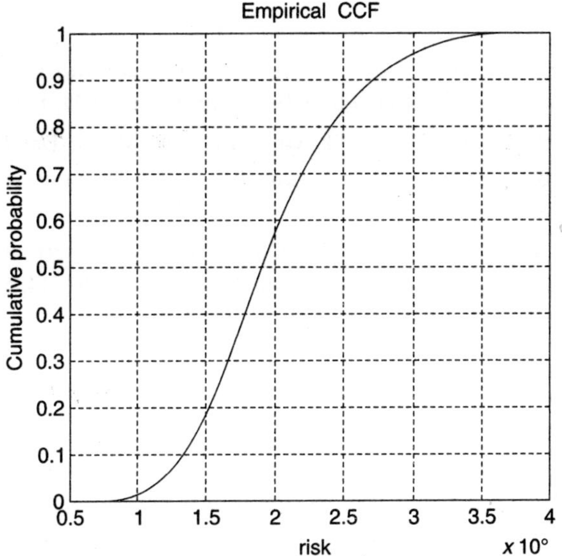

Fig. 5(a) Inhalation risk for stability class *F* (using *p*-box method)

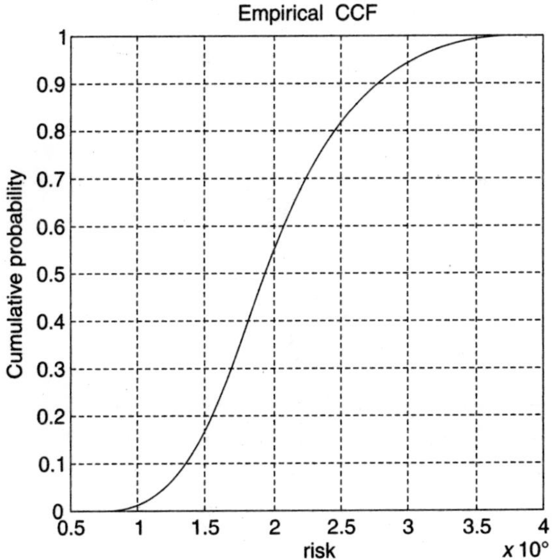

Fig. 5(b) Inhalation risk for stability class *F* (using DMC method)

References

1. Abrahamsson, M., 2002: *Uncertainty in Quantitative Risk Analysis-Characterization and Methods of Treatment*; Report 1024, Lunds University.

2. Ferson, S., Kreinovich, Vladik, Ginzburg, Lev, Myers, Davis, S. and Sentz, Kari, 2003: *Constructing Probability Boxes and Dempster-Shafer Structures*; SAND2002-4015.

3. Ferson, S., and Tucker, W.T., 2006: *Sensitivity analysis using probability bounding*; Reliability Engineering and System Safety; 91, 1435-1442.

4. Rao, K.D., Kushwaha, H.S., Verma, A.K., and Ajit, S., 2009: *Uncertainty Analysis Based on Probability Bounds (P-Box) Approach in Probabilistic Safety Assessment*; Risk Analysis, **29**(5), 662-675.

5. Rugen, P., Callahan, B., 1996: *An overview of Monte Carlo: A fifty years perspective*; Human Heath and Ecological Risk assessment; **2**(4), 671-680.

6. Tucker, W.T., and Ferson, S., 2003: *Probability bounds analysis in environmental risk assessments*; Applied Biomathematics Setauket; New York 11733.

Rough Sets, Fuzzy Sets and Soft Computing
Editor: S. Bhattacharya Halder

Continuity in RNA Space

Chandra Kanta Phukan[*] **and Tazid Ali**[#]
Department of Mathematics, Dibrugarh University
Dibrugarh, Assam
E-mail: chandra_kanta25@yahoo.com[*]*; tazidali@yahoo.com*[#]

ABSTRACT

The notion of proximity evolved in the space of RNA phenotypes help to clarify the biological processes underlying the evolutionary change of living organism. We know that phenotype space (P) is one of the best fields to investigate continuous and discontinuous change in evolutionary biology. The genotype space (G) is determined by physical processes, namely mutation and recombination, acting on genes of all living organisms. On the other hand phenotypes are the physical realization of the process. Fontana, Stadler, Stadler, and Wagner argued that *continuous* changes in the RNA sequences can lead to apparent jumps at the phenotypic level, so that evolution is not solely influenced by natural selection and genetic drift, but is also directed by internal dynamics. From an abstract point of view, an evolutionary trajectory is a function from time axis R into the phenotype space (here RNA space) with the topology implied by the accessibility relationships between RNA shapes. In this paper we have structured the RNA space as a fuzzy pre-topological space and thereby introduced a notion of continuity of evolutionary transition. Finally we have discussed a necessary and sufficient condition for an evolutionary map to be continuous.

Keywords: Genotype, Phenotype, RNA, Fuzzy Pre-topology, Continuity

1. INTRODUCTION

Genotype and phenotype are two aspects of the same molecule, genotype is internally coded, inheritable information possessed by all living organisms, while the phenotype is the physical realization of that information. For example, the collection of genes responsible for eye colour in a particular individual is a genotype. The observable eye coloration in the individual is the corresponding phenotype. The idea of phenotype demonstrated on the ribonucleic acid (RNA) is ideally suited to explore the concepts of evolution in a wide context and it is also simple in computational matter. There is a unique RNA shape which is associated to each genotype sequence; as a result of this we get the canonical map from the set of genotype to phenotype ($f : G \rightarrow P$). This canonical map in not one-to-one since, multiple genotype sequences can result in the same RNA shape. To distinguish continuous from discontinuous evolutionary change, a relation of nearness between phenotypes is needed. Such a relation is based on the probability as well as possibility of one phenotype being accessible from another through changes in the genotype. Various aspects are applied to explore the accessibility notion on the set of RNA sequences. Fontana and Schuster(1998) and Cupal et al.(2000) equipped the phenotypes with topology based on probability of mutation. Stadler et al.(2001) suggested that the set of phenotypes P with a pretopology defined in terms of the GP-map $f : G \rightarrow P$ can describe the evolutionary mechanism. They further argue that "continuous" changes in the RNA sequences can lead to apparent "jumps" at the phenotypic level, so that evolution is not solely influenced by natural selection and genetic drift, but is also directed by internal dynamics. In order to analyze the "jumps" in the shapes of RNA sequences, the authors equip the genotype set with a *pretopology*. A probabilistic version of pretopology was also suggested by Stadler and Stadler (2006). Mynard and Seal (2010) improved the model by considering probabilistic atopology. They also proposed an alternative model which is intrinsically non-topological.

In order to answer the question whether RNA evolution is continuous or not, we must be precise regarding the following points:

In what sense are we using the term continuity?

How this continuity be characterized by the notion of topology?

What topology is most natural in the phenotype space as well as the time axis.

Fontana and Schuster (1998) have argued that accessibility is not a yes/no attribute but rather a continuous likelihood. It is turned into a binary attribute by choosing a likelihood threshold below which one phenotype is regarded as not accessible from another. This leads to a range of (pre)topologies with various degrees of continuity/discontinuity. Moreover in Ali ((2011),it is argued that the uncertainty involved in the process of mutation/transition is not purely probabilistic. It is a graded concept. So more than probability theory it is fuzzy theory that is appropriate in this situation (Ali and Phukan, 2011).

2. FUZZY SET AND FUZZY TOPOLOGY

In this section we discuss some preliminaries on fuzzy sets (Klir and Yuan, 2003) and fuzzy topology(Palaniappan) that will be required in the sequel. Further we propose some new notions.

Fuzzy (sub)sets are generalization of classical (sub)sets. In a classical subset an element of the universal set either belongs to the subset or not which can be identified by the corresponding characteristic function $\chi_A : X \to \{0, 1\}$ such that $\chi_A(x) = 1$, if $x \in A$ and $\chi_A(x) = 0$, if $x \notin A$. Here the boundary of a fuzzy subset is not precisely defined and so an element of the universal set belongs to the fuzzy subset with some level of membership which can be identified by the membership function μ_A. In short fuzzy set expresses the concept of graded membership.

Mathematically a fuzzy set A of a universal set X is a function

$$\mu_A : X \to [0, 1]$$

For each $x \in X$, $\mu_A(x)$ is called the membership grade of x in A. For convenience the fuzzy subset as well as the corresponding membership function is represented by μ.

Given two fuzzy sets A and B, their standard intersection $A \cap B$, standard union $A \cup B$ and standard complement A^c are defined for all $x \in X$ by the equations

$$\mu_{A \cap B}(x) = \min [\mu_A (x), \mu_B(x)]$$

$$\mu_{A \cap B}(x) = \max [\mu_A(x), \mu_B(x)]$$

$$\mu_A{}^c(x) = 1 - \mu_A(x)$$

For infinite collection of fuzzy subsets, min and max are respectively replaced by *infimum* and *supremum*.

A fuzzy set A is said to be contained in a fuzzy set B, denoted by $A \subseteq B$, if

$$\mu_A(x) \leq \mu_B(x) \ \forall \ x \in X.$$

Definition 2.1: A fuzzy pretopology on a set X is described by an application Cl of I^X into I^X, which verifies:

$$P1 : Cl (\phi) = \phi,$$

$$P2 : Cl (A) \supset A \text{ for every } A \in I^X,$$

(X, Cl) is then said to be a fuzzy pretopological space and Cl is called fuzzy closure (or adherence) (Badard, 1981).

Definition 2.2: Neighbourhood systems: Let X be a arbitrary set. For each $x \in X$ a collection of fuzzy neighbourhood systems $\mathcal{N}(x)$ on X is a function $\mathcal{N} : X \to I^X$ (where I^X is the set of all fuzzy subsets of X) such that for all $x \in X$

(N1) $N(x) \geq 0$ for all $N \in \mathcal{N}(x)$.

(N2) $N_1, N_2 \in \mathcal{N}(x)$ implies $N_1 \cap N_2 \in \mathcal{N}(x)$.

(N3) $N_1 \in \mathcal{N}(x)$ and $N_1 \subset N$ implies $N \in \mathcal{N}(x)$.

The Ns are fuzzy neighbourhoods, $\mathcal{N}(x)$ is a fuzzy neighbourhood system for element x, and the pair $(X, \mathcal{N})$ is called a pretopological space. We speak of a neighbourhood basis, if only (N1) and a weakened version of (N2) i.e., (N2') $N_1, N_2 \in \mathcal{N}(x)$ implies that there is a $N_3 \in \mathcal{N}(x)$ such that $N_3 \subseteq N_1 \cap N_2$ are satisfied.

Definition 2.3: A family of fuzzy preneighbourhoods at the point $x \in X$ is a family $\mathcal{B}(x)$ of fuzzy subsets μ_V which verify $\mu_V(x) = 1$ (Badard,1981).

Definition 2.4: Let φ be an application of I^X in to $[0, 1]$, φ is said to be a degree of non-vacuity (it associates to every fuzzy subset a number which represents the fact that it is more or less void) if it verifies:

(i) $\varphi(\phi) = 0$,

(ii) $\varphi(A) = 1$ if there exists x such that $\mu_A(x) = 1$,

(iii) $A \supset B$ implies $\varphi(A) \supset \varphi(B)$.

In particular $\varphi(A) = sup_{x \in X} \mu_A(x)$ is the degree of non-vacuity.

If we define a fuzzy subset $\overline{\mu}$ of X such that $\overline{\mu}(x) = inf_{v \in B(x)} \varphi(v \wedge \mu)$, then we get a *fuzzy closure* operator on X. This gives an connection between preneighbourhoods and fuzzy closure operator (Badard, 1981).

In the fuzzy pretopological setting, we propose the following:

Definition 2.5: Let $g : (X, N) \rightarrow (Y, M)$ be an arbitrary function between two fuzzy pretopological space. Then g is said to be continuous in x if for every fuzzy preneighbourhood M of $g(x)$ there is a fuzzy preneighourhood N of x such that $g(N) \subseteq M$.

Definition 2.6: (Fuzzy pretopology on real line): We equip R with a fuzzy pretopology which is given by a system of neighbourhood basis $N(x)$ consisting of all triangular fuzzy sets with core x such that for any two members μ_1, μ_2, either $\mu_1 \subseteq \mu_2$ or $\mu_2 \subseteq \mu_1$ and supp. $\mu_1 \neq$ supp. μ_2.

3. PHENOTYPIC NEARNESS WITH FUZZY PRETOPOLOGY

Different types of accessibility notion such as accessibility topology, shadow pre-topology, probabilistic atopology etc. are discussed in (Stadler et al., 2001; Mynard and Seal, 2010) to equip the phenotype space. As accessibility of phenotypes is a graded concept so in this paper we equip the phenotype set with a fuzzy pre-topology induced by the system of preneighbourhood defined for each shape of the phenotype space. If we fixed a phenotype (shape), say, α, the accessibility of each phenotype to the shape α, leads to a meaningful notion of fuzzy preneighbourhood. Keeping this aspect in mind, we consider for each phenotype $\alpha \in P$ as a fuzzy subset of P with membership functions define as,

$$v_\alpha : P \to [0, 1] \tag{1}$$

where $v_\alpha(\beta)$ is the accessibility of β from α.

We assume that any shape is accessible from itself with a grade of one. Then the fuzzy subset v_α corresponding to the shape α can be regarded as a fuzzy pre-neighbourhood of α. Then the preneighbourhood system $\mathbf{B}(\alpha)$ generates a pretopology on the set P through the closure operator

$$Cl(\mu)\,(x) = inf_{\,\gamma \in\, \mathbf{B}(\alpha)}\,\varphi(\gamma \wedge \mu) \tag{2}$$

where, μ is an arbitrary fuzzy subset on P and $\mathbf{B}(\alpha)$ is the set of all pre-neighbourhoods of α. We should note the fact that for each $\alpha \in P$, $\mathbf{B}(\alpha)$ consist of only one member, viz, the fuzzy set of Equation 1.

4. EVOLUTIONARY TRAJECTORY AND CONTINUITY

An evolutionary trajectory is a map ϕ from the time axis into phenotype space. Here as the map is uniquely defined, the continuity of the map is solely based on the neighbourhood structure of time axis and the phenotype set. As we already have a fuzzy preneighbourhood as well as pretopological structure on the phenotype space, the question that arises is what neighbourhood structure should be considered in the time axis T. The time axis is usually described as well-known topological space, namely the real line with its standard (usual) topology. In the case of computer simulations and samples from the fossil record, which intrinsically represent time in the form of discrete steps, it is more natural to use the pretopology corresponding to the directed infinite path graph. In Stadler et al.(2001) time axis T is taken to be discrete (the set of natural numbers) equipped with natural pretopology.

Time flows continuously, however mutation/transformation takes place at discrete level. Suppose at instant t_0 the state of a shape is α. The shape changes to β at instant t_1, i.e., in the period $[t_0, t_1)$ the shape does not change. Again at instant t_2 the shape changes to β. Thus transition is taking place at discrete level while time is flowing continuously. So we prefer to take the time axis as positive real line $[0, \infty)$. Further as P is equipped with fuzzy pretopology we feel it is natural to equip $[0, \infty)$ with the fuzzy pretopology as defined in Section 2. It may be noted that by replacing $[0, \infty)$ with the real line (with fuzzy pretopology) will not affect the continuity of the evolutionary map. So we can just consider the evolutionary trajectory as a map $\phi : R \to P$ where R and P have the relevant fuzzy pretopologies. However as all our practical studies will always involve finite time period we can simply restrict ourselves to some finite interval say, $[a, b]$.

Cupal et a.(2000) defined trajectory as

Definition 4.1: Let V be a finite topological space and let $[u , v] \subset R$ be a finite interval, $u = t_0 < t_1 < \ldots < t_m < t_{m+1} = v$, $m \geq 0$. A trajectory is a function $\phi : [u, v] \to V$ with the following properties:

(i) ϕ is constant on the intervals $I_0 = [u , t_1)$, $I_m = (t_m , v]$, and $I_k = (t_k , t_{k+1})$ for $1 \leq k < m$.

(ii) $\phi(I_k) \neq \phi(I_{k-1})$ for $1 \leq k \leq m$.

(iii) $\phi(t_k)$ equals $\phi(I_k)$ or $\phi(I_{k-1})$ for $1 \leq k \leq m$.

Thus a trajectory follows a sequence

$$\{x_i \in V, 0 \leq I \leq m\} \text{ where } \phi(I_k) = x_k \text{ for } 0 \leq i \leq m.$$

A trajectory can be decomposed into parts that have only one intermediate point, which is called a *transition* (Cupal et al.(2000).

Definition 4.2: Let V be a finite topological space, $x \neq y \in V$, $t_0 < t_2 \in R$. A transition from x to y is a function $\theta : [t_0 , t_2] \to \{x , y\}$ (with the induced topology) such that $\theta([t_0 , t_1)) = \{x\}$ and $\theta((t_1 , t_2]) = \{y\}$ for some $t_1 \in (t_0, t_2)$

Replacing topology by fuzzy pretopology we proposed the following definition:

Definition 4.3: Let V be a finite fuzzy pretopological space, $x \neq y \in V, t_0 < t_2 \in R$. A transition from x to y is a function $\theta : [t_0, t_2] \to \{x, y\}$ (with the induced fuzzy pretopology) such that $\theta([t_0 , t_1)) = \{x\}$ and $\theta((t_1 , t_2]) = \{y\}$ for some $t_1 \in (t_0 , t_2)$.

We prove the following result

Result 4.1: Let P be the phenotype space equipped with the fuzzy pretopology discussed above and $[t_0, t_2]$ is endowed with fuzzy pretopology inherited from the fuzzy pretopology on R. Suppose $\theta(t_1) = \alpha$. Then the transition from α to β is continuous iff $v_\alpha (\beta) > 0$.

Proof: We have $\theta : [t_0, t_2] \to Y = \{\alpha, \beta\}$. We are to discuss continuity of θ at the point $t_1 \in [t_0, t_2]$. Since $\theta(t_1) = \alpha$, we consider the fuzzy preneighbourhood v_α of α. θ is continuous at t_1 iff there exist a fuzzy neighbourhood of t_1, say, μ_1, such that

$$\theta(\mu_1) \subseteq v_\alpha.$$

Equivalently,

$$\theta(\mu_1)(\gamma) \leq v_\alpha(\gamma), \ \gamma \in \{\alpha, \beta\} \qquad (2)$$

If $\gamma = \alpha$,

$$\theta(\mu_1)(\alpha) = sup\{\mu_1(t) : \theta(t) = \alpha\}$$
$$= sup\{\mu_1[t_0, t_1)\} = \mu_1(t_1) = 1 = v_\alpha(\alpha).$$

Hence Eqn. 2 is true.

Again, if $\gamma = \beta$

$$\theta(\mu_1)(\beta) = sup\{\mu_1(t) : \theta(t) = \beta\}$$
$$= sup\{\mu_1(t_1, t_2]\} = \mu_1(t_1)$$

Now,

$$\theta(\mu_1)(\beta) \leq v_\alpha(\beta) \text{ iff } \mu_1(t_1) \leq v_\alpha(\beta)$$

Let $v_\alpha(\beta) = \kappa$. Then θ is continuous at t_1 iff $\mu_1(t_1) \leq \kappa$. By the definition of fuzzy neighbourhood basis of t_1 we observe that it is always possible to choose a fuzzy neighbourhood μ_1 of t_1 such that $\mu_1(t_1) \leq \kappa$ so long as $\kappa > 0$. Hence θ is continuous at t_1 iff $\kappa > 0$, i.e., iff $v_\alpha(\beta) > 0$.

Similarly we have the following result

Result 4.2: Let P be the phenotype space equipped with the fuzzy pretopology discussed above and $[t_0, t_2]$ is endowed with fuzzy pretopology inherited from the fuzzy pretopology on R. Suppose $\theta(t_1) = \beta$. Then the transition from α to β is continuous iff $v_\beta(\alpha) > 0$.

5. CONCLUSION

In this paper we have attempted to obtain a topological model of RNA space so as to discuss continuity of evolutionary trajectory. Several researchers have considered different models in RNA set for discussing accessibility notion. Many of such models require choosing a threshold value for discussing continuity notion. Here, we try to model the situation in the fuzzy pretopological setting by considering each shape in phenotype space to be a fuzzy preneighbourhood and construct a fuzzy pretopology on the set of phenotypes. Our model gets rid of any threshold value. To have a meaningful discussion of continuity we equipped the time axis with fuzzy pretopology which is generated by fuzzy neighbourhood basis. We have further discussed a necessary and sufficient condition for continuity of the trajectory map.

Acknowledgement

The second author would like to express his gratitude to the funding Agency Department of Science and Technology, Govt. of India, for providing financial support under a Research Project(Ref. No. SR/S4/MS:686/10-1 dated 06/05/11).

References

1. Ali, T., 2011: *Modelling RNA Evolution*, Proceeding of WASET conferences Paris; 27-29, 1136-1141.

2. Ali, T., and Phukan, C.K., 2011: *Fuzzy proximity in phenotype space*, Int. J. Adv. Comp. Math. Sc; **2**, 117-125.

3. Badard, R., 1981: *Fuzzy pretopological spaces and their representation*, J. Math. Anal. Appl., **81**, 378-390.

4. Cupal, J., Kopp, S., and Stadler, P., 2000: *RNA shape space topology*; Alife **6**, 3-23.

5. Fontana, W., and Schuster, P., 1998: *Shaping space: The possible and the attainable in RNA genotype-phenotype mapping*; J. Theor. Biol.; **194**, 491-515.

6. Klir, G.J., and Yuan, B., 2003: *Fuzzy Sets and Fuzzy Logic: Theory and Application*, Prentice Hall of India Private Limited.

7. Mynard, F., and Seal, G.J., 2010: *Phenotype spaces*; J. Math. Biol.; **60**, 247-266.

8. Palaniappan, N., 2005: *Fuzzy Topology*; Narosa Publication House, India.

9. Stadler, B., Stadler, P., Wagner, G., and Fontana, W., 2001: *The topology of the possible: Formal spaces underlying patterns of evolutionary change*; J. Theor. Biol.; **213**, 241-274.

10. Stadler, P., and Stadler, B., 2006: *Genotype phenotype maps*; Biol. Theory; **3**, 268-279.

A New Fuzzy Operator in Enhancing Images

Sharmistha Bhattacharya (Halder)[*] and Md. Atikul Islam[#]
Department of Mathematics, Tripura University
Suryamaninagar, Agartala, Tripura
E-mail: halder_731@rediffmail.com[*]; atik.math@yahoo.com[#]

ABSTRACT

Machine vision has become an integral part of today's electronic industry. Digital images capturing over-brightness, low-contrast and low-illumination conditions often cause serious problems in pattern recognition systems. This paper uses over-bright, over-contrast, low-contrast and also in low-illumination real digital images and document images to demonstrate the effectiveness of the new fuzzy operator called Fuzzy Image Intensification Operator in reducing background noise and increasing the readability of text of document image by contrast enhancement. Experiments were performed using gray-scale and colour images.

1. INTRODUCTION

Image enhancement methods were categorized by Gonzalez and Woods [04] into two broad classes: Transform domain methods and spatial domain methods. The techniques in the first category are based on modifying the frequency transform of an image. However, computing a two dimensional transform for a large array (an image) is a very time consuming task even the fast transformation techniques [08] are not suitable for real time processing. The techniques in the second category directly operate on the pixels. Contrast enhancement is one of the important image enhancement techniques in spatial domain. Besides two popular methods: histogram equalization and histogram specifications [04], we may mention a few important spatial domain methods such as an iterative histogram modification of gray images [03], an efficient adaptive neighbourhood histogram equalization [11] and Gabor's technique [10]. Fuzzy set [14] offers a problems solving tool between the precision of classical mathematics and the inherent imprecision of the real world. In 1991, Zimmermann [07] initiated that the imprecision in an image contained within colour value can be handled using fuzzy sets. In the fuzzy framework of image enhancement and smoothing, two contributions merit an elaboration. The first one deals with 'IF..THEN.. ELSE' fuzzy rules [13] for image enhancement. The second one proposes a rule based filtering [02] in which different filter classes are devised on the basis of compatibility with the neighbourhood. An image can be considered as an array of fuzzy singletons [12] having a membership value that denotes the degree of some image property in the range [0,1]. In 1997, Hanmandlu et al. [05] have proposed a Gaussian type of fuzzification function that contains a single fuzzifier and a new intensification operator called NINT. Fuzzifier is obtained by maximizing the fuzzy contrast and the parameter is obtained by minimizing the entropy. Cheng and Huijuan [01] in 2000 have proposed a fuzzy logic approach to contrast enhancement. Then Hanmandlu, Jha and Sharma [06] in 2003 described colour image enhancement by fuzzy intensification. Leung [09] in 2005 presented a method based on Generalized Fuzzy Operator, pre-processed with HE and POSHE for low -contrast and low- illumination images.

We propose a new approach for image enhancement based on the Fuzzy Image Intensification Operator (FIO). This method gives better performance in image enhancement of extremely over-bright, over-contrast, low- contrast and also in low-illumination condition.

2. PRELIMINARIES

Definition 2.1 [12]. An image I of size $m \times n$ and L gray levels k ranging between 0 to $L - 1$ can be consider as an array of fuzzy singletones, each having a value of membership denoting its degree of brightness relative to some brightness levels. For an image I, we can write in the notion of fuzzy sets as,

$$I = \{\langle I_k(i, j), \mu_{I_k}(i, j)\rangle | \ i = 1, 2, \ldots, m; j = 1, 2, \ldots, n\} \tag{1}$$

where $I_k(i, j)$ is the values of I at position (i, j) at any gray levels k, $\mu_{I_k}(i, j)$ denotes the degree of brightness possessed by the gray level intensity $I_k(i, j)$ of the $(i, j)^{\text{th}}$ pixel.

3. FUZZY IMAGE INTENSIFICATION OPERATOR (FIO)

We define Fuzzy Intensification Operator (FIO) for enhancement of extremely over-bright, over-contrast, low-contrast and also low-illumination real digital images and document images. A mathematical expression of such operator for an image G at any gray level or colour level k may be denoted as, FIO: $[0, 1] \rightarrow [0, 1]$ and expressed as

$$\text{FIO}\left(\mu_{G_K}(i, j)\right) = \mu'_{G_K}(i, j)$$
$$= 1 - e^{-\alpha \mu_{G_K}^2(i, j)}, \text{ if } 0 \leq \mu_{G_K}(i, j) \leq 1 \tag{2}$$

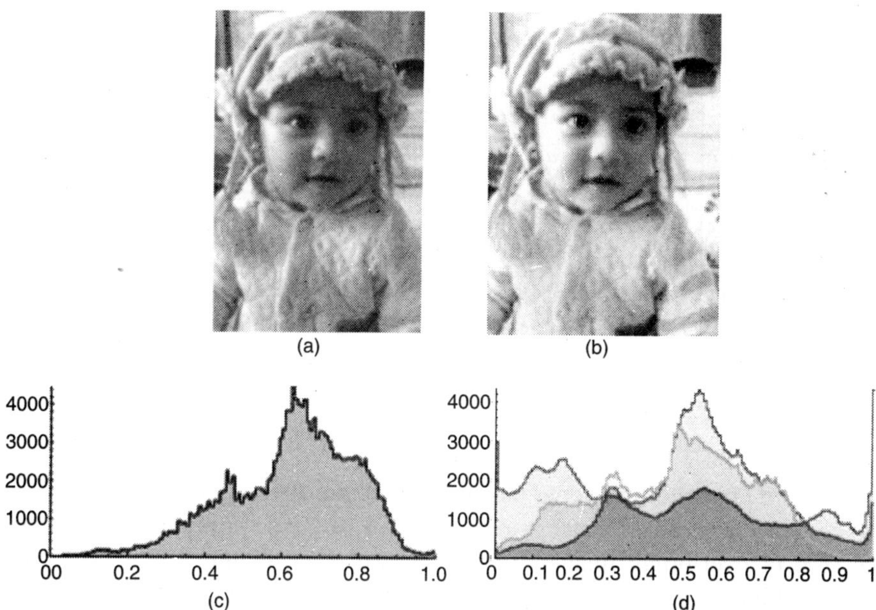

Fig. 1 (a) Gray-scale image; (b) colour(RGB) image; (c) fuzzy membership histogram of (a); (d) fuzzy membership histogram of (b).

where $\mu_{G_K}(i, j)$ denotes the degree of brightness possessed by the gray level or colour level intensity $G_K(i, j)$ of the $(i, j)^{th}$ pixel, $\mu'_{G_K}(i, j)$ is the modified membership value of $\mu_{G_K}(i, j)$ by FIO and α is an intensification parameter which can range from 1 to infinity.

Then we generate new gray level or colour level $\mu'_{G_K}(i, j)$ for $G_K(i, j)$ by the following way,

$$(i, j) = (L - 1)\left(1 - \left(\mu'_{G_K}(i, j)^{\frac{\sigma}{\tau}}\right)\right) \tag{3}$$

where L is the highest gray level of the image, $\sigma > 0$ and $\tau \geq 1$ are arbitrary parameters.

Property 3.1: When $\alpha \to \infty$

$$\mu'_{G_K}(i, j) = \begin{cases} 1, & \text{if } \mu_{G_K}(i, j) \neq 0 \\ 0, & \text{if } \mu_{G_K}(i, j) = 0 \end{cases} \tag{4}$$

Property 3.2: When $\alpha \geq 1$

$$\mu'_{G_K}(i, j) = \begin{cases} (0, 1], & \text{if } \mu_{G_K}(i, j) \neq 0 \\ 0, & \text{if } \mu_{G_K}(i, j) = 0 \end{cases} \tag{5}$$

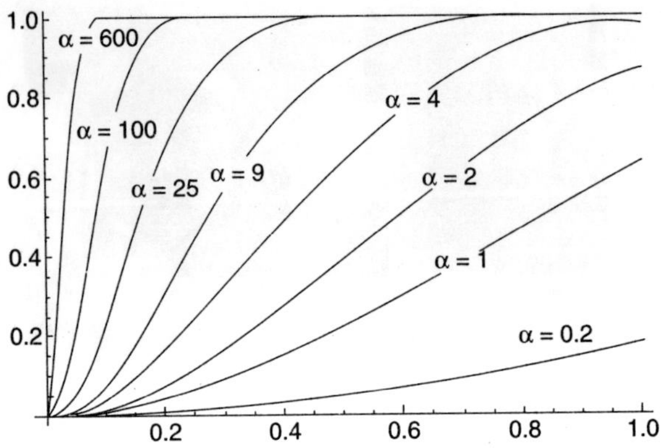

Fig. 2 Graph of FIO for α = 0.2, 1, 2, 4, 9, 25, 100, 600 from bottom to top respectively.

The FIO operator distribute membership values from lower luminance (for the case of low-contrast images) to the interval [0,1] and upper luminance (for the case of over-bright images) to the interval [0,1]. Figure 2 represents the graph of FIO for different values of α. It is easy to see from the Fig. 2 that, as the value of parameter α increases degree of brightness of images increases and as α decreases degree of brightness decreases.

4. RESULT AND DISCUSSION

Enhancement algorithms are used to enhance contrast, to reduce over-brightness and to remove fuzziness from the image. Enhancement improves the quality of the image and generally provides a clearer image for a human observer. The images taken for analysis are shown in Fig. 3 and Fig. 4.

5. CONCLUSION

In this paper we have proposed image enhancement method based on Fuzzy Image Intensification Operator (FIO). Enhancement algorithm is used to reduce over-contrast, over-brightness of image and to improve the contrast of low-contrast, low-illumination images. Generally provides a clearer image for a human observer. However more experimental results are required in order to establish a reliable comparison.

Acknowledgement

I would like to thank UGC, New Delhi for the financial support to doing this work under the scheme of Maulana Azad National Fellowship for Minority students to pursue M.Phil/Ph.D Degree 2010-11.

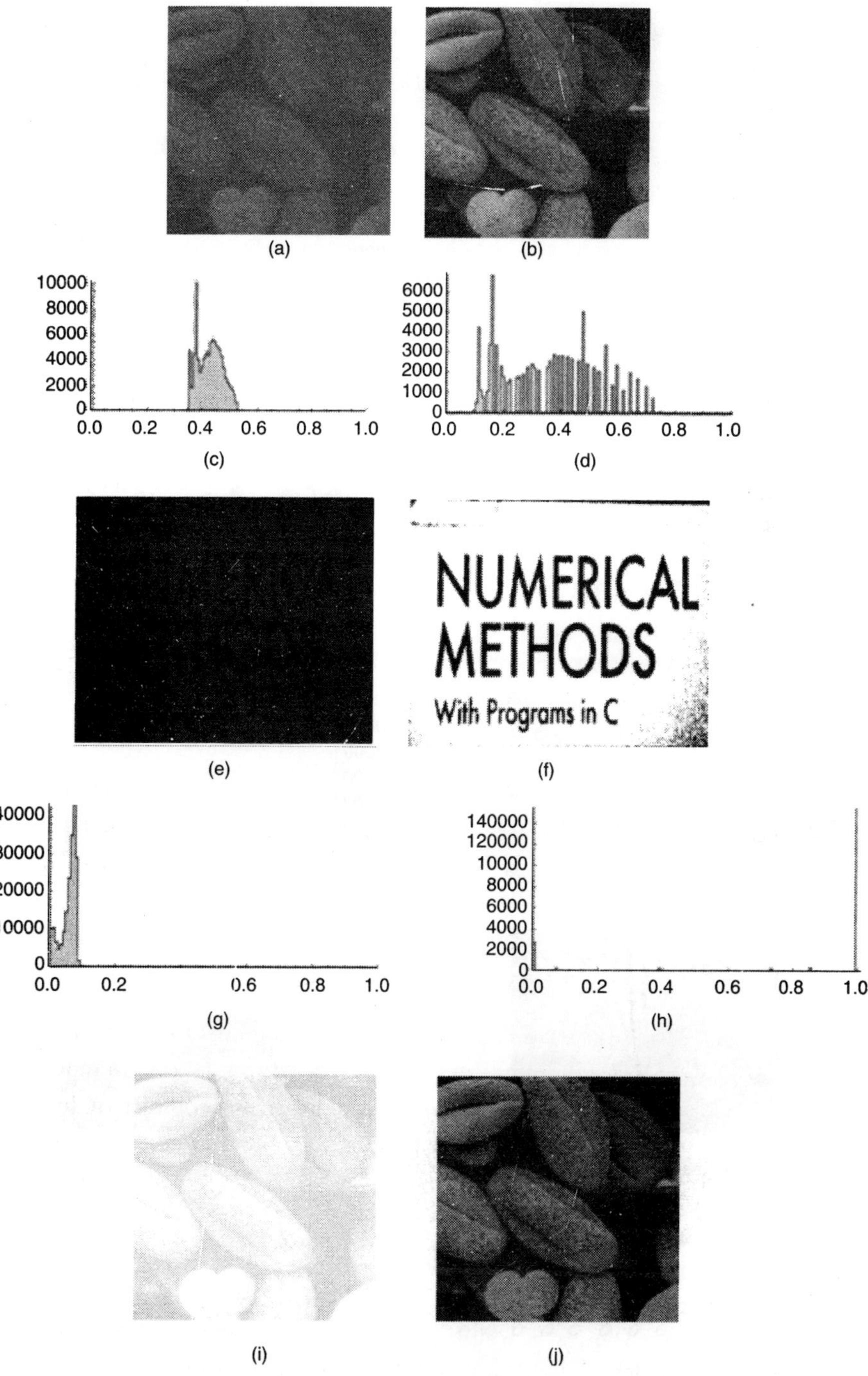

Fig. 3 (a),(e),(i) Original image; (b),(f),(j) By the proposed method; (c), (g) fuzzy membership histogram before enhancement; (d), (h) fuzzy membership histogram after enhancement.

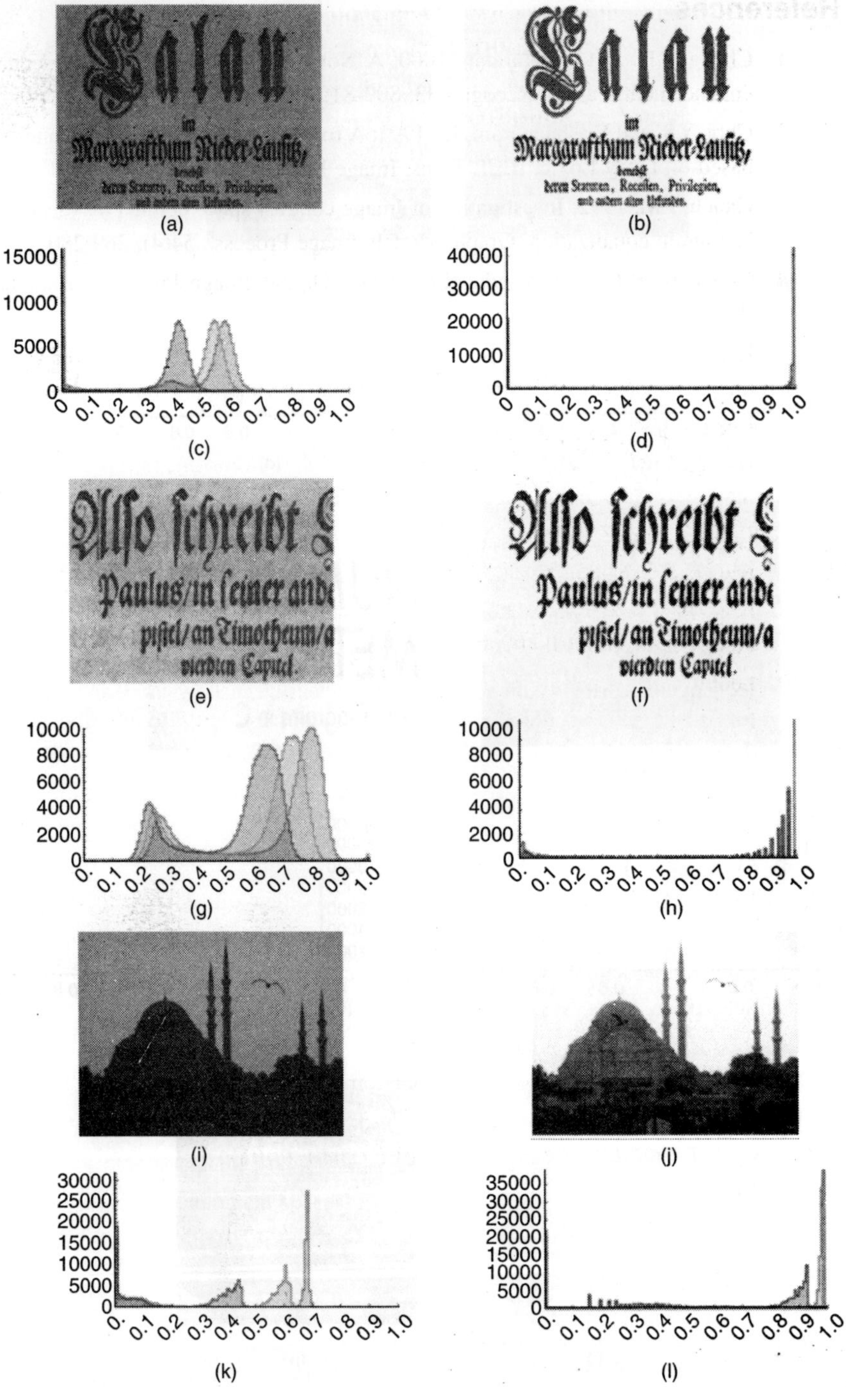

Fig. 4 (a), (e), (i) Original image; (b), (f), (j) By the proposed method; (c), (g), (k) fuzzy membership histogram before enhancement; (d), (h), (l) fuzzy membership histogram after enhancement.

References

1. Cheng H.D. and Xu Huijuan, 2000: A Novel fuzzy logic approach to contrast enhancement, Pattern Recogn., 33, 809-819.

2. Choi, Y.S. and Krishnapuram, R., 1997: A robust approach to image enhancement based on fuzzy Logic, IEEE Trans. Image Process., 6(6), 808-825.

3. Gauch, J.M. 1992: Investigation of Image contrast space defined by variation of histogram equalization, Graph. Models Image Process., 54(4), 269-280.

4. Gonzalez, R.C. and Woods, R.E., 1992: Digital Image Processing, Addison-Wesley, Reading, MA.

5. Hanmandlu, M., Tandon, S.N., and Mir, A.H., 1997: A new fuzzy logic based image enhancement, 34th Rocky Mountain Symposium on bioengineering, Dayton, Ohio, USA., 590-595.

6. Hanmandlu, M., JhaD., and Sharma, R., 2003, Color image enhancement by fuzzy Intensification, Pattern Recogn. Lett., 24, 81-87.

7. Jimmermann, H.J., 1991: Fuzzy Set Theory and its Applications, second ed. Kluwer Academic Publishers, Dordrecht.

8. Lee,. S., 1980: Digital Image Enhancement and noise filtering, IEEE Trans. Pattern Anal. Machine Intell., 2,165-168.

9. Leung, C.-C., Chan, K.-S., Chan, H.-M., and Tsui W.-K. 2005: A new approach for image enhancement applied to low-contrast–low-illumination IC and document images, Pattern Recogn. Lett., 26, 769-778.

10. Lindenbaum, M., Fischer M., and Bruckestein, A., 1994: On Gabor's Contribution to Image Enhancement, Pattern Recogn., 7, 1-8.

11. Mukherjee, P., and Chatterji, B.N., 1995: Note: Adaptive Neighbourhood Extended Contrast Enhancement and its Modifications, Graph. Models Image Process., 57(3), 254-265.

12. Pal S.K. and King R.A., 1981: Image enhancement using smoothing with Fuzzy Sets, IEEE Trans. Sys. Man Cybern., SMC-11, 494-501.

13. Russo, F., and Ramponi, G., 1965: A fuzzy operator for the enhancement of blurred and noisy images, IEEE Trans. Image Process. 4(8), 1169-1174.

14. Zadeh, L.A., 1973: A, Outline of a new approach to the analysis of complex systems and decision Processes, IEEE Trans. Sys. Man. Cybern. SMC-3, 29-44.

Rough Sets, Fuzzy Sets and Soft Computing

Editor: S. Bhattacharya Halder

A Fuzzy Approach to Select an Expert from Decision-Makers based on Fuzzy Clustering and Fuzzy Preference Relations for Decision-Maker Relationship Management

Akoijam Bimolnanda Singh[1*] and Thoudam Biren Singh[2#]

[1]*Research Scholar, Department of Mathematics*
Manipur University, Canchipur, Imphal
[2]*Principal, Oriental College, Imphal*
E-mail: akoijam79@gmail.com[*] ; biren_thoudam@yahoo.com[#]

ABSTRACT

In terms of logistical system and administration, expert candidate (EC) selection from a group of decision-makers is a common problem faced by logistic managers. Decision-makers commonly take relevant attributes into consideration for selection of services or products. The attribute assessment of a service or product is often presented by linguistic data sequences. To select an expert candidate, first, we require to partition these linguistic data sequences of decision-makers' assessment on a service or product into clusters and based on these clusters, a proper fuzzy preference relation method is essential and proposed in this paper. In the clustering method, the linguistic data sequences are presented by fuzzy data sequences and a fuzzy compatible relation is constructed and then a fuzzy equivalence relation is derived. Based on the fuzzy equivalence relation, the linguistic data sequences are easily classified into clusters. For each cluster, a fuzzy preference relation is constructed to find an expert candidate. Finally, from the expert candidates so obtained, an overall expert candidate is selected through fuzzy preference relation.

Keywords: Clustering, Decision-Maker relationship management, Fuzzy compatible relation, Fuzzy equivalence relation, Fuzzy preference relation.

1. INTRODUCTION

In system and administration management for the selection of an indicated expert, the manager often has to assess related decisions of the decision-makers (*DMs*).

The decision-makers have to assess related attributes of the service or product of the system. Commonly, the related attributes are so many and their importance degrees are varied for different decision-makers and attribute ratings will stand for *DMs'* assessments that are presented with linguistic terms such as very poor (*VP*), poor (*P*), medium poor (*MP*), fair (*F*), medium good (*MG*), good (*G*), very good (*VG*) and around (*AR*), about (*AB*) and approximately (*AP*) depending on the context considered. Thus, a *DM*'s assessment on several attributes is expressed with a linguistic data sequence and many *DMs'* assessments on the attributes are given with linguistic data sequences. In order to discover essential knowledge, linguistic data sequences of *DMs'* preferences have to be processed by using the technique of clustering. In order to select an expert, firstly, we are to discover essential knowledge from linguistic data sequences of *DMs'* preferences. To process the linguistic data sequence of a *DM*'s preference, the technique of clustering is an important technique (Bandyopadhyay and Maulik, 2002; Hirano, Sun and Tsumoto, 2004; Miyamoto, 2003; Wang and Lee, 2008; Wu and Yang, 2002). For a given system of service or product, clustering methods can partition linguistic data sequences of *DMs'* preferences into clusters. The clusters respectively represent preferences of different *DM* groups in the service or product. Secondly, we employ a fuzzy ranking procedure based on a fuzzy preference relation to find out a preferred *DM* from each *DM* group. Our problem is to select an overall expert from among the preferred *DMs*.

In the past, lots of clustering methods (Bandyopadhyay and Maulik, 2002; Hirano, Sun and Tsumoto, 2004; Khan and Ahmad, 2004; Kuo, Chang and Chien, 2004; Kuo, Liao and Tu, 2005; Lee, 2001; Miyamoto, 2003; Wang and Lee, 2008; Wu and Yang, 2002) and compatible preference relations were proposed. However, in this paper, a clustering method has to partition linguistic data sequences into clusters to present the preferences of different *DM* groups. The linguistic terms are transferred into fuzzy numbers in the clustering method. Then, the method will construct a fuzzy binary relation between any two fuzzy sequences. The fuzzy binary relation so constructed satisfies compatible relation. We utilize the max-min transitive closure (Lee, 2001) as the mechanism to transform fuzzy compatible relation into fuzzy equivalence relation. Then fuzzy data sequences are partitioned into clusters by the equivalence relation. One cluster stands for a *DM* group having similar preferences up to a certain degree to the service or product on attributes. For finding out the preferred *DM* in each *DM* group, we use the ranking procedure by utilizing the fuzzy preference relation. The same procedure will be performed for the selection of the most preferred *DM* which is termed as overall expert in the paper.

The organization of the paper is as follows: In Section 2, we introduce the basic definitions and notations of fuzzy numbers, linguistic variables and fuzzy relations. In Section 3, we define a fuzzy clustering method based on fuzzy max-min transitive closure of fuzzy preference relation and in Section 4, a fuzzy ranking procedure based on fuzzy preference relation for selection of an overall expert. In Section 5, a numerical example is discussed.

2. MATHEMATICAL PRELIMINARIES

In this section, some fuzzy mathematical theories are stated below. We review some basic notions of fuzzy sets, fuzzy numbers and fuzzy relations (Klir and Yuan, 2002; Zadeh, 1971; Zimmerman, 1991).

Definition 2.1: Let U be a universal set. A fuzzy set $\tilde{A}$ of U is defined with a membership function $\mu_{\tilde{A}} : U \to [0, 1]$ where $\mu_{\tilde{A}}(x)$ indicates the degree of membership of x in $\tilde{A}$.

Definition 2.2: A fuzzy set $\tilde{A}$ is said to be normal iff $\sup\limits_{x \in U} \mu_{\tilde{A}}(x) = 1$ where U is the universal set.

Definition 2.3: A fuzzy set $\tilde{A}$ of a universal set U is said to be convex iff $\mu_{\tilde{A}}(\lambda x + (1 - \lambda)y) \geq (\mu_{\tilde{A}}(x) \wedge \mu_{\tilde{A}}(y))$ where $\wedge$ denotes the minimum operator.

Definition 2.4: A fuzzy set $\tilde{A}$ is a fuzzy number if it is both convex and normal on U.

Definition 2.5: A fuzzy number $\tilde{A}$ is called a triangular fuzzy number if for any three real numbers a, b, c such that $a < b < c$, the membership function is defined as:

$$\mu_{\tilde{A}}(x) = \begin{cases} 0 & \text{for } x < a, x > c \\ \dfrac{x - a}{b - a} & \text{for } a < x < b \\ 1 & \text{for } x = b \\ \dfrac{c - x}{c - b} & \text{for } b < x < c \end{cases}$$

It is denoted by $\tilde{A} = (a, b, c)$. The set of all triangular fuzzy numbers is denoted by F_s.

Definition 2.6: For any $\tilde{A} = (a\ b\ c) \in F_s$, the α-level set of $\tilde{A}$, $\tilde{A}(\alpha) = [l(\alpha), u(\alpha)]$ is a closed interval where $l(\alpha) = a + \alpha(b - a)$, $u(\alpha) = c - \alpha(c - b)$.

Next, we review the fuzzy compatible (semi-equivalence) relation and fuzzy equivalence relation on fuzzy sets.

Definition 2.7: Let X and Y be universal sets. A fuzzy binary relation $\tilde{R}$ on $X \times Y$ is defined as the set of ordered pairs:

$$\tilde{R} = \{((x, y), \mu_{\tilde{R}}(x, y)) : (x, y) \in X \times Y\}$$

where $\mu_{\tilde{A}} : X \times Y \to [0, 1]$ is the membership function. In particular, the relation $\tilde{R}$ is called a fuzzy binary relation on X when $X = Y$.

Definition 2.8: Let $\tilde{R}$ be a binary relation on S where S is a set of fuzzy vectors. The following conditions may hold for $\tilde{R}$. Denoting $\mu_{\tilde{R}} (\tilde{x}, \tilde{y}) = r(\tilde{x}, \tilde{y})$ for all $\tilde{x}, \tilde{y} \in S$, we have:

$\tilde{R}$ is reflex if $r(\tilde{x}, \tilde{x}) = 1 \; \forall \; \tilde{x} \in S$

$\tilde{R}$ is symmetric if $r(\tilde{x}, \tilde{y}) = r(\tilde{y}, \tilde{x}) \; \forall \; \tilde{x}, \tilde{y} \in S$

$\tilde{R}$ is transitive if $r(\tilde{x}, \tilde{z}) \geq \vee_{\tilde{y} \in S} (r(\tilde{x}, \tilde{y}) \wedge r(\tilde{x}, \tilde{z})) \; \forall \; \tilde{x}, \tilde{y}, \tilde{z} \in S$

where $\vee$ and $\wedge$ denote the maximum operator and minimum operator respectively. If $\tilde{R}$ is reflexive, symmetric and transitive, $\tilde{R}$ is an equivalence relation on S.

Definition 2.9: Let $\tilde{R}$ be a fuzzy binary relation on S. A non empty set defined by $\tilde{R}_\lambda = \{(\tilde{x}, \tilde{y}) : r(\tilde{x}, \tilde{y}) \geq \lambda \; \forall \; \tilde{x}, \tilde{y} \in S, 0 \leq \lambda \leq 1\}$ is called a λ-level set of the fuzzy relation $\tilde{R}$ on S.

Lemma 2.1: Let $\tilde{R}$ be a fuzzy equivalence relation on S. Then $\tilde{R}_\lambda$ is an equivalence relation on S.

Proof: By definition of $\tilde{R}_\lambda$, $\tilde{R}_\lambda$ is an equivalence relation on S because it satisfies the following three conditions:

Reflexive: $(\tilde{x}, \tilde{x}) \in \tilde{R}_\lambda$ for $r(\tilde{x}, \tilde{x}) = 1 \geq \lambda \; \forall \; \tilde{x} \in S$

Symmetric: $(\tilde{x}, \tilde{y}) \in \tilde{R}_\lambda$ implies $(\tilde{y}, \tilde{x}) \in \tilde{R}_\lambda$ for

$$r(\tilde{x}, \tilde{y}) = r(\tilde{y}, \tilde{x}) \geq \lambda \; \forall \; \tilde{x}, \tilde{y} \in S.$$

Transitive: $(\tilde{x}, \tilde{y}) = \in \tilde{R}_\lambda, (\tilde{y}, \tilde{z}) \in \tilde{R}_\lambda$ implies $(\tilde{x}, \tilde{z}) \in \tilde{R}_\lambda$ for

$$r(\tilde{x}, \tilde{z}) \geq \vee_{\tilde{y} \in S} (r(\tilde{x}, \tilde{y}) \wedge r(\tilde{y}, \tilde{z})) \; \forall \; \tilde{x}, \tilde{y}, \tilde{z} \in S$$

Definition 2.10: An $n \times n$ fuzzy relation matrix $\tilde{A} = [r_a(\tilde{x}_i, \tilde{x}_j)]_{n \times n}$ is a matrix where $0 \leq r_a (\tilde{x}_i, \tilde{x}_j) \leq 1$ indicates the relation between two fuzzy numbers $\tilde{x}_i$ and $\tilde{x}_j$ and $0 \leq i \leq n, 0 \leq j \leq n$. It is reflexive if $r_a (\tilde{x}_i, \tilde{x}_j) = 1$ for $0 \leq i \leq n$ and symmetric if $r_a (\tilde{x}_i, \tilde{x}_j) = r_a (\tilde{x}_j, \tilde{x}_i)$. The matrix $\tilde{A}$ is called max-min transitive if $r_a (\tilde{x}_i, \tilde{x}_j) \geq \vee_k (r_a (\tilde{x}_i, \tilde{x}_k) \wedge r_a (\tilde{x}_k, \tilde{x}_j))$.

Definition 2.11: A fuzzy semi-equivalence matrix is a fuzzy relation matrix that is both reflexive and symmetric.

Definition 2.12: A fuzzy semi-equivalence matrix which satisfies max-min transitive is called a fuzzy equivalence matrix.

An $n \times n$ fuzzy relation matrix $\tilde{B} = [r_b\,(\tilde{x}_i, \tilde{x}_j)]_{n \times n}$ includes the matrix $\tilde{A}$ and is denoted by $\tilde{A} \subseteq \tilde{B}$ where $r_a\,(\tilde{x}_i, \tilde{x}_j) \le r_b\,(\tilde{x}_i, \tilde{x}_j)$ for all i, j.

Definition 2.13: (Wang, 2010). A fuzzy equivalence matrix $\tilde{B}$ is called the equivalence closure of a fuzzy semi-equivalence matrix $\tilde{A}$ if $\tilde{A} \subseteq \tilde{B}$ and is included by any fuzzy equivalence matrix that includes $\tilde{A}$.

Definition 2.14: A partition of a set S means a family of disjoint subsets of S, say $S_1, S_2, ..., S_n$ such that:

$$S_1 \cup S_2 \cup, ... \cup S_n \text{ and } S_i \cap S_j = \phi,\ i \ne j.$$

Definition 2.15: Let $P_{\tilde{R}_\lambda}$ be a partition of S by a fuzzy equivalence relation $\tilde{R}_\lambda$. Then, define the elements $\tilde{x}, \tilde{y} \in S$ to be in a subset S_i of S where $(\tilde{x}, \tilde{y}) \in \tilde{R}_\lambda$ and $S_i \in P_{\tilde{R}_\lambda}$.

3. FUZZY SEMI-EQUIVALENCE RELATION, EQUIVALENCE RELATION AND FUZZY CLUSTER

In this section, a fuzzy relation is proposed to represent a binary relation between any two fuzzy sequences.

Let $\tilde{X}_i = \{\tilde{X}_{i1}, \tilde{X}_{i2}, ..., \tilde{X}_{in}\}$ denote the ith sequence consisting of n fuzzy numbers $\tilde{X}_{ij}$ ($j = 1, 2, 3, ..., n$). Suppose D to indicate a data matrix compose of m fuzzy sequences *i.e.*

$$D = [\tilde{X}_{ij}]_{m \times n} \tag{3.1}$$

where i is the index of sequence and j is index of fuzzy number.

Following Wang (2010), for a given data matrix D, the fuzzy binary relation between any two fuzzy sequences is derived as follows:

Let $\tilde{X}_{ij} = (x_{ijl}, x_{ijh}, x_{ijr})$ be a triangular fuzzy number, $i = 1, 2, ..., m$; $j = 1, 2, ..., n$. We assume that $\tilde{T}_j^+ = t_{jl}^+, t_{jh}^+, t_{jr}^+$ and $\tilde{T}_j^- = t_{jl}^-, t_{jh}^-, t_{jr}^-$ respectively, represent upper and lower boundaries of jth fuzzy number of all sequences in the data matrix D where:

$$t_{jl}^+ = \max_i\ \{x_{ijl}\}$$

$$t_{jl}^+ = \max_i\ \{x_{ijh}\}$$

$$t_{jr}^+ = \max_i\ \{x_{ijr}\}$$

$$t_{jl}^- = \min_i\ \{x_{ijl}\}$$

$$t_{jh}^- = \min_i \{x_{ijh}\}$$

$$t_{jr}^- = \min_i \{x_{ijr}\}$$

and $i = 1, 2, ..., m; j = 1, 2, ..., n$.

Then, we define a fuzzy binary relation, $\tilde{R} = \{((i, q), \mu_{\tilde{R}}(i, q))\}$ as the binary relation between fuzzy sequences $\tilde{X}_i$ and $\tilde{X}_q$. The membership $\mu_{\tilde{R}}(i, q)$ between fuzzy sequences $\tilde{X}_i$ and $\tilde{X}_q$ will present the similarity of the two fuzzy data sequences. Let

$$\mu_{\tilde{R}}(i, q) = \left[\frac{1}{n} \sum_{j=1}^{n} \left(\frac{d\left(T_j^+, T_j^-\right) - \left| d\left(\tilde{X}_{ij}, \tilde{X}_{qj}\right)\right|}{d\left(T_j^+, T_j^-\right)} \right) \right] \tag{3.2}$$

where

$$d\left(\tilde{T}_j^+, \tilde{T}_j^-\right) = \int_0^1 \left(\left(\tilde{T}_j^+ - \tilde{T}_j^-\right)_\alpha^L + \left(\tilde{T}_j^+ - \tilde{T}_j^-\right)_\alpha^U \right) d\alpha$$

$$= \frac{\left(t_{jl}^+ - t_{jl}^-\right) + 2\left(t_{jh}^+ - t_{jh}^-\right) + \left(t_{jr}^+ - t_{jr}^-\right)}{2}$$

and

$$d\left(\tilde{X}_{ij}, \tilde{X}_{ij}\right) = \int_0^1 \left(\tilde{X}_{ij} - \tilde{X}_{qj}\right)_\alpha^L + \left(\tilde{X}_{ij} - \tilde{X}_{qj}\right)_\alpha^U d\alpha$$

$$= \frac{(x_{ijl} - x_{qjr}) + 2(x_{ijh} - x_{qjh}) + (x_{ijr} - x_{qil})}{2}$$

Lemma 3.1: The fuzzy binary relation $\tilde{R}$ is a fuzzy semi-equivalence relation.

Proof: Reflexive : $\qquad \mu_{\tilde{R}}(i, i) = 1$.

Symmetric : $\qquad \mu_{\tilde{R}}(i, q) = \mu_{\tilde{R}}(q, i)$.

Since $\tilde{R}$ satisfies reflexive and symmetric laws, $\tilde{R}$ is a fuzzy semi-equivalence relation. Following Lemma 3.1, $\tilde{R}$ is merely a fuzzy semi-equivalence relation, not a fuzzy equivalence relation. By utilizing the fuzzy semi equivalence relation to partition the set of fuzzy numbers may overlap (Mingyi, Danning and Ying, 2010). To resolve the overlapping, the fuzzy semi-equivalence relation has to be transformed into a fuzzy equivalence relation by a mechanism. In the paper, the max-min transitive closure is the mechanism. The max min transitive closure $\tilde{R}^*$ of $\tilde{R}$ is displayed as follows:

$$\mu_{\widetilde{R}}^{2}(i, q) = \max_{k} \min [\mu_{\widetilde{R}}(i, k), \mu_{\widetilde{R}}(k, q)]$$

$$\cdot$$
$$\cdot$$
$$\cdot$$

$$\mu_{\widetilde{R}}^{n}(i, q) = \max_{k} \min \left[\mu_{\widetilde{R}}^{n-1}(i, k), \mu_{\widetilde{R}}^{n-1}(k, q)\right]$$

and

$$\mu_{\widetilde{R}}^{*}(i, q) = \max_{k} \mu_{\widetilde{R}}^{k}(i, q) \; \forall \; i, q$$

From the above equations, $\widetilde{R}^{*}$ and $\widetilde{R}^{n}$ can be revised as follows:

$$\mu_{\widetilde{R}}^{n}(i, q) \geq \mu_{\widetilde{R}}^{*}(i, q) \quad \text{i.e.} \quad \widetilde{R}^{*} \in \widetilde{R}^{n}.$$

Now, $\widetilde{R}^{*}$ based on $\widetilde{R}$ and satisfying transitive law is a fuzzy equivalence relation. With $\widetilde{R}^{*}$ and a threshold value of λ, fuzzy numbers of data set are partitioned into clusters. Therefore, we must determine the proper threshold value λ after $\widetilde{R}^{*}_{\lambda}$ being derived. Moreover, $\widetilde{R}^{n}$ can be substituted for $\widetilde{R}^{*}$ in partitioning the fuzzy numbers.

By definition 2.8, we have

$$\widetilde{R}^{*}_{\lambda} = \{(i, q) : \mu_{\widetilde{R}}(i, q) \geq \lambda\}, \quad 0 \leq \lambda \leq 1.$$

According to $\widetilde{R}^{*}_{\lambda}$, the partition rule is shown below:

If $\mu_{\widetilde{R}}^{*}(i, q) \geq \lambda$, then i^{th} fuzzy sequence and q^{th} fuzzy sequence are in the same cluster; otherwise the two fuzzy sequences respectively belong to two different clusters. It is also noted that $\widetilde{R}^{*}_{\lambda}$ will be an equivalence relation because $\widetilde{R}^{*}$ is a fuzzy equivalence relation. When $\lambda = 0$, $\widetilde{R}^{*}_{\lambda}$ partitions n fuzzy numbers into one cluster. On the contrary, when $\lambda = 1$, $\widetilde{R}^{*}_{\lambda}$ commonly partitions n fuzzy numbers into n clusters. In the following, a validation index is defined to determine the proper value of λ. In (Wang and Lee, 2008), they assumed that $c(\lambda)$ was the number of clusters of n fuzzy numbers partitioned by $\widetilde{R}^{*}_{\lambda}$. To determine a proper value of λ, their validation index was $\lambda - c(\lambda)/n$. Commonly a good partition has a high intra-cluster relation and a low inter-cluster relation. This means that if there is only one cluster, there is no inter-cluster relation and the clusters' number increases as the value of the inter-cluster relation rises. Thus, the inter-cluster relation can be represented by $c(\lambda)/n$ and λ represents the intra-cluster relation. A good partition can be determined by large value of validation index

$\lambda - c(\lambda)/n$ i.e. λ is large and $c(\lambda)/n$ is small. In view of the possible difference of variation scales in λ and $c(\lambda)/n$, a validation index was proposed by Wang (2010) which was in similar variation scales of the intra-cluster relation and inter-cluster relation. Wang (2010) discussed cohesion within clusters and coupling between clusters. Based on the investigation of Wang (2010), we propose the validation index, Val(Ind) of partition $P_{\tilde{R}^*_\lambda}$ (for different λ^*) as given below:

$$\text{Val(Ind)} = \sum_{1}^{n} \sum_{q = 1 (i \neq q)}^{n} [a(i, q) \times \mu_{\tilde{R}}(i, q) + (1 - a(i, q)) \times (1 - \mu_{\tilde{R}}^{*}(i, q))]$$

$$(3.3)$$

$$\text{where } a(i, q) = \begin{cases} 1 \text{ if } i, q \in S_1 \text{ and } S_1 \in P_{\tilde{R}^*_\lambda} \\ 0 \text{ if } i \in S_1 \text{ and } q \in S_2, S_1, S_2 \in P_{\tilde{R}^*_\lambda}, S_1 \neq S_2 \end{cases}$$

From Eqn. (3.3), it is observed that the value of the Val(Ind) is maximum when there is no inter-cluster relation for the relation matrix $\tilde{R}^*$ where $\mu_{\tilde{R}}^{*}(i, q) \in [0.5, 1]$, for all $i, q = 1, 2, 3, ..., n$. Thus, a good partition may not be given by the maximum value of Val(Ind). In order to solve such problems, we consider a method by introducing the concept of intra-cluster.

Definition 3.1: Suppose that a data set be partitioned with respect to with a threshold value into clusters C_i $(i = 1, 2, ..., n)$. Then, a cluster C_i is called an intra-cluster if there are at least two data in C_i.

We propose that a good partition exists only when there are maximum numbers of intra-clusters.

In this paper, $\tilde{R}^n$ will substitute for $\tilde{R}^*$ based on $\tilde{R}^* \subseteq \tilde{R}^n$ for simplification. Different validation indices are determined for those partitions having maximum number of intra-clusters and the best partition is obtained corresponding to the greatest validation index among the partitions having maximum number of intra-clusters.

In the next section, we will consider the selection process for the selection of preferred sequence in a cluster. The method is an extension from the one discussed in (Chen, 2001).

4. FUZZY RANKING PROCEDURE

Suppose $C = \{C_1, C_2, ..., C_k\}$ be the set which consists of clusters of a cluster arrangement having maximum number of intra-clusters with maximum validation index obtained from the fuzzy sequences given by Eqn. (3.1). Without loss of generality, assume $C_s \in C, 1 \leq s \leq k,$ is an intra-cluster with:

$$C_s = \left\{ \tilde{X}_1, \tilde{X}_2, ..., \tilde{X}_p \right\} \text{ where } 1 \leq p \leq m.$$

Then
$$\tilde{X}_1 = \left(\tilde{X}_{11}, \tilde{X}_{12}, ..., \tilde{X}_{1n} \right)$$
$$\tilde{X}_2 = \left(\tilde{X}_{21}, \tilde{X}_{22}, ..., \tilde{X}_{2n} \right)$$

$$\tilde{X}_p = \left(\tilde{X}_{p1}, \tilde{X}_{p2}, ..., \tilde{X}_{pn} \right).$$

Then cluster C_s can be concisely expressed in matrix format as

$$\tilde{D}' = \begin{bmatrix} \tilde{X}_{11} & \tilde{X}_{12} & \tilde{X}_{13} & & \tilde{X}_{1n} \\ \tilde{X}_{21} & \tilde{X}_{22} & \tilde{X}_{23} & \cdots & \tilde{X}_{2n} \\ \vdots & \vdots & \vdots & \vdots & \vdots \\ \tilde{X}_{p1} & \tilde{X}_{p2} & \tilde{X}_{p3} & \cdots & \tilde{X}_{pn} \end{bmatrix} \tag{4.1}$$

where each fuzzy number can be described by triangular fuzzy number $\tilde{X}_{ij} = (a_{ij}, b_{ij}, c_{ij})$

In order to preserve the property that the ranges of the triangular fuzzy numbers in Eqn. (4.1) belong to [0, 1], we can normalize $\tilde{D}'$ and can obtain the normalized fuzzy matrix denoted by $\tilde{R}$ as:

$$\tilde{R} = [\tilde{r}_{ij}]_{p \times n}$$

$$\tilde{r}_{ij} = \left(\frac{a_{ij}}{c^*_j}, \frac{b_{ij}}{c^*_j}, \frac{c_{ij}}{c^*_j} \right) \tag{4.2}$$

$$c^*_j = \max_i c_{ij}$$

Considering the importance of each criterion, we calculate the fuzzy evaluation value of each fuzzy sequence as:

$$\tilde{P}_i = \sum_{j=1}^{n} \tilde{r}_{ij}, \quad i = 1, 2, ..., p. \tag{4.3}$$

where $\tilde{P}$ is the fuzzy evaluation value of ith sequence. After the calculation of the fuzzy evaluation value of each sequence in the cluster C_s the pair wise comparison of the preference relationship between $\tilde{X}_i$ and $\tilde{X}_j$ can be established as stated in the following.

The above fuzzy evaluation values of $\tilde{P}_i$ and $\tilde{P}_j$ are triangular fuzzy numbers and hence the difference between $\tilde{P}_i$ and $\tilde{P}_j$ is also a triangular fuzzy number. It can be calculated as

$$\tilde{Z}_{ij} = \tilde{P}_i \, (-) \, \tilde{P}_j \tag{4.4}$$

$$\tilde{Z}_{ij}(\alpha) = [z_{ijl}(\alpha), z_{iju}(\alpha)] \tag{4.5}$$

where $\tilde{Z}_{ij}(\alpha)$ is the α-level set of $\tilde{Z}_{ij}$ and

$$\tilde{P}_i(\alpha) = [p_{il}(\alpha), p_{lu}(\alpha)]$$

$$\tilde{P}_j(\alpha) = [p_{jl}(\alpha), p_{ju}(\alpha)]$$

$$z_{ijl}(\alpha) = p_{il}(\alpha) - p_{ju}(\alpha)$$

$$z_{iju}(\alpha) = p_{iu}(\alpha) - p_{jl}(\alpha)$$

$z_{ijl}(\alpha) > 0$ for $\alpha \in [0, 1]$, then $\tilde{X}_i$ is absolutely preferred to $\tilde{X}_j$. If $z_{ijl}(\alpha) < 0$ for $\alpha \in [0, 1]$, then $\tilde{X}_i$ is not absolutely preferred to $\tilde{X}_j$. If $z_{ijl}(\alpha) < 0$ and $z_{iju}(\alpha) > 0$ for some α, we define e_{ij} as the fuzzy preference relation between $\tilde{X}_i$ and $\tilde{X}_j$ to represent the degree of preference of $\tilde{X}_i$ over $\tilde{X}_j$. The e_{ij} is defined as:

$$e_{ij} = \frac{I_1}{I}, \quad I > 0 \tag{4.6}$$

where

$$I_1 = \int_{x > 0} \mu_{zij}(x) \, dx,$$

$$I_2 = \int_{x < 0} \mu_{zij}(x) \, dx$$

$$I = I_1 + I_2,$$

$\mu_{z_{ij}}(x)$ is the membership functional of $\tilde{P}_i \, (-) \, \tilde{P}_j$.

Obviously according to the definition of e_{ij}, we can see that $e_{ij} + e_{ji} = 1$. Therefore, $e_{ij} > 0.5$ indicates that is $\tilde{X}_i$ preferred to $\tilde{X}_j$. If $e_{ij} = 0.5$, then $\tilde{X}_i$ and $\tilde{X}_j$ are equally preferred and if $e_{ij} < 0.5$, $\tilde{X}_j$ is preferred to $\tilde{X}_i$. Using the fuzzy preference relation, we can construct a fuzzy preference relation matrix as:

$$E = [e_{ij}]_{k \times k}$$

From the fuzzy preference matrix E, the fuzzy strict preference matrix E^s can be defined as:

$$E^s = \left[e_{ij}^s \right]_{k \times k} \tag{4.7}$$

where
$$e_{ij}^s = \begin{cases} e_{ij} - e_{ji} & \text{when } e_{ij} > e_{ji} \\ 0 & \text{otherwise} \end{cases}$$

Then, the non-dominated degree of each $\tilde{X}_i$ ($i = 1, 2, ..., k$) can be calculated by using the fuzzy strict preference relation matrix as:

$$\mu^{ND}(\tilde{X}_i) = \min_{\substack{j \in C_k \\ I \neq J}} \left\{ 1 - e_{ij}^s \right\} = 1 - \max_{\substack{j \in C_k \\ j \neq i}} e_{ji}^s \qquad (4.8)$$

where $\mu^{ND}(\tilde{X}_i)$ is the non-dominated degree of each $\tilde{X}_i$ and C_k is the cluster of fuzzy sequences.

The ranking procedure can be described as follows:

(i) Set $k = 0$ and $C_k = \{\tilde{X}_1, \tilde{X}_2, ..., \tilde{X}_n\}$.

(ii) Select the decision-maker which have the highest non-dominated degree $\tilde{X}_h$ (say) where

$$\mu^{ND}(\tilde{X}_h) = \max_i \{\mu^{ND}(\tilde{X}_i)\}.$$

(iii) Set the ranking for $\tilde{X}_h$, say $r(\tilde{X}_h) = k + 1$ and declare $\tilde{X}_h$ is the preferred fuzzy sequence in the cluster C_k. Similarly, we can define the preferred fuzzy sequence for each cluster. By repeating the above process, we can define the most preferred fuzzy sequence from among the preferred fuzzy sequence obtained from each cluster of the service sector.

5. NUMERICAL EXAMPLE

Here we consider a problem the content of which is extended from the problem in (Ross, 1995) to illustrate the proposed methods for decision making problem with fuzzy data for selection of most preferred *DM*. Suppose that a manager of Environmental Protection Agency (*EPA*) is faced with the challenge of immediate need to remediate soil and groundwater of a particular site where a carcinogen, trichloroethylene (*TCE*) has been detected at a level higher than the *EPA* maximum contaminant levels (*MCLs*). The manager has decided to remediate soil and ground water by using the remediation technique: bioremediation of soil with pump and treat and air stripping. Ten environmental professional engineers (*DMs*) were employed to assess the requirements for the implementation of the technique. The assessments were performed based on four criteria:

(i) Cost (O_1, it wants a reasonable cost).

(ii) Effectiveness (O_2, capacity to reduce the contaminant concentration).

(iii) Duration (O_3).

(iv) Speed of implementation (O_4).

The fuzzy data sequences based on the four criteria obtained from the assessments by the *DMs* are defined in the following table/matrix format (Table 5.1) as follows:

Table 5.1 Assessment by DMs

	O_1	O_2	O_3	O_4
D_1	(4.5, 4.86, 5.5)	(2.8, 3.31, 4)	(1.2, 1.45, 1.6)	(0, 0.22, 0.5)
D_2	(4.2, 5.21, 5.9)	(2.9, 3.65, 4.5)	(1, 1.42, 1.8)	(0, 0.25,0.5)
D_3	(4.5, 5.01, 5.5)	(2.9, 3.39, 3.8)	(0.9, 1.55, 2)	(0.1, 0.27, 0.6)
D_4	(4.3, 5.07, 5.6)	(2.9, 3.46, 4.3)	(1.1, 1.42, 1.7)	(0, 0.2, 0.5)
D_5	(4.9, 5.85, 6.8)	(1.9, 2.65, 3.2)	(3.4, 4.14, 4.6)	(0.9, 1.27, 1.6)
D_6	(5.6, 6.26, 6.9)	(2.4, 2.85, 3.3)	(3.4, 4.49, 5.1)	(0.9, 1.41, 1.9)
D_7	(5.3, 5.83, 6.8)	(2.2, 2.75, 3.5)	(3.8, 4.27, 4.8)	(0.9, 1.34, 1.7)
D_8	(5.5, 6.63, 7.8)	(2.6, 2.96, 3.4)	(4.7, 5.5, 6.8)	(1.5, 1.93, 2.4)
D_9	(5.3, 6.74, 8)	(2.5, 3.04, 3.9)	(4.7, 5.62, 6.5)	(1.3, 1.94, 2.5)
D_{10}	(5.7, 6.45, 7)	(2.4, 3.03, 3.5)	(5, 5.33, 6)	(1.7, 2.17, 2.6)

From the above data matrix, using Eqn. (4.2), we construct the fuzzy semi-equivalence relation matrix $\tilde{R}$ as follows:

$$\tilde{R} = \begin{vmatrix} 1 & .79 & .94 & .88 & .18 & .15 & .19 & .05 & .06 & .06 \\ .79 & 1 & .80 & .88 & .14 & .44 & .15 & .03 & .04 & .04 \\ .94 & .80 & 1 & .88 & .19 & .17 & .21 & .06 & .08 & .07 \\ .88 & .88 & .88 & 1 & .15 & .22 & .16 & .04 & .05 & .05 \\ .18 & .14 & .19 & .15 & 1 & .72 & .84 & .40 & .35 & .41 \\ .15 & .44 & .17 & .22 & .72 & 1 & .86 & .60 & .55 & .62 \\ .19 & .15 & .21 & .16 & .84 & .86 & 1 & .50 & .45 & .51 \\ .05 & .03 & .06 & .04 & .40 & .60 & .50 & 1 & 90 & .85 \\ .06 & .04 & .08 & .05 & .35 & .55 & .45 & .90 & .1 & .78 \\ .06 & .04 & .07 & .05 & .41 & .62 & .51 & .85 & .78 & 1 \end{vmatrix}$$

Performing max-min transitive closure mechanism, after two operations we obtain the max-min transitive closure matrix, $\tilde{R}^*$ of $\tilde{R}$ as shown below:

$$\tilde{R} = \begin{vmatrix} 1 & .88 & .94 & .88 & .44 & .44 & .44 & .44 & .44 & .44 \\ .88 & 1 & .88 & .88 & .44 & .44 & .44 & .44 & .44 & .44 \\ .94 & .88 & 1 & .88 & .44 & .44 & .44 & .44 & .44 & .44 \\ .88 & .88 & .88 & 1 & .44 & .44 & .44 & .44 & .44 & .44 \\ .44 & .44 & .44 & .44 & 1 & .84 & .84 & .62 & .62 & .62 \\ .44 & .44 & .44 & .44 & .84 & 1 & .86 & .62 & .62 & .62 \\ .44 & .44 & .44 & .44 & .84 & .86 & 1 & .62 & .62 & .62 \\ .44 & .44 & .44 & .44 & .62 & .62 & .62 & 1 & .90 & .85 \\ .44 & .44 & .44 & .44 & .62 & .62 & .62 & .90 & 1 & .85 \\ .44 & .44 & .44 & .44 & .62 & .62 & .62 & .85 & .85 & 1 \end{vmatrix}$$

For different values of λ, we can partition the group of the decision makers into clusters. The cluster arrangements with the validation indices are obtained as follows (Table 5.2)

Table 5.2 Partition of DMs into clusters

λ	Clustering arrangement	Val(Ind)
1	D_1, D_2, D_3, D_4, D_5, D_6, D_7, D_8, D_9, D_{10}	36.76
.94	$\{D_1, D_3\}$, D_2, D_4, D_5, D_6, D_7, D_8, D_9, D_{10}	38.52
.90	$\{D_1, D_3\}$, D_2, D_4, D_5, D_6, D_7, $\{D_8, D_9\}$, D_{10}	40.12
.88	$\{D_1, D_2, D_3, D_4\}$, D_5, D_6, D_7, $\{D_8, D_9\}$, D_{10}	47.72
.86	$\{D_1, D_2, D_3, D_4\}$, D_5, $\{D_6, D_7\}$ $\{D_8, D_9\}$, D_{10}	49.16
.85	$\{D_1, D_2, D_3, D_4\}$, D_5, $\{D_6, D_7\}$, $\{D_8, D_9, D_{10}\}$	51.96
.84	$\{D_1, D_2, D_3, D_4\}$, $\{D_5, D_6, D_7\}$, $\{D_8, D_9, D_{10}\}$	54.68
.62	$\{D_1, D_2, D_3, D_4\}$, $\{D_5, D_6, D_7, D_8, D_9, D_{10}\}$	59
.44	$\{D_1, D_2, D_3, D_4, D_5, D_6, D_7, D_8, D_9, D_{10}\}$	53.24

From the validation index column, we obtain the optimum cluster arrangement as given below:

$$\{D_1, D_2, D_3, D_4\}, \{D_5, D_6, D_7\}, \{D_8, D_9, D_{10}\}$$

From the first cluster $\{D_1, D_2, D_3, D_4\}$, using the various steps as explained in the equations [4.3-4.8] for selection of the preferred *DM*. we obtained the followings:

$$\tilde{P}_1 = (1.98 \quad 2.63 \quad 3.44)$$

$$\tilde{P}_2 = (1.85 \quad 2.81 \quad 3.73)$$

$$\tilde{P}_3 = (2.01 \quad 2.81 \quad 3.77)$$

$$\tilde{P}_4 = (1.91 \quad 2.65 \quad 3.57)$$

$$E = \begin{bmatrix} .5 & .42 & .39 & .48 \\ .58 & .5 & .55 & .55 \\ .61 & .53 & .59 & .59 \\ .52 & .45 & .41 & .5 \end{bmatrix}, E_s = \begin{bmatrix} 0 & 0 & 0 & 0 \\ .16 & 0 & 0 & .10 \\ .22 & .06 & 0 & .18 \\ .04 & 0 & 0 & 0 \end{bmatrix},$$

$$\mu^{ND}(D_1) = .78, \ \mu^{ND}(D_2) = .94, \ \mu^{ND}(D_3) = 1, \ \mu^{ND}(D_4) = .82.$$

Therefore, D_3 is the preferred decision maker in the first cluster. Similarly, D_6 and D_9 (or D_{10}) are respectively, the preferred decision makers of the second and third clusters. By repeating the same process, among the three preferred DMs, D_9 is found to be the most preferred *DM*.

6. CONCLUSION

In the paper, we have developed the selection of the most preferred *DM* from a group of decision makers in such a way that firstly, the *DMs'* are partitioned into groups having similar opinions; secondly, from each group a leader in the name of preferred *DM* is selected and then the most preferred *DM* is selected from among the preferred *DMs*. An example is given as an application of the proposed method. The proposed method highlights the working of the most problems faced in the real world.

References

1. Bandyopadhyay, S. and Maulik, U., 2002:. *An evolutionary technique based on K-means algorithm for optimal clustering.* Information Sciences, **146**, 221, 237.

2. Chen-Tung, Chen, 2001:. *A fuzzy approach to select the location of the distribution center.* Fuzzy Sets and Systems, **118**, 65-73.

3. Hirano, S., Sun, X. and Tsumoto, S., 2004: *Comparison of clustering methods for clinical databases.* Information Sciences, **159**, 155-165.

4. Khan, S.S. and Ahmad, A., 2004: *Cluster centre initialization algorithm for K-means clustering.* Pattern Recognition Letters, **25**, 1293-1302.

5. Klir, J., and Yuan, B., 2002:. *Fuzzy Sets And Fuzzy Logic Theory And Application,* Prentice Hall Of India Pvt. Ltd. New Delhi.

6. Kuo, R.J., Chang, K., and Chien, S.Y., 2004: *Integration of self-organizing feature maps and genetic algorithm based clustering method for market segmentation.* Journal of Organizational Computing and Electronic Commerce, **14**, 43-60.

7. Kuo, R.J., Liao, J.L., and Tu, C., 2005: *Integration of ART2 neural network and genetic K-means algorithm for analysing web browsing paths in electronic commerce.* Decision Support Systems, **40**, 355-374.

8. Lee, H.S., 2001: *An optimal algorithm for computing the max-min transitive closure of a fuzzy similarity matrix.* Fuzzy Sets and Systems, **123**, 129-136.

9. Mingyi, Z., Danning, L., and Ying, Z., 2010: *An approach to generation of decision rules based on a fuzzy semi-equivalence relation.* Journal of Universal Computer Sciences, **16**, 140-158.

10. Miyamoto, S., 2003, *Information clustering based on Fuzzy multi-sets.* Information Processing and Management, **39**, 195-213.

11. Ross, J., 1995, *Fuzzy Logic With Engineering Applications,* McGraw-Hill Inc International Edition, New York.

12. Wang, Y.J., and Lee, H.S., 2008: *A clustering method to identify representative financial ratios.* Information Sciences, **178**, 1087-1097.

13. Wang, Y.J., 2010: *A clustering method based on fuzzy equivalence relation for customer relationship management*. Expert Systems With Applications, **37**, 6421-6428.

14. Wu, K.L., and Yang, M.S., 2002: *Alternative c-means clustering algorithms*. Pattern Recognition, **35**, 2267-2278.

15. Zadeh, L.A., 1971: *Similarity Relation and Fuzzy Orderings*, Information Sciences, **3**, 177-200.

16. Zimmerman, H.J. 1991: *Fuzzy Set Theory- and Its Applications*, 2nd Edition, Allied Publishers, New Delhi.

Rough Sets, Fuzzy Sets and Soft Computing
Editor: S. Bhattacharya Halder

Concept Lattice for Rough Approximation

Dipankar Rana and Sankar Kumar Roy[*]
*Department of Applied Mathematics with Oceanology and Computer
Programming, Vidyasagar University
Midnapore, West Bengal
E-mail: sankroy2006@gmail.com*[*]

ABSTRACT

The theory of concept lattice is an efficient tool for knowledge representation and knowledge discovery.It is applied to many fields successfully. However, in many real life applications the problem under investigation cannot be described by formal concepts. Such concepts are called the non-definable concepts. The hierarchical structure of formal concept (called concept lattice) represents a structural information which obtained automatically from the input data table. We deal with the problem in which how further additional information be supplied to utilize the basic object attribute data table. In this paper, we provide rough concept lattice to incorporate the rough set into the concept lattice by using equivalence relation. Numerical examples are presented to illustrate the paper.

Keywords: Rough Set, Formal Concept Lattice, Equivalence Relation, Lattice.

1. INTRODUCTION

The formal concept analysis (FCA)is a mathematical framework, developed by Wille[1] and his colleagues at Darmstadt University, which is very useful for representation and analysis of data. The concept lattice is also called Galois lattice, was proposed by Wille in 1982 [1]. A concept lattice is an ordered hierarchy that is defined by a binary relationship between objects and attributes in a data set. As an efficient tool of data analysis and knowledge processing, the concept lattice has been applied in many fields, such as knowledge engineering, data mining, information searches, and software engineering [2]. Most of the researchers are concentrated on their attention to the concept lattice and defined on such topics as: construction of the concept lattice, pruning of the concept lattice, acquisition of rules, relationship between the concept lattice and rough set [3] and approximation. The basic formal concept analysis deals with input data in the form of a table with rows corresponding to objects and columns corresponding to attributes. The data table is formally represented by a formal context, which is a triplet (A, B, I) where A and B are sets and I is subset of $A \times B$ (i.e, $I \subseteq A \times B$) and defined a binary relation between A and B. The elements of A are called objects while the elements of B are called attributes or simply considered as the characteristics of objects. For a object and b characteristic, $(a, b) \in I$ or aIb shall indicate the following: a object owns the b attribute. Let us assume that (A, B, I) is a formal context. Using two operations, a lower and an upper approximations, we can describe every subset of the universe. The concept of the lower and upper approximations in rough set theory are fundamental to the examination of granularity in knowledge.

In this paper, we discuss the properties of concept lattice under rough set environment. FCA and rough set theory are two kinds of complementary mathematical tools for data analysis and data processing [6], [7]. Up to now, many efforts have been made to combine these two theories [6], [7], [9] in which the concept lattices based on rough set theory, including the attribute oriented concept lattice [8] and the object oriented concept lattice [6], [7] are perspective concept lattices for knowledge representation and knowledge discovery. In this paper, we provide rough concept lattice to incorporate rough set into the concept lattice by using equivalence relation.

Definition 1.1   If $X \subseteq A$, $Y \subseteq B$, then two operators α and β can be defined as $\alpha : 2^A \to 2^B$, $\alpha(Y) = \{a \in A: aIb, \forall \in B\}$, $\beta(x) = \{b \in B: aIb, \forall a \in A\}$. $\beta(X)$ will take us to the set of attributes that are common in the entire objects in the set X. Similarly $\alpha(Y)$ will take us to the attribute set of A that owns the entire attributes of Y. In other word $\beta(X)$, shall give the maximum object set that it hired by the entire objects in X while $\alpha(Y)$ still give the maximum object set that it owns by the entire objects in Y. (β, α) forms a Galois connection between 2^A and 2^B.

Definition 1.2   Let $K = (A, B, I)$ be a formal context, $X \to P(A)$ and $Y \to P(B)$, where $P(A)$ and $P(B)$ are the power sets of A and B respectively. (X, Y) is called a concept, if $\alpha(X) = Y$ and $\beta(Y) = X$ hold for X and Y, where X is called the extent of the concept and Y is called the intent of the concept.

Definition 1.3   Let (A, B, I) be a formal context and $L(K)$ denotes the set of all concepts in the formal context. If there exists an attribute set D such that $L(A, B, D) \approx L(A, B, I)$, then D is called a consistent set of (A, B, I). And further, if $\forall d \in D, L(A, D - \{d\}, D - \{d\}) L(A, B, I)$. Then D is called a reduct of (A, B, I) The intersection of all the reducts is called the core of (A, B, I).

Definition 1.4   For the formal context $K = (A, B, I)$, let $H_1 = (X_1, Y_1)$ and $H_2 = (X_2, Y_2)$ be two elements of $L(K)$ If there exists $H_1 \leq H_2$, $Y_2 \leq Y_1$, then $\leq$

is a partial order of $L(K)$, which produce a lattice structure in $Latt(K)$, called concept lattice of formal context $K = (A, B, I)$ also denoted by $L(K)$. Table 1 is a formal context, and Figure 1 shows its Hasse diagram.

Lemma 1.1: Let (A, B, I) be a context. Then the following assertions hold:

- $X_1 \leq X_2$ implies $\beta(A_1) \supseteq (A_2)$ for every $X_1, X_2 \subseteq A$, and $Y_1 \subseteq Y_2$ implies $\alpha(Y_1) \supseteq \alpha(Y_2)$ for every $Y_1, Y_2 \subseteq B$.

- $X \subseteq \alpha(\beta(X))$ and $\beta(X) = \beta(\alpha(\beta(X)))$ for all $X \subseteq A$, and $Y \subseteq \beta(\alpha(Y))$ and $\alpha(Y) = \alpha(\beta(\alpha(X)))$ for all $Y \subseteq B$.

2. FUNDAMENTAL THEOREM OF FCA

The fundamental theorem of FCA states that the set of all formal concepts on a given context with the Ordering $(X_1, Y_1) \leq (X_2, Y_2)$ if and only if $X_1 \subseteq X_2$ is a complete lattice called the concept lattice, in which the infimum and supremum are given by

$$\bigwedge_{i \in J} (X_j, Y_j) = \left(\bigcap_{i \in J} X_j, \beta \left(\alpha \left(\bigcup_{i \in J} Y_j \right) \right) \right)$$

$$= \left(\bigcap_{i \in J} X_j, \beta \left(\bigcap_{i \in J} X_j \right) \right)$$

$$\bigvee_{i \in J} (X_j, Y_j) = \left(\alpha \left(\beta \left(\bigcup_{i \in J} X_j \right) \right), \bigcap_{i \in J} Y_j \right)$$

$$= \left(\alpha \left(\bigcap_{i \in J} X_j \right), \bigcap_{i \in J} X_J \right)$$

Theorem 2.1 For two elements $H_1 = (X_1, Y_1)$ and $H_2 = (X_2, Y_2)$ of concept lattice $Latt(K) = (L(K), \cap, \cup)$ if $\alpha\,(X_1 \cap X_2) = Y_1 \cap Y_2$ and $\beta(Y_1 \cap Y_2) = X_1 \cup X_2$ then $(L(K), \cap, \cup)$ is a distributive lattice.

Proof: Since

$$(X_1, Y_1) \cup (X_2, Y_2) = (X_1 \cup X_2, Y_1 \cap Y_2)$$

and $\quad (X_1, Y_1) \cap (X_2, Y_2) = (X_1 \cap X_2, Y_1 \cup Y_2).$

Therefore,

$$H_1 \cap (H_2 \cap H_3) = (X_1, Y_1) \cap (X_2, Y_2) \cup (X_3, Y_3)$$

$$= (X_1, Y_1) \cap (X_2 \cup X_3, Y_2 \cap Y_3)$$

$$= (X_1 \cap (X_2 \cup X_3), Y_1 \cup (Y_2 \cap Y_3))$$

$$= ((X_1 \cap X_2) \cup (X_1 \cap X_3), (Y_1 \cup Y_2) \cap (Y_1 \cup Y_3))$$

and

$$(H_1 \cap H_2) \cup (H_1 \cup H_3) = (X_1 \cap H_2, Y_1 \cap Y_2) \cup (X_1 \cap X_3, Y_1 \cap Y_3)$$
$$= ((X_1 \cap X_2) \cup (X_1 \cup X_3), (Y_1 \cup Y_2) \cap (Y_1 \cup Y_3))$$

i.e., $H_1 \cap (H_2 \cup H_3)$

$$= (H_1 \cup H_2) \cap (H_1 \cup H_3).$$

Similarly, we can prove

$$H_1 \cup (H_2 \cap H_3) = (H_1 \cap H_2) \cap (H_1 \cup H_3).$$

Thus $\quad\quad Latt(K) = (L(K), \cap, \cup)$

is a distributive lattice.

Theorem 2.2: Let $Latt(K)$ be a concept lattice in formal context $K = (A, B, I)$. Let $H_1 = (X_1, Y_1)$ and $H_2 = (X_2, Y_2)$ be elements of $L(K)$. The following propositions are equivalent. (i) $H_1 \leq H_2$ (ii) $H_1 \cap H_2 = H_1$; $H_1 \cup H_2 = H_2$.

Proof: Suppose that (i) is true. Since $Y_2 \subseteq Y_1$ and $X_1 \subseteq H_2$, we have $(X_1, Y_1) \cap (X_2, Y_2) = (X_1 \cap X_2, \alpha(X_1 \cap X_2)) = (X_1, \alpha(X_1)) = (X_1, Y_1) = H_1$ and $H_1 \cup H_2 = (X_1, Y_1) \cup (X_2, Y_2) = (\beta(Y_1 \cap Y_2), Y_1 \cap Y_2)) = (\beta(Y_2), Y_2) = (X_2, Y_2) = H_2$,

Hence, (ii) is true.

Suppose that (ii) is true. Since $H_1 \cap H_2 = H_1 \leftrightarrow (X_1, Y_1) = (X_1 \cap X_2, \alpha(X_1 \cap X_2))$, from definition, it follows that $(Y_1 \cup Y_2) \cap \alpha(X_1 \cap X_2) = Y_1$ and $Y_1 \cup Y_2 \subseteq Y_1$. Thus $Y_2 \subseteq Y_1$, i.e., $H_1 \leq H_2$. Hence (i) is true.

Example 1: Table 1 shows a formal context (A, B, I), in which $A = \{1, 2, 3, 4\}$ and $B = \{a, b, c, d, e\}$.

The concepts are $(\{1\}, \{a, b, d, e\})$; $(\{2, 4\}, \{a, b, c\})$; $(\{1, 3\}, \{d\})$; $(\{1, 2, 4\}, \{a, b\})$; (A, ϕ). The concept lattice is shown in Fig. 1.

Table 1

	a	b	c	d	e
1	1	1	0	1	1
2	1	1	1	0	0
3	0	0	0	1	0
4	1	1	1	0	0

Table 2

	a	b	c	d	e
1	1	1	0	1	1
2	1	1	1	0	0
3	0	0	0	1	0
4	1	1	1	0	0

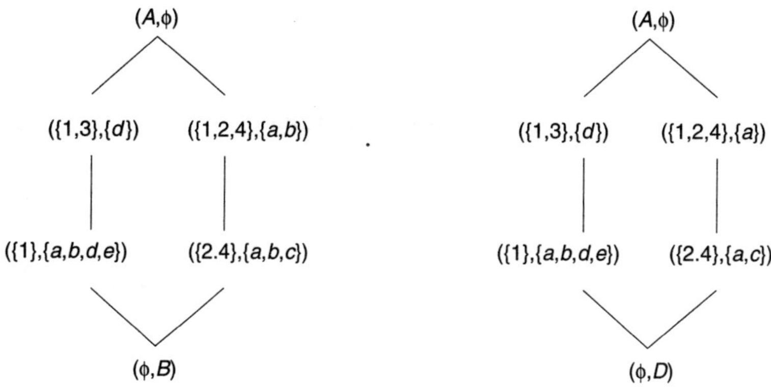

Fig. 1 Latt (*A, B, I*) in example 1   **Fig. 2** Latt (*A, B, D*) in example 1

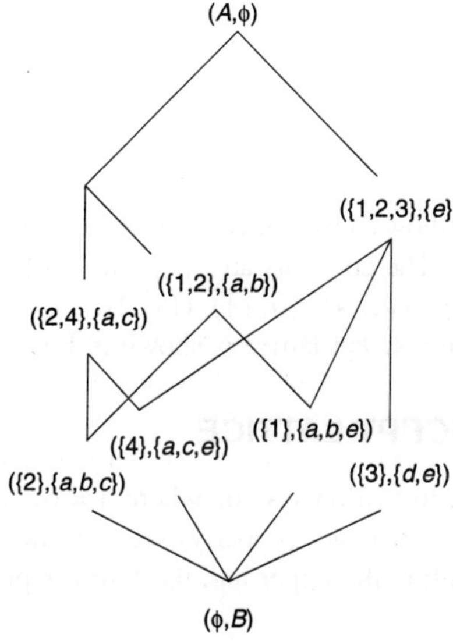

Fig. 3 Concept lattice for Table 2

If we consider another example then it will able to understand how far the approximation causes for concept approximation in rough set. For (*A, B, I*) formal context we have the following results:

Proposition 2.3: The core of the formal context is a reduct, There is only on reduct in the formal context.

Proposition 2.4: For $a \in B$ is an unnecessary attribute, $A - \{a\}$ is a consistent set.

Proposition 2.5: For $a \in B$ is an element of the core, $A - \{a\}$ is not a consistent set.

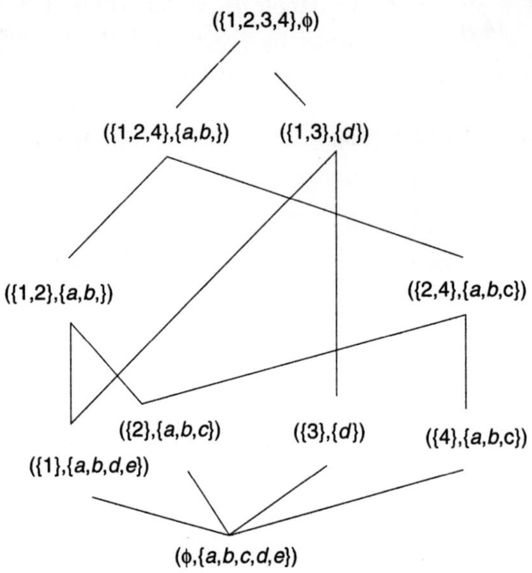

Fig. 4 Rough concept lattice for through formal context for Table 1

Example 2: Table 2 shows a formal context (A, B, I), in which $A = \{1, 2, 3, 4\}$ and $B = \{a, b, c, d, e\}$. The concepts are $(\{1\}, \{a, b, e\})$; $(\{4\}, \{a, c, e\})$; $(\{2\}, \{a, b, c\})$; $(\{3\}, \{d, e\})$; $(\{2, 4\}, \{a, c\})$; $(\{1, 2\}, \{a, b\})$; $\{1, 2, 4\}, \{a\})$; $(\{1, 3, 4\}, \{e\})$; (A, ϕ). The concept lattice is shown in Fig. 3.

3. ROUGH CONCEPT LATTICE

Let (A, B, I, E) be an information system, where $A = \{a_1, a_2, \ldots, a_m\}$ is an object set, $B = \{x_1, x_2, \ldots, x_n\}$ attribute (property) set, E is an equivalence relation on X, $\forall X \in A$, we can define the upper and the lower approximations of X about E as follows:

$$X\!\downarrow_E = \cup\{M \in A/E: M \subseteq X\}$$

$$= \{x \in A: EA(x) \subseteq X\} \tag{1}$$

$$X\!\uparrow_E = \cup\{M \in A/E: M \cap X \neq \phi\}$$

$$= \{x \in A: E_x(x) \cap X \neq \phi\} \tag{2}$$

Then $X\!\downarrow_E$, $X\!\uparrow_E$ are called E-lower and E-upper approximations of X respectively. If $X\!\downarrow_E = X\!\uparrow_E$, we say that X is definable, otherwise, X is rough. Similarly, we can define the upper and lower approximations of attributes set $Y \subseteq B$ about an equivalence relation on B.

Definition: For all $Y \subseteq B$, we denote $E_Y = \{(a_i, a_j) \in X \times X: f_p (aj), p \in Y\}$, where $f_p : A \rightarrow \{0, 1\}$ is defined by $f_p (ai) = 1$ if and only if the object a_i possesses property p, $(p \in A, a_i \in A)$, and E_Y is an equivalence relation, and E_Y can generate a partition of A, $X(Y) = \{Y(x) : x \in A\} = X/E_Y$, where $Y(x) = \{y \in A : yE_Yx\} = \{y \in A : f_p (y) = f_p (x), a_p \in Y\}$, that is, $Y(x) = \wedge \{(a_p, f_p (x)) : a_p \in Y\}$, $\cup Y(x) = \wedge(\wedge\{(a_p, f_p (x)) : a_p \in Y\})$.

Example 3: The above Table 1 is rough formal context (A, B, I, E) where $A = \{1, 2, 3, 4\}$, $B = \{a, b, c, d, e\}$ rough set theory $X = \{1, 2\} \subseteq A$, $Y = \{a, e\} \subseteq B$ then $f_a (1) = f_a (2) = f_a (4) = 1; f_a (3) = 0.$ $f_b (1) = f_b (2) = f_b (4) = 1; f_b (3) = 1.$ $f_c (2) = f_c (4) = 1; f_c (1) = f_c (3) = 0.$ $f_d (1) (1) = 1; f_d (1) (2) = f_d (1) (4) = 0.$ $f_e (1) = 1; f_e (2) = f_e (3) = f_e (4) = 0.$

Then the partition of X is $B/X(1) = \{\{a, b, d, e\}, \{c\}\}$; $B/X(2) = \{\{a, b, c\}, \{d, e\}\}$. So $B/X = \{\{a, b\}, \{d, e\}, \{c\}, \phi\}$; and the partition of Y is $A/Y(a) = \{\{1, 2, 4\}, \{3\}\}$. $A/Y(e) = \{\{1\}, \{2, 3, 4\}\}$. So $A/Y = \{\{1\}, \{2, 4\}, \{3\}, \phi\}$. Under $B/X = E_X$ the lower approximation of $Y = Y\uparrow_{E_X} = \phi$, the upper approximation of $Y = Y\downarrow_{E_X} = \{a, b, d, e\}$. Under $B/Y = E_Y$ the lower approximation of $X = X\downarrow_{E_Y} = \{1\}$, the upper approximation of $X = Y\uparrow_{E_Y} = \{1, 2, 4\}$. Here $f_p : A \rightarrow \{0, 1\}$, set $\{0, 1\}$ can extent to $[0, 1]$, and the condition of equivalence relation can also substitute for other relations, for example, the similar relation (that is, it satisfies reflexivity, symmetry, and transitivity). In essential, those functions are the same.

The information system (A, B, I, E) which have lower approximation set, upper approximation set and partition is called a rough formal context. $\forall a \in A, b \in B$, object a has attribute B then $(a, b) \in I$, aIb. For the rough formal context in Table 1 the Hasse diagram is shown in Fig. 4. One can find the extension of their meet by the set intersection of their extensions, and the intension of their join by the set intersection of their intensions. One cannot find directly the intension of their meet and the extension of their join by simply applying set-theoretic operators.

4. CONCLUSION

This paper presents the approach to approximating concepts in the framework of the formal concept analysis. The main focus is already shown how rough set theory can be employed as an approach to the problem of knowledge extraction. The approach shows how to approximate single set of objects, single set of features, and non-definable concepts. We use both the set of objects and the set of features for approximating non-definable concepts, which results in the fact that non-definable concepts with the same set of objects have different and more accurate concept approximations. Rough Lattice combines the advantages of concept lattice and rough set, so it is widely used in Information Retrieval, Data Mining, Software Engineering and other fields [7].

References

1. Wille, R., 1982: Restructring Lattice Theory: *An Approach Based on Hierarchies of Concepts, in Order Sets* (Ed. Rival, 1.), Dordrecht. Boston: Reidelpp 445-470.

2. Oosthuizen, G.D., 1996: *The Application of Concept Lattice to Machine Learning*, Technical Report, University of Pretoria, South Africa.

3. Oosthuizen, G.D. 1994: *Rough Set and Concept Lattices, in Rough Sets, and Fuzzy Sets and knowledge Discovery* (RSKD'93) (Ed. Ziarko, w.p.), London: Springer Verlag, pp 24-31.

4. Rana, D. and Roy, S.K., 2011: *Rough Set Approach on Lattice*, Journal of Uncertain Systems, vol 5(1): pp 72-80.

5. Wu, Q. and Liu, Z., 2009: *Real Formal Concept Analysis Based on Grey-Rough Set Theory*, Knowledge-Based Systems, vol 22(1): pp 38-45.

6. Yao, Y.Y., 2004: *Concept Lattices in Rough Set Theory*, In: Proceedings of 23rd International Meeting of the North American, Fuzzy Information Processing Society.

7. Yao, Y.Y., 2004: *A Comparative Study of Formal Concept Analysis and Rough Set Theory in Data Analysis, Rough Sets and Current Trends in Computing*, In: Proceedings of 3rd International Conference, RSCT04.

8. Gediga, G. and Duntsch, I., 2002: *Modal-Style Operators in Qualitative Data Analysis*, in: Proceedings of the 2002 IEEE International Conference on Data Mining, pp 155-162.

9. Hu, K., Sui, Y., Lu, Y., Wang, J. and Shi, C., 2001: *Concept Approximation in Concept Lattice, Knowledge Discovery and Data Mining*, In: Proceedings of the 5th Pacific-Asia Conference, PAKDD 2001, in:, Lecture Notes in Computer Science, vol. 2035: pp. 167-173.

Functional Dependences in Intuitionistic Fuzzy Rough Relational Databases

Chhaya Gangwal[1*] and R.N. Bhaumik[2#]
[1]Ph.D Scholar
[2]Emeritus Fellow (UGC) and Professor of Mathematics (Rtd.)
Department of Mathematics, Tripura University
Suryamaninagar, Tripura
E-mail: c_hhaya@hotmail.com[*]; bhaumik_r_n@yahoo.co.in[#]

ABSTRACT

Intuitionistic fuzzy (IF) rough set theory is emerging as a powerful tool for dealing with imprecision and uncertain information in relational database theory. Measures of similarity between IF sets and also between fuzzy rough sets have gained attention to the researchers for their wide applications in real life problems. In this paper, a new approach of similarity measure on IF rough sets based on the set-theoretic approach is proposed. The proposed measure can provide a useful way for measuring the degree of similarity between IF rough sets. Then the notion of similar equality for comparing the data values in IF rough relations is defined. Based on the proposed similarity measure, the paper focuses on the issues of IF rough functional dependencies. There are analogies to Armstrong's Axiom for classical functional dependency because Armstrong's axioms are the most important base of the theory of relational databases. The Armstrong's axioms with IF rough functional dependencies are checked and few results on the Armstrong's axioms of IF rough nature are verified.

1. INTRODUCTION

The relational data model [5] has been extensively studied for over three decades. This data model basically handles precise and exact data in an information source. However, many real life applications involve imprecise and inexact data. It is well known that since Zadeh[12] introduced fuzzy set in 1965, a lot of new theories treating imprecision and uncertainty have been proposed. IF sets originated by Atanassov [1] are an intuitively straightforward extension of Zadeh's fuzzy sets. While a fuzzy set gives a degree to which an element belongs to a set, an IF set gives both a membership degree and non-membership degree. In 1982, Pawlak[10] introduced the notion of a rough set as a new mathematical tool to deal with the approximation of a concept in the context of inexact, noisy or incomplete information. Since rough sets and IF sets both capture particular facets of the same notion-imprecision. Combining IF sets with rough sets, Chakrabarty et. al. [3] proposed IF rough sets.

Measures of similarity between fuzzy sets, as an important content in fuzzy mathematics, have gained attention from researchers for their wide applications in real world. Based on similarity measures pattern recognition, machine learning, decision making and market prediction, many measures of similarity between some fuzzy sets have been proposed and researched in recent years. Binyamin [2] introduced similarity measures of IF sets and Qi & Zhang [11, 13] introduced similarity measures for measuring the degree of similarity between fuzzy rough sets.

In this paper, the classical relational data model is extended to deal with IF rough information. Our first objective is to extend relational databases to include IF rough domains by suitably defining the IF rough functional dependency based on the new notion of similarity measure.

The rest of this paper is organized as follows. In Section II, some basic concept related to databases and IF rough sets is presented. In Section III, a new similarity measure between IF rough sets based on the set-theoretic approach is proposed. In Section IV, the concept of IF rough functional dependencies between two sets of attributes is introduced. In Section V, Armstrong's axioms with the notion of IF rough functional dependencies is verified. The conclusions are discussed in Section VI.

2. PRELIMINARIES

In this section, some basic concepts related to the classical relational data model and the IF rough set theory is given.

2.1 Relational Database Model [5]

We assume that the readers are familiar with the basic concepts of the relational data model. In the logical design of a relational data model, the integrity constraints play an important role. Among the integrity constraints, data dependency is important. There are various types of data dependencies, viz. functional dependency, join dependency, multivalued dependency etc. The single most important concept in relational schema design is that of functional dependency. It is a constraint between two sets of attributes X and Y where $X \subseteq R$, $Y \subseteq R$ and relation schema $R = \{A_1, A_2, ..., A_n\}$. A relation r of the relation schema $R = \{A_1, A_2, ..., A_n\}$ denoted by $r(R)$, is thus a set of tuples t_1, t_2, ..., t_n. A functional dependency $X \to Y$ specifies a constraint such that for any two tuples t_1 and t_2 in $r(R)$ whenever $t_1[X] = t_2[X]$ implies $t_1[Y] = t_2[Y]$. Thus

X functionally determines in R iff, whenever two tuples of $r(R)$ agree on their X-values, they must necessarily agree on their Y-values.

IF Rough Set [7]

Definition 2.2.1: Let X be a rough set in (U, S). An IF rough set $A = (\underline{A}, \overline{A})$ in X can be defined by a pair of mapping $\mu_{\underline{A}} : \underline{X} \to [0, 1]$, $v_{\underline{A}} : \underline{X} \to [0, 1]$ where $0 \leq \mu_{\underline{A}} + v_{\underline{A}} \leq 1$, $\mu_{\overline{A}} : \overline{X} \to [0, 1]$, $v_{\overline{A}} : \overline{X} \to [0, 1]$ and $0 \leq \mu_{\overline{A}} + v_{\overline{A}} \leq 1$ with the property that $\mu_{\underline{A}}(x) \leq \mu_{\overline{A}}(x)$ and $v_{\underline{A}}(x) \geq v_{\overline{A}}(x)$. Then, an IF rough set A in X is denoted by $A = \{\langle x, \mu_{\underline{A}}(x), \mu_{\overline{A}}(x), v_{\underline{A}}(x), v_{\overline{A}}(x)\rangle | \ \forall x \in X\}$.

2.3 IF Rough Relation [7]

Definition 2.3.1: Let $P(D_i)$ is the power set of D_i. An IF rough relation R is a subset of the product set $P(D_1) \times P(D_2) \times ... P(D_m) \times D_{[\mu, v]}$, where $D_{[\mu, v]}$ is the domain for membership and non membership value of the closed interval $[0, 1]$.

The Similarity Measures

Many measures of similarity between fuzzy sets, IF sets and rough sets have been proposed and researched in recent years.

Based on the set-theoretic approach, Pap pis and Karacapilidis [9] defined the similarity measure between fuzzy values A and B with fuzzy values a and b as follows:

$$M_p (A, B) = \frac{|A \cap B|}{|A \cup B|} = \frac{\sum\limits_{i=1}^{n} (a_i \wedge b_i)}{\sum\limits_{i=1}^{n} (a_i \vee b_i)} \tag{1}$$

In[4], Chen defined a similarity measure between two IF sets with IF values x and y (from the point of set-theoretic view) as follows:

$$M_C (x, y) = \frac{(\min (\mu(x), \mu(y)) + \min (v(x), v(y)) + \min (\pi(x), \pi(y)))}{(\max (\mu(x), \mu(y)) + \max (v(x), v(y)) + \max (\pi(x), \pi(y)))} \tag{2}$$

Zhang Chengyi et. al. in [13] also give a similarity measure between fuzzy rough values as follows: Let A be a fuzzy rough set in X, $x = \langle \mu_{\underline{A}}(x), \mu_{\overline{A}}(x)\rangle$, $y = \langle \mu_{\underline{A}}(y), \mu_{\overline{A}}(y)\rangle$ be the fuzzy rough values in A. The degree of similarity between the fuzzy rough values x and y can be evaluated by the function M_Z.

$$M_Z (x, y) = 1 - \frac{1}{2} \left(|\mu_{\underline{A}}(x) - \mu_{\underline{A}}(y)| + |\mu_{\overline{A}}(x) - \mu_{\overline{A}}(y)|\right) \tag{3}$$

Qi and Chengyi [11] introduced the following definition of similarity measures between the fuzzy rough (FR) values as follows:

Definition 2.4.1: Let $A \in FR(X)$, $M: A \times A \to [0, 1]$, $(x, y) \to M(x, y)$. Then M is called the similarity measure on A and $M(x, y)$ is called the similarity degree between the fuzzy rough values $x = \langle \mu_{\underline{A}}(x), \mu_{\overline{A}}(x) \rangle$ and $y = \langle \mu_{\underline{A}}(y), \mu_{\overline{A}}(y) \rangle$, if M satisfies the following conditions:

1. $M(x, y) = M(y, x)$
2. $M(x, y) \in [0, 1]$
3. $\forall x \in X, M(x, y) = M(x, z) \Rightarrow M(y, z) = 1$
4. $M(x, y) = M(x^c, y^c)$

3. THE SIMILARITY MEASURE BETWEEN IF ROUGH SETS AND ITS ELEMENTS

In this section, a new similarity measure between IF rough values based on set-theoretic approach is defined.

Definition 3.1: Let A bean IF rough set in X and $x = \langle \mu_A(x), \mu_{\overline{A}}(x), \nu_A(x), \nu_{\overline{A}}(x) \rangle$, $y = \langle \mu_A(y), \mu_{\overline{A}}(y), \nu_A(y), \nu_{\overline{A}}(y) \rangle$ be two IF rough values in A. The degree of similarity between the IF rough values x and y can be defined by the function M as follows:

$$M(x, y) = \frac{\left| \mu_{\underline{A}}(x) \wedge \mu_{\underline{A}}(y) + \mu_{\overline{A}}(x) \wedge \mu_{\overline{A}}(y) + \nu_{\underline{A}}(x) \wedge \nu_{\underline{A}}(y) + \nu_{\overline{A}}(x) \wedge \nu_{\overline{A}}(y) \right|}{\left| \mu_{\underline{A}}(x) \vee \mu_{\underline{A}}(y) + \mu_{\overline{A}}(x) \vee \mu_{\overline{A}}(y) + \nu_{\underline{A}}(x) \vee \nu_{\underline{A}}(y) + \nu_{\overline{A}}(x) \vee \nu_{\overline{A}}(y) \right|} \tag{4}$$

where $M(x, y) \in [0, 1]$.

The large the value of $M(x, y)$, the more the similarity between the IF rough values x and y.

Example 3.2: Let x and y be two IF rough values, where $x = [0.5, 0.8, 0.5, 0.2]$ and $y = [0.4, 0.7, 0.6, 0.3]$. By applying Eqn. (4), the degree of similarity between x and y can be evaluated as follows:

$M(x, y) = (\min(0.5, 0.4) + \min(0.8, 0.7) + \min(0.5, 0.6) + \min(0.2, 0.3)/ (\max(0.5, 0.4) + \max(0.8, 0.7) + \max(0.5, 0.6) + \max(0.2, 0.3)) = 0.8$

Example 3.3: Let x and y be two IF rough values, where $x = [0.5, 0.8, 0.5, 0.2]$ and $y = [0.0, 0.0, 0.0, 0.0,]$. By applying Eqn. (4), the degree of similarity between x and y can be evaluated as follows:

$M(x, y) = (\min(0.5, 0.0) + \min(0.8, 0.0) + \min(0.5, 0.0) + \min(0.2, 0.0)/ (\max(0.5, 0.0) + \max(0.8, 0.0) + \max(0.5, 0.0) + \max(0.2, 0.0)) = 0.$

Example 3.4: Let x and y be two IF rough values, where $x = y = [0.5, 0.8, 0.5, 0.2]$. By applying Eqn. (4), the degree of similarity between x and y can be evaluated as follows:

$M(x, y) = (\min (0.5, 0.5) + \min (0.8, 0.8) + \min (0.5, 0.5) + \min (0.2, 0.2)/$
$(\max (0.5, 0.5) + \max (0.8, 0.8) + \max (0.5, 0.5) + \max (0.2, 0.2)) = 1.$

That is, if the IF rough values x and y are equal, then $M(x, y) = 1$.

Theorem 3.5: The following statements are true:

1. The similarity degree is bounded, i.e., $0 \leq M(x, y) \leq 1$.
2. If the IF rough values x and y are equal, then $M(x, y) = 1$.
3. The similarity degree between the zero IF rough value and any non-zero IF rough value is equal to 0.
4. The similarity measure is commutative, i.e. $M(x, y) = M(y, x)$.

Now, the similarity measure between two given IF rough sets is generalized.

Let A and B be two IF rough sets in the universe of discourse $U = \{u_1, u_2, ..., u_n\}$, where

$$A = \left[\mu_{\underline{A}}(u_1), \mu_{\overline{A}}(u_1), v_{\underline{A}}(u_1), v_{\overline{A}}(u_1) \right]/u_1 + \dots$$

$$+ \left[\mu_{\underline{A}}(u_n), \mu_{\overline{A}}(u_n), v_{\underline{A}}(u_n), v_{\overline{A}}(u_n) \right]/u_n$$

and
$$B = \left[\mu_{\underline{B}}(u_1), \mu_{\overline{B}}(u_1), v_{\underline{B}}(u_1), v_{\overline{B}}(u_1) \right]/u_1 + \dots$$

$$+ \left[\mu_{\underline{B}}(u_n), \mu_{\overline{B}}(u_n), v_{\underline{B}}(u_n), v_{\overline{B}}(u_n) \right]/u_n$$

Then, based on the Eqn. (1) and Eqn. (4), the degree of similarity between the IF rough sets A and B can be defined as follows:

$$T(A, B) = \frac{1}{n} \sum_{i=1}^{n} M \left(\left[\mu_{\underline{A}}(u_i), \mu_{\overline{A}}(u_i), v_{\underline{A}}(u_i), v_{\overline{A}}(u_i) \right], \right.$$

$$\left. \left[\mu_{\underline{B}}(u_i), \mu_{\overline{B}}(u_i), v_{\underline{B}}(u_i), v_{\overline{B}}(u_i) \right] \right)$$

$$= \frac{1}{n} \sum_{i=1}^{n} \frac{\begin{array}{l} [\min (\mu_{\underline{A}}(u_i), \mu_{\underline{B}}(u_i)) + \min (v_{\underline{A}}(u_i), v_{\underline{B}}(u_i)) \\ + \min (\mu_{\overline{A}}(u_i), \mu_{\overline{B}}(u_i)) + \min (v_{\overline{A}}(u_i), v_{\overline{B}}(u_i))] \end{array}}{\begin{array}{l} [\max (\mu_{\underline{A}}(u_i), \mu_{\underline{B}}(u_i)) + \max (v_{\underline{A}}(u_i), v_{\underline{B}}(u_i)) \\ + \max (\mu_{\overline{A}}(u_i), \mu_{\overline{B}}(u_i)) + \max (v_{\overline{A}}(u_i), v_{\overline{B}}(u_i))] \end{array}} \tag{5}$$

Here $T(A, B) \in [0, 1]$. The larger the value of $T(A, B)$, the more similarity between the IF rough sets A and B.

From Eq. (5), we obtain the following Theorems for IF rough sets for $T(A, B)$.

Theorem 3.6: The following statements are true:

1. The similarity degree is bounded, i.e., $0 \leq T(A, B) \leq 1$.

2. If the IF rough sets A and B are equal, then $T(A, B) = 1$.

3. If $A \cap B$ is a zero IF rough set, then $T(A, B) = 0$.

4. The similarity measure is commutative, i.e., $T(A, B) = T(B, A)$.

Example 3.7:   Let A and B be two IF rough sets of the universe of discourse $U = \{u_1, u_2, u_3, u_4, u_5\}$, where

$$A = [0.7, 0.9, 0.3, 0.1]/u_1 + [0.4, 0.7, 0.6, 0.3]/u_2$$
$$+ [0.4, 0.8, 0.6, 0.2]/u_3 + [0.3, 0.6, 0.7, 0.4]/u_4$$
$$+ [0.8, 0.9, 0.2, 0.1]/u_5$$

and $$B = [0.6, 0.7, 0.4, 0.3]/u_1 + [0.7, 0.8, 0.3, 0.2]/u_2$$
$$+ [0.6, 0.7, 0.4, 0.3]/u_3 + [0.5, 0.9, 0.5, 0.1]/u_4$$
$$+ [0.6, 0.8, 0.4, 0.2]/u_5.$$

By applying Eq. (5), the degree of similarity between A and B can be evaluated as follows:

$$T(A, B) = \frac{1}{5} \sum_{i=1}^{5} \frac{\begin{array}{l}[\min(\mu_A(u_i), \mu_B(u_i)) + \min(\nu_A(u_i), \nu_B(u_i)) \\ + \min(\mu_{\overline{A}}(u_i), \mu_{\overline{B}}(u_i) + \min(\nu_{\overline{A}}(u_i), \nu_{\overline{B}}(u_i))]\end{array}}{\begin{array}{l}[\max(\mu_A(u_i), \mu_B(u_i)) + \max(\nu_A(u_i), \nu_B(u_i)) \\ + \max(\mu_{\overline{A}}(u_i), \mu_{\overline{B}}(u_i) + \max(\nu_{\overline{A}}(u_i), \nu_{\overline{B}}(u_i))]\end{array}}$$

$$= \frac{1}{5} \left(\begin{array}{c} \dfrac{(0.6 + 0.3) + (0.7 + 0.1)}{(0.7 + 0.4) + (0.9 + 0.3)} + \dfrac{(0.4 + 0.3) + (0.7 + 0.2)}{(0.7 + 0.6) + (0.8 + 0.3)} \\[2ex] + \dfrac{(0.4 + 0.4) + (0.7 + 0.2)}{(0.6 + 0.6) + (0.8 + 0.3)} + \dfrac{(0.3 + 0.5) + (0.6 + 0.1)}{(0.5 + 0.7) + (0.9 + 0.4)} \\[2ex] + \dfrac{(0.6 + 0.2) + (0.8 + 0.1)}{(0.8 + 0.4) + (0.9 + 0.2)} \end{array} \right)$$

$$= 0.139$$

It indicates that the degree of similarity between the IF rough sets A and B is equal to 0.139.

4. IF ROUGH FUNCTIONAL DEPENDENCY

In this section, the notion of similar equality (S_{EQ}) of IF rough relations, for comparing IF rough relations are introduced. Next the definition of IF rough functional dependency is defined. Lastly, a new notion of IF rough functional dependencies in relational databases is conducted to verify the so-called Armstrong's axioms of IF rough nature.

4.1 Similar Equality of IF Rough Relations

Similar Equality (S_{EQ}) of IF rough relations defined below can be used as an IF rough similarity measure to compare elements of a given domain. Suppose t_p and t_q are two tuples in a relation r over the scheme R.

4.1.1 Similar Equality of Tuples

Definition 4.1.1.1: The similar equality of two IF rough tuples t_p and t_q on the attribute A_i in an IF rough relations is given by:

$$S_{EQ}(t_p[A_i], t_q[A_i]) = M(t_p[A_i], t_q[A_i]).$$

The degree of similar equality of two IF rough tuples t_p and t_q on the attributes $X = \{A_1, ..., A_n\}$ $(X \subseteq R)$ in an IF rough relation is

$$S_{EQ}(t_p[X], t_q[X]) = S_{EQ}(t_p[A_1, ..., A_n], t_q[A_1, ..., A_n])$$

$$= \min\{S_{EQ}(t_p[A_1], t_q[A_1]), ...,$$

$$S_{EQ}(t_p[A_n], t_q[A_n])\}.$$

Theorem 4.1.1.2: The following statements of the properties of $S_{EQ}(t_p[X], t_q[X])$ are true:

1. The similar equality is bounded: $0 \leq S_{EQ}(t_p[X], t_q[X]) \leq 1$;
2. $S_{EQ}(t_p[X], t_q[X]) = 1$ iff all IF rough sets $t_p[A_s]$ and $t_q[A_s]$ $(i \leq s \leq j)$ are equal (i.e. $t_p[A_s] = t_q[A_s](i \leq s \leq j)$;
3. $S_{EQ}(t_p[X], t_q[X]) = 0$ iff $\exists A_i \in X$, $S_{EQ}(t_p[A_i], t_q[A_i]) = 0$, if and only if, $\exists A_i \in X$.
4. The similar equality is commutative: $S_{EQ}(t_p[X], t_q[X]) = S_{EQ}(t_q[X], t_p[X])$.

Informally, an IF rough functional dependency captures the semantics of the fact that, for given two tuples, Y values should not be less similar than X values. We now give the following definition of an IF rough functional dependency.

4.2 IF Rough Functional Dependency

Definition 4.2.1: Given a relation r over a relation schema $R = \{A_1, A_2, ..., A_m\}$ where $\text{Dom}(A_i)$, $i = 1, ..., m$, are sets of IF rough sets, an IF rough functional dependency $X \rightarrow_R Y$ where $X, Y \subseteq R$ holds over r, for all tuples t_p, t_q in r, we have $S_{EQ}(t_p[X], t_q[X]) \leq S_{EQ}(t_p[Y], t_q[Y])$.

5. VALIDATION OF ARMSTRONG'S AXIOMS WITH IF ROUGH FUNCTIONAL DEPENDENCY

Armstrong's axioms are the most important base of the theory of relational databases. In this section we check up whether Armstrong's axioms are also true with IF rough functional dependencies. So we study the Armstrong's axioms with IF rough functional dependencies and verify few results on the Armstrong's axioms of IF rough nature. The Armstrong's axioms dealing with IF rough functional dependencies are called IF rough Armstrong's Axioms.

Let U be the set of attributes with relation scheme $R(A_1, A_2, ..., A_m)$ and F be an IF rough functional dependency. Let X, Y and Z are subsets of the relation scheme R. We obtain mainly three Armstrong's axioms on IF rough functional dependency and checked whether it is true or not. Here IFRFD is the shorthand notation for the IF rough functional dependency.

IFRFD 1: Reflexive rule: If $Y \subseteq X \subseteq U$ holds, then $X \to_R Y$ holds in F.

Let t_p and t_q are two IF rough tuples in an IF rough relation. Since $Y \subseteq X \subseteq U$, from definition 4.1.1.1, whenever $S_{EQ}(t_p[X], t_q[X])$ is true, the result $S_{EQ}(t_p[Y], t_q[Y])$ is also then obviously true. Hence $X \to_R Y$ holds in F.

IFRFD2: Transitivity rule: If $X \to_R Y$, $Y \to_R Z$ holds in F, then $X \to_R Z$ holds in F.

Assume that both the IF rough functional dependence is $X \to_R Y$, $Y \to_R Z$ hold in the relation r. Therefore whenever $S_{EQ}(t_p[X], t_q[X])$ is true, $S_{EQ}(t_p[Y], t_q[Y])$ is also true. But whenever $S_{EQ}(t_p[Y], t_q[Y])$ is true, $S_{EQ}(t_p[Z], t_q[Z])$ is also true. Combining these two results we see that whenever $S_{EQ}(t_p[X], t_q[X])$ is true, the result $S_{EQ}(t_p[Z], t_q[Z])$ is also true. Hence $X \to_R Z$ holds in F.

IFRFD3: Augmentation rule: If $X \to_R Y$ holds in F, and $Z \subseteq U$, then $XZ \to_R YZ$ holds in F.

Let us consider $S_{EQ}(t_p[X], t_q[X])$	(i)
Therefore we have $S_{EQ}(t_p[Y], t_q[Y])$	(ii)
Now suppose that $S_{EQ}(t_p[XZ], t_q[XZ])$	(iii)
From (i) and (iii) we get $S_{EQ}(t_p[Z], t_q[Z])$	(iv)
From (ii) and (iv) we get $S_{EQ}(t_p[YZ], t_q[YZ])$	(v)

From (iii) and (v) we deduce that $XZ \to_R YZ$ is included in F. For IF rough functional dependency, we obtain *following* additional inference rules:

IFRFD4: Union rule: If $X \to_R Y$, $X \to_R Z$ holds, then $X \to_R YZ$ holds in F.

From IFRFD1, we know that $X \to_R Y$	(i)
and $X \to_R Z$	(ii)
From (i) we can write $X \to_R XY$	(iii)
From (ii) we can write $XY \to_R YZ$	(iv)
From (iii) and (iv) we get $X \to_R YZ$.	

IFRFD5: Decomposition rule: If $X \to_R YZ$ holds, then $X \to_R Y$ and $X \to_R Z$ holds in F.

Because $\quad X \to_R YZ$ $\hfill$ (i)

Now $Y \subset YZ$. Therefore by IFRFD1, we have $YZ \to_R Y$ $\hfill$ (ii)

Applying IFRFD2 on (i) and (ii) we get $X \to_R Y$.

IFRD6: Pseudotransitive rule: If $X \to_R Y$ and $WY \to_R Z$ holds, then $WX \to_R Z$ holds in F.

6. CONCLUSION

In this paper, we incorporate the notion of IF rough relational data model, with an objective to provide a generalized approach for treating imprecise data. We propose a new similarity measure between IF rough sets. We apply Similar Equality in IF rough relations. Based on the concept of similar equality of attribute values in IF rough relations, we develop the notion of IF rough functional dependency, which is a simple and natural generalization of classical or fuzzy rough functional dependencies. In this paper we have also worked on the validation of Armstrong's axioms with IF rough functional dependencies. Armstrong's axioms are the most important base of the theory of relational databases. It is true that the results which are true on crisp environment may not be true on fuzzy, IF or IF rough environment. So, we have studied the Armstrong's axioms with IF rough functional dependency and verified few results on the Armstrong's axioms of IF rough nature. The Armstrong's axioms dealing with IF rough are called by IF rough-Armstrong's-axioms.

References

1. Atanassov, K., 1986: *Intutionistic fuzzy sets;* Fuzzy Sets and Systems; **64**, 287-96.

2. Binyamin, Y., 2011: *A new similarity measure on intuitionistic fuzzy sets*; World academy of science; **78**, 36-40.

3. Chakrabarty, K., Gedeon, T., and Koczy, L., 1998: *Intuitionistic fuzzy rough sets*; In: Proceeding of the Fourth Joint Conference of Information Sciences, Durham, NC, JCTS; 211-214.

4. Chen, S.M., 1995: *A comparison of similarity measures of fuzzy values*; Fuzzy sets and systems; **72**, 79-89.

5. Codd, F.E., 1970: *A relational model of data for large shared data banks*; Comm. ACM.; **13**, 377-387.

6. Dutta, A.K., Al-adhaileh, Mosleh, H., and Biswas, R., 2009: *A method of intuitionistic fuzzy functional dependencies in relational databases*; European journal of scientific research; **29**, 415-425

7. Gangwal, C., and Bhaumik, R.N., 2012: *Intuitionistic fuzzy rough relational database model*; International Journal of Database Theory and Application; **5(3)**, 91-102.

8. Hamouz, S., and Biswas, R., 2006: *Fuzzy functional dependencies in relational databases*; International journal of computational cognition; **4,** 39-43.

9. Pappis, P., and Karacapilidis, N.I., 1993, *A comparative assessment of measures of similarity of fuzzy values*; Fuzzy Sets and Systems; **56**, 171-174.

10. Pawlak, Z., 1982, *Rough Sets*; Int. J. Inf. Comp. Sci.; **11(5)**, 341 – 356.

11. Qi, N., and Chengyi, Z., 2008: *A new similarity measure on fuzzy rough sets*; International journal of pure and applied mathematics; **47**, 89-100.

12. Zadeh, L.A., 1965: *Fuzzy sets*; Information and control; **8**, 338 – 353.

13. Zhang, C., Dang, P., and Fu H., 2004: *On measures of similarity between fuzzy rough sets*; International Journal of Pure and Applied Mathematics; **10(4)**, 451-458.

Rough Sets, Fuzzy Sets and Soft Computing
Editor: S. Bhattacharya Halder

ISO-Compactness in Smooth L-Fuzzy Topological Spaces

Jaydip Bhattacharya
Iswar Chandra Vidyasagar College
Belonia, South Tripura
E-mail: jay_bhatta@rediffmail.com

ABSTRACT

The aim of this paper is to introduce the definition of iso-compactness in smooth L-fuzzy topological spaces, where L is a fuzzy lattice ie. a completely distributive lattice. We define this concept for arbitrary L-fuzzy sets and obtain some properties like union, invariance, productivity etc.

Keywords: Smooth L-fuzzy compactness, Smooth L-fuzzy countably compactness, smooth L-fuzzy isocompactness, smooth L-fuzzy perfect mapping, smooth L-fuzzy continuous mapping.

AMS Subject Classification: 54 A 40

1. INTRODUCTION

The concept of fuzzy topology was first defined by Chang [4] in 1968 and later redefined in a somewhat different way by R. Lowen [11] in 1976 and by R. Hutton [7] in 1980. A.P. Sostak [14] in 1985, introduced the fundamental concept of a fuzzy topological structure, as an extension of both classical topology and fuzzy topology, in the sense that not only objects are fuzzified, but also the axiomatic. He gave some rules and showed how such an extension can be realized. After that, several authors viz. Ramadan [12, 13], Chottopadhyay et. al. [5, 6], and R. Srivastava [15] have re-introduced the same definition and studied smooth fuzzy topological spaces being aware of Sostak's work.

The notion of compactness is one of the most important concept in general topology. Many papers on fuzzy compactness have been published and various kinds of fuzzy compactness have been presented and studied. Among these compactness, the fuzzy compactness in L-fuzzy topological spaces introduced by Kudri possesses more satisfactory properties than others. Aygun et. al. [1] introduced the notion of L-fuzzy compactness in smooth L-fuzzy topological spaces as a generalization of L-fuzzy compactness introduced by Kudri [9]. In his paper, "The compactness of countably compact spaces", P. Bacon [2], studied about the class of those spaces in which closed, countably compact subsets are always compact and these spaces are known as iso-compact spaces in general topology. In 2001, Bhaumik and Bhattacharya [3] introduced the concept of iso-compactness in L-fuzzy topological spaces. In this paper we introduce a new class of spaces, called smooth L-fuzzy isocompact spaces and studied some of its properties.

2. PRELIMINARIES

Throughout this paper X and Y will be non-empty ordinary sets and $L = L$ ($\leq$, $\vee$, $\wedge$, $'$) will denote a fuzzy lattice, i.e. a completely distributive lattice with a smallest element 0 and a largest element 1 ($0 \neq 1$), and with and order reversing involution $a \to a'$ ($a \in L$). An L-fuzzy subset on X is a mapping λ: $X \to L$, and the family of L-fuzzy subsets on X is denoted by L^X. X is called the carrier domain of each L-fuzzy subset on X. We shall denote by L^X the lattice of L-fuzzy subsets of X, by λ' the complement of λ and if $A \subset x$ by χ_A the characteristic function of A in X.

Definition 2.1 [1]:   An element p of L is called prime if and only if $p \neq 1$ and whenever $a, b \in L$ with $a \wedge b \leq p$ then $a \leq p$ or $b \leq p$. The set of all prime elements of L will be denoted by $\mathrm{pr}(L)$.

Definition 2.2 [1]:   An element α of L is called union-irreducible or coprime if and only if whenever $a, b \in L$ with $\alpha \leq a \leq b$ then $\alpha \leq a$ or $\alpha \leq b$. The set of all nonzero union-irreducible elements of L will be denoted by $M(L)$. It is obvious that $p \in \mathrm{pr}(L)$ if and only if $p' \in M(L)$.

Definition 2.3 [10]:   Let (X, τ) be an L-fuzzy topological space and $\lambda \in L^X$. The L-fuzzy subset λ is said to be compact (countably compact) if and only if for every $p \in \mathrm{pr}(L)$ and every (countable) collection $(\gamma_i)_{i \in J}$ of open L-fuzzy sets with $(\vee_{i \in J} \gamma_i)(x) \leq p$ for all $x \in X$ with $\lambda(x) \geq p'$, there exists a finite subset F of J such that $(\vee_{i \in F} \gamma_i)(x) \leq p$ for all $x \in X$ with $\lambda(x) \geq p'$.

If the L-fuzzy set λ is the whole space X, then the L-fuzzy topological space (X, τ) is (countable) compact.

Definition 2.4 [3]:   Let (X, τ) be an L-fuzzy topological space and $\lambda \in L^X$. The L-fuzzy subset λ is said to be L-fuzzy isocompact if every countably compact and closed L-fuzzy subset of λ is L-fuzzy compact.

If λ is the whole space, then L-fuzzy topological space (X, τ) is isocompact. It can be seen that every compact L-fuzzy topological space is L-fuzzy isocompact.

Definition 2.5 [14]: A smooth L-fuzzy topology on X is a map $\tau\colon L^X \to L$ satisfying the following three axioms:

    (i) $\tau(0) = \tau(1) = 1$,

    (ii) $\tau(\lambda \wedge \mu) \geq \tau(\lambda) \wedge \tau(\mu)$ for every $\lambda, \mu \in L^X$,

    (iii) $\tau(\vee_{i \in I} \lambda_i) \geq \wedge_{i \in I} \tau(\lambda_i)$ for every family $(\lambda_i)_{i \in I}$ in L^X.

The pair (X, τ) is called a smooth L-fuzzy topological space. For every $\lambda \in L^X$, $\tau(\lambda)$ is called the degree of openness of the fuzzy subset λ.

Definition 2.6 [14]: Let (X, τ) be a smooth L-fuzzy topological space. The map $\phi\colon L^X \to L$ defind by $\phi(\mu) = \tau(\mu')$ for every $\mu \in L^X$ is called the degree of closedness on X.

Definition 2.7 [12]: Let (X, τ) and (Y, τ') be smooth L-fuzzy topological spaces. A function $f\colon (X, \tau) \to (Y, \tau')$ is called

    (i) smooth continuous if and only if $\forall \lambda \in L^Y$, $\tau(f^{-1}(\lambda)) \geq \tau'(\lambda)$ where $f^{-1}(\lambda)(x) = \lambda(f(x))$.

    (ii) smooth open if and only if $\forall \lambda \in L^X$, $\tau'(f(\lambda)) \geq \tau(\lambda)$ where $f(\lambda)(y) = \vee\{\lambda(x)\colon x \in X \text{ with } f(x) = y\}$.

    (iii) smooth closed if and only if $\forall \mu \in L^X$, $\phi'(f(\mu)) \geq \phi(\mu)$.

    (iv) smooth perfect [8] iff f is smooth continuous, smooth closed and for each $y \in Y$, $f^{-1}(y)$ is smooth compact L-fuzzy subset in (X, τ).

Definition 2.8 [13]: Let (X, τ) be a smooth L-fuzzy topological space and $\lambda \in L^X$. The L-fuzzy subset λ is said to be smooth compact (countably compact) if and only if for every $p \in pr(L)$ and every (countable) collection $(\gamma_i)_{i \in J}$ of L-fuzzy sets with $\tau(\gamma_i) \leq p$, $\forall i \in I$ and $(\vee_{i \in J} \gamma_i)(x) \leq p$ for all $x \in X$ with $\lambda(x) \geq p'$, there exists a finite subset F of J such that $(\vee_{i \in F} \gamma_i)(x) \leq p$ for all $x \in X$ with $\lambda(x) \geq p'$.

If the L-fuzzy set λ is the whole space X, then the L-fuzzy topological space (X, τ) is smooth (countable) compact.

Theorem 2.9 [13]: Let $f\colon (X, \tau) \to (Y, \tau')$ be a smooth continuous mapping and let λ be a smooth compact L-fuzzy subset of (X, τ). Then $f(\lambda)$ is a smooth compact L-fuzzy subset of (Y, τ').

3. SMOOTH ISOCOMPACT SPACES

In this section, a new class of smooth L-fuzzy topological spaces, called smooth isocompact spaces is studied.

Definition 3.1: Let (X, τ) be a smooth L-fuzzy topological space and $\lambda \in L^X$. The L-fuzzy subset λ is said to be smooth isocompact if and only if every closed countably compact subset of λ is compact. If λ is the whole space, then the L-fuzzy topological space (X, τ) is also smooth isocompact.

Clearly, every smooth compact space is smooth isocompact. Now we discuss some properties of smooth isocompact spaces.

Theorem 3.2: If an L-fuzzy topological space (X, τ) is the union of a countable collection of smooth closed and smooth isocompact L-fuzzy subsets, then (X, τ) is smooth L-fuzzy isocompact.

Proof: Suppose $X = \vee \mu_i$, where each μ_i is smooth closed and smooth isocompact L-fuzzy subset of X and let β be a smooth closed and smooth countably compact L-fuzzy subset of X. Let $p \in pr(L)$ and let$\{\gamma_i\}_{i \in J}$ be a family of smooth open L-fuzzy sets with $(\vee_{i \in J} (\gamma_i))(x) \leq p$ for all $x \in X$ such that $\beta(x) \geq p'$.

For each i, $\beta \wedge \mu_i$ is smooth closed, smooth countably compact L-fuzzy subset of μ_i. Since each μ_i is smooth isocompact then $\beta \wedge \mu_i$ is smooth compact. Thus there exist a finite subset F of J with $(\vee_{i \in F} (\gamma_i))(x) \leq p$ for all $x \in X$ such that $(\beta \wedge \mu_i) (x) \geq p'$ i.e., $\beta(x) \geq p'$. Hence β is a smooth compact L-fuzzy subset, which implies that X is smooth isocompact.

Theorem 3.3: Let $f \colon (X, \tau) \to (Y, \tau')$ be a smooth continuous mapping from smooth isocomapact space (X, τ) onto an L-fuzzy topological space (Y, τ'). Then $(Y, \tau')$is smooth isocompact.

Proof: Let β be a smooth closed and smooth countably compact L-fuzzy subset of (Y, τ'). Since f is smooth continuous map then $f^{-1} (\beta)$ is closed and countably compact L-fuzzy subset of (X, τ). By smooth isocompactness of (X, τ), $f^{-1} (\beta)$ is smooth compact. Since f is onto smooth continuous then $ff^{-1} (\beta) = \beta$ is smooth compact L-fuzzy subset in (Y, τ'). Hence (Y, τ') is smooth isocompact.

Theorem 3.4: Let $f \colon (X, \tau) \to (Y, \tau')$ be a smooth perfect map from an L-fuzzy topological space (X, τ) onto an smooth isocomopact space (Y, τ'). Then (X, τ) is smooth isocompact.

Proof: Let β be a smooth closed and smooth countably compact L-fuzzy subset of (X, τ). Since f is smooth closed, $f(\beta)$ is also smooth closed. $f(\beta)$ is also smooth countably compact L-fuzzy subset of (Y, τ') as f is smooth continuous. Since (Y, τ') is smooth isocompact $f(\beta)$ is smooth compact. By smooth perfectness of f, the inverse image $f^{-1} f(\beta)$ of smooth compact $f(\beta)$ is smooth compact L-fuzzy subset. But $\beta \leq f^{-1} f(\beta)$ and is smooth closed. So β is smooth compact subset of (X, τ). Hence (X, τ) is smooth isocompact.

Definition 3.5 [8]: An L-fuzzy topological space (X, τ) is called fully stratified if for each $p \in L$, the L-fuzzy set which takes constant value p at each point $x \in X$ belongs to τ.

Theorem 3.6 [8]: If (X, τ) be a compact L-fuzzy topological space and (Y, τ') be a fully stratified L-fuzzy topological space, then the projection mapping $P_Y \colon X \times Y \to Y$ is L-fuzzy perfect.

Theorem 3.7: The product of a smooth compact L-fuzzy topological space and a fully stratified smooth L-fuzy isocompact space is smooth L-fuzzy isocompact.

Proof: Let (X, τ) be a smooth compact L-fuzzy topological space and (Y, τ') be a fully stratified smooth isocompact L-fuzzy topological space and consider the projection map $P_Y \colon X \times Y \to Y$.

Let β be smooth countably compact, smooth closed L-fuzzy subset of $X \times Y$. $P_Y(\beta)$ is smooth countably compact L-fuzzy subset, P_Y being smooth L-fuzzy continuous. From 3.6, P_Y is smooth L-fuzzy perfect and so is smooth L-fuzzy closed. Thus $P_Y(\beta)$ is smooth countably compact and smooth L-fuzzy closed subset of (Y, τ'). By smooth L-fuzzy isocompactness of (Y, τ'), $P_Y(\beta)$ is smooth L-fuzzy compact subset of (Y, τ'). Thus $X \times P_Y(\beta)$ is smooth L-fuzzy compact. Since β is smooth L-fuzzy closed subset of $X \times P_Y(\beta) \le X \times Y$, β is smooth compact L-fuzzy subset of $X \times Y$. Hence $X \times Y$ is smooth L-fuzzy isocompact.

Definition 3.8: An smooth L-fuzzy topological space (X, τ) is called smooth hereditarily isocompact if every sub space of it is smooth isocompact.

Theorem 3.9: Let (X, τ) be a fully stratified smooth isocompact L-fuzzy topological space and (Y, τ') be a hereditarily smooth isocompact L-fuzzy topological space. Then $X \times Y$ is smooth L-fuzzy isocompact.

Proof: Let (X, τ) be a fully stratified smooth isocompact L-fuzzy topological space and (Y, τ') be a smooth hereditarily isocompact L-fuzzy topological space. Let us consider the projection map $P_Y \colon X \times Y \to Y$. Let β be a smooth countably compact, smooth closed L-fuzzy subset of $X \times Y$. Then $P_Y(\beta)$ is smooth countably compact L-fuzzy subset of (Y, τ'). Since (Y, τ') is smooth hereditarily isocompact L-fuzzy topological space then $P_Y(\beta)$ is smooth L-fuzzy compact. Thus from 3.7, $X \times P_Y(\beta)$ is smooth L-fuzzy isocompact. Since β is a smooth countably compact and smooth closed subset of $X \times P_Y(\beta) \le X \times Y$, β is smooth compact L-fuzzy subset of $X \times Y$. Hence $X \times Y$ is smooth L-fuzzy isocompact.

The concept of closed-complete space and the relation between closed-complete space and isocompact space in L-fuzzy topological space are investigated by Bhaumik and Bhattacharya[3].

Definition 3.10: An smooth L-fuzzy topological space (X, τ) is said to be smooth closed-complete iff every smooth L-fuzzy closed ultrafilter of X with countable intersection property is fixed.

The following theorem can be deduced on the line of Theorem 2.10[3].

Theorem 3.12: Every smooth L-fuzzy closed complete space is smooth isocompact.

The definition of Hausdorffness is considered here due to Warner and Mclean[16].

Definition 3.13: A smooth Hausdorff L-fuzzy topological space (X, τ) is said to be smooth L-fuzzy almost realcompact if for any given smooth open ultra filter $\mathcal{U}$ in X with $\wedge_{n \in N}$ cl $U_n \neq \phi$ for each smooth countable subfamily $\{U_n\}_{n \in N}$, of U , $\wedge\{$cl $U: U \in$ U$\} \neq \phi$.

Definition 3.14 [13]: A smooth L-fuzzy topological space (X, τ) is said to be regular iff for every $p \in pr(L)$ and each $\lambda \in L^X$ satisfying $\tau(\lambda) \leq p$ can be written as $\lambda = \vee\{\beta \in L^X: \tau(\beta) \geq \tau(\lambda), \beta \leq \lambda\}$.

It can be easily proved that the smooth closed L-fuzzy subset of a smooth regular almost real compact space is smooth almost real compact.

Relating this concept with that of smooth iso-compactness, we have the following theorem:

Theorem 3.15: Every smooth regular, almost realcompact L-fuzzy topological space is smooth isocompact.

Proof: Let (X, τ) be a smooth regular and smooth almost realcompact L-fuzzy topological space and β be a smooth countably compact, smooth closed L-fuzzy subset of X. Since smooth closed L-fuzzy subset of a smooth regular almost realcompact space is smooth almost real compact, β is smooth almost real compact. Let ζ be a smooth closed L-fuzzy filter in β. Consider $\mathcal{U} = \{\lambda : \lambda$ is smooth open L-fuzzy subset in β and $\lambda \leq \gamma, \gamma \in \zeta\}$. Then $\mathcal{U}$ is a smooth filter which is contained in some open smooth ultra L-fuzzy filter $\mathfrak{M}$. Since β be a smooth countably compact, for any countable subfamily $\mathcal{F}$ of $\mathfrak{M}$, $\wedge \{$cl $\lambda : \lambda \in$ F $\}$ $\neq \phi$. But by smooth almost real compactness of β, infimum of the whole family $\mathfrak{M}$ is non-empty, i.e., there exist a smooth L-fuzzy point x_p in $\wedge \{$cl $\lambda : \lambda \in$ M $\}$. Suppose there exists an element λ of ζ that does not contain the L-fuzzy point x_p. Since β is smooth regular, there exists a smooth open L-fuzzy subset γ containing in λ such that cl λ does not contain x_p which gives a contradiction. It follows that $x_p \in \zeta$ and β is smooth compact. Hence (X, τ) is a smooth isocompact space.

References

1. Aygün, H., Warner, M.W., and Kudri, S.R.T., 1997: On smooth L-fuzzy topological spaces, J. Fuzzy Math. 5(2), 321-338.

2. Bacon, P., 1970: On compactness of countably compact spaces, Pac. J. Math. 32, 587- 592.

3. Bhaumik, R.N., and Bhattacharya, J., 2001: Fuzzy compactness in L-fuzzy top spaces, Proc. Nat. Conf. on Rec. Dev. in Math. & Appl, Assam Univ. 104-109.

4. Chang, C.L., 1968: Fuzzy topological spaces, J. Math. Anal. Appl. 24, 182-190.

5. Chattopadhyay, K.C., Hazra, R.N., and Samanta, S.K., 1993: Gradation of openness, Fuzzy topology, Fuzzy Sets and Systems 49, 207-212.

6. Chattopadhyay K.C., and Samanta, S.K., 1993: Fuzzy topology: Fuzzy closure operator, fuzzy compactness and fuzzy connectedness, Fuzzy Sets and Systems 54, 207-212.

7. Hutton, R., 1980: Products of fuzzy topological spaces, Top. and its Appl. 14, 59-67.

8. Jie, Z., 1995: A characterization of fuzzy compactness and its applications, Fuzzy Sets and Systems, 73, 89-96.

9. Kudri, S.R.T., 1994: Compactness in L-Fuzzy topological spaces, Fuzzy Sets and Systems 67, 329-336.

10. Kudri, S.R.T., and Warner, M.W., 1995: Some good L-fuzzy compactness-related concepts and their properties-I, Fuzzy Sets and Systems 76, 141-155.

11. Lowen, R., 1976: Fuzzy topological spaces and fuzzy compactness, J. Math. Anal. Appl., 56, 621-633.

12. Ramadan, A.A., 1992: Smooth topological spaces, Fuzzy Sets and Systems 48, 371-375.

13. Ramadan, A.A., and Abbas, S.E., 2001: On L-smooth compactness, J. Fuzzy Math., 9(19): 59-73.

14. Sostak, A.P., 1985: On a fuzzy topological structure, Suppl. Rend. Circ. Mat. Palermo Ser. II, 11, 89-103.

15. Srivastava, R., 1994: On separation axioms in newly defined fuzzy topology, Fuzzy Sets and Systems 62, 341-346.

16. Warner, M.W., and Mclean, R.G., 1993: On compact Hausdorff L-fuzzy spaces, Fuzzy Sets and Systems, 56, 103-110.

Rough Sets, Fuzzy Sets and Soft Computing
Editor: S. Bhattacharya Halder

Fuzzy Pairwise γ-Regular Space and Fuzzy Pairwise γ-Normal Space

Anjan Mukherjee[1*] and Dipankar Dee[2#]

[1]*Department of Mathematics, Tripura University Surjyamaninagar*
Agartala, Tripura
[2]*Department of Mathematics, Ramthakur College*
Badharghat, Agartala, Tripura
E-mail: anjan2002_m@yahoo.co.in[*] ; dipankardee@yahoo.co.in[#]

ABSTRACT

In 1990, M.N. Mukherjee and S.P. Sinha introduced the concepts of fuzzy regular, fuzzy almost regular and fuzzy semi regular space in fuzzy topological space (X, T). S. Sampat Kumar generalized these concepts in fuzzy bitopological space in 2000. The aim of this paper is to introduce fuzzy pairwise γ-regular space, fuzzy pairwise γ normal space and study their basic properties.

Keywords: (τ_i, τ_j)-fuzzy γ open set, fuzzy pairwise γ-continuous mappings, fuzzy regular space, fuzzy almost regular space, fuzzy pairwise γ-regular space.

AMS Subject Classifications: 54A40; 54E55.

1. INTRODUCTION

I.M. Hanafy [5] introduced the concept of fuzzy-γ-open sets and fuzzy-γ-continuity in 1999. After this F.S. Mahamoud and others [11] generalized these notions from fuzzy topological space to fuzzy bitopological spaces. S. Sampat Kumar studied these concepts of fuzzy regular space [15] in a fuzzy bitopological space in 2000. In Section 2, we introduce fuzzy pairwise γ regular space, fuzzy pairwise γ normal space and study the basic properties of these concepts.

Now, we give some basic definitions and results in order to study fuzzy pairwise γ regular space and fuzzy pairwise γ normal space.

A system (X, T_1, T_2) consisting of a set X with two fuzzy topological spaces T_1 and T_2 on X is called fuzzy bitopological space (fbts, in short). Throughout this paper indices i, j take values in $\{1, 2\}$ with $i \neq j$.

In short, we use $\tau_i - fo$ and $\tau_j - fc$ for τ_i-fuzzy open and τ_j-fuzzy closed respectively and τ_i-cl and τ_j-int for τ_i-closure and τ_j-interior respectively.

Definition 1.1 [14]: Let μ be a fuzzy set of a fuzzy bitopological space X. Then μ is called

(a) *a* (τ_i, τ_j)-fuzzy semi open $[(\tau_i, \tau_j)$-fso]set if
$$\mu \leq \tau_j\text{-cl}(\tau_i\text{-(int}(\mu)$$

(b) *a* (τ_i, τ_j)-fuzzy semi closed $[(\tau_i, \tau_j)$-fsc]set if
$$\mu \geq \tau_j\text{-int}(\tau_i\text{-(cl }(\mu)$$

(c) *a* (τ_i, τ_j)-fuzzy pre-open $[(\tau_i, \tau_j)$-fpo]set if
$$\mu \leq \tau_i\text{-int}(\tau_j\text{-(cl}(\mu)$$

(d) *a* (τ_i, τ_j)-fuzzy pre-closed $[(\tau_i, \tau_j)$-fpc] set if
$$\mu \geq \tau_i\text{-cl }(\tau_j\text{-(int}(\mu)$$

Definition 1.2 [14]: Let μ be a fuzzy set of a fuzzy bitopological space X. Then μ is called

(a) *a* (τ_i, τ_j)-fuzzy semi pre-open $[(\tau_i, \tau_j)$-fspo]set if
$$\mu \leq \tau_j\text{-cl}(\tau_i\text{-(int}(\tau_j\text{-cl }(\mu))$$

(b) *a* (τ_i, τ_j)-fuzzy semi pre-closed $[(\tau_i, \tau_j)$-fspc]set if
$$\mu \geq \tau_j\text{-int}(\tau_i\text{-(cl}(\tau_j\text{-int}(\mu))$$

(c) *a* (τ_i, τ_j)-fuzzy strongly semi open $[(\tau_i, \tau_j)$-fsso]set if
$$\mu \leq \tau_i\text{-int}(\tau_j\text{-(cl}(\tau_i\text{-int}(\mu))$$

(d) *a* (τ_i, τ_j)-fuzzy strongly semi closed$[(\tau_i, \tau_j)$-fssc]set if
$$\mu \geq \tau_i\text{-cl}(\tau_j\text{-(int}(\tau_i\text{-cl}(\mu))$$

Definition 1.3 [4]: Let μ be a fuzzy set of a fbts X. Then μ is called *a* (τ_i, τ_j)-fuzzy γ-open (resp. (τ_i, τ_j)-fuzzy γ-closed) set, briefly $[(\tau_i, \tau_j)$-fγo(resp. (τ_i, τ_j)-ϕγc] of X if $\mu \leq \tau_i\text{-int}(\tau_j\text{-cl }(\mu)) \vee \tau_j\text{-cl}(\tau_i\text{-int }(\mu))$. (resp. if $\mu \geq \tau_i\text{-cl}(\tau_j\text{-int }(\mu)) \wedge \tau_j\text{-int}(\tau_i\text{-cl }(\mu))$).

Remark 1.4: From the above definition, it is clear that (τ_i, τ_j)-fγo set is weaker than the concepts of (τ_i, τ_j)-fso set or (τ_i, τ_j)-fpo set and stronger than the concept of (τ_i, τ_j)-fspo set.

It is clear from their definitions that every (τ_i, τ_j)-fso set is (τ_i, τ_j)-fγo set and also (τ_i, τ_j)-fpo is (τ_i, τ_j)-fγo set and every (τ_i, τ_j)-fγo set is (τ_i, τ_j)-fspo set. But the converse need not be true in general.

Definition 1.5 [11]: Let μ be a fuzzy set of fbts (X, τ_1, τ_2). Then the (τ_i, τ_j)-γ-interior $[(\tau_i, \tau_j)$-γ-int in short] and (τ_i, τ_j)-γ-closure $[(\tau_i, \tau_j)$-γ-cl in short] of μ are defined respectively as follows:

$$(\tau_i, \tau_j)\text{-γ-int}(\mu) = \vee\{O \mid O \text{ is } (\tau_i, \tau_j)\text{-γ }o \text{ and } O \leq \mu\}$$

$$(\tau_i, \tau_j)\text{-γ-cl}(\mu) = \wedge\{C \mid C \text{ is } (\tau_i, \tau_j)\text{-γ }c \text{ and } C \geq \mu\}.$$

Now, from the above definitions, we have shown the following diagrams:

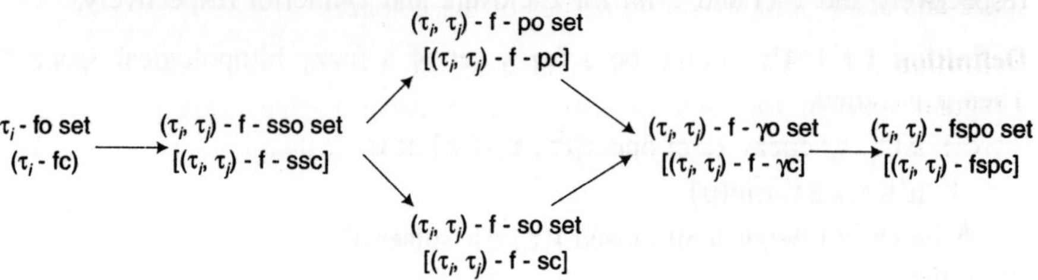

The converse need not be true in general.

Result 1.6:   Let μ be a fuzzy set of fbts (X, τ_1, τ_2). Then.

(a) $(\tau_i, \tau_j)\text{-}\gamma\text{-int}(\mu^c) = [(\tau_i, \tau_j)\text{-}\gamma\text{cl }(\mu)]^c$

(b) $(\tau_i, \tau_j)\text{-}\gamma\text{-cl}(\mu^c) = (\tau_i, \tau_j)\text{-}\gamma o(\mu)]^c$

Definition 1.7 [14]:   Let $f : (X, \tau_1, \tau_2). \to (Y, \sigma_1, \sigma_2)$ be a mapping. Then f is called

(i) a fuzzy pairwise semi continuous [fpsc] mapping if $f^{-1}(\alpha)$ is a (τ_i, τ_j)-fso set of X for each σ_i-fo set α of Y,

(ii) a fuzzy pairwise precontinuous [fppc] mapping if $f^{-1}(\alpha)$ is a (τ_i, τ_j)-fpo set of X for each σ_i-fo set α of Y,

Definition 1.8 [10]:   Let $f : (X, \tau_1, \tau_2) \to (Y, \sigma_1, \sigma_2)$ be a mapping. Then f is called a fuzzy pairwise γ - continuous $[fp\ \gamma\ c]$ mapping if $f^{-1}(\alpha)$ is a (τ_i, τ_j)-fγo set of X for each σ_i-fo set α of Y.

From definitions it is observed that every fpsc is a fpγc mapping and every fppc is a fpγc mapping. But the converses are not true in general.

Theorem 1.9 [8]:   Let $f : (X, \tau_1, \tau_2) \to (Y, \sigma_1, \sigma_2)$ be a mapping. Then the following are equivalent:

1. f is fpγc,

2. The inverse image of each σ_i-fc set of Y is a (τ_i, τ_j)-fγo set of X,

3. $f((\tau_i, \tau_j)\text{-}\gamma\,\text{cl}\mu) \le \sigma_i\text{-cl}(f(\mu))$ for each fuzzy set μ of X.

4. $(\tau_i, \tau_j)\text{-}\gamma\,\text{cl}(f^{-1}(\alpha)) \le f^{-1}(\sigma_i\,\text{-cl}\ \alpha)$ for each fuzzy set α of Y,

5. $f^{-1}(\sigma_i\,\text{-int}\ \alpha) \le (\tau_i, \tau_j)\text{-}\gamma\ \text{int}(f^{-1}(\alpha))$ for each fuzzy set α of Y.

2. FUZZY PAIRWISE γ-REGULAR SPACES

Sampat Kumar introduced the concept of fuzzy pairwise regular space, fuzzy pairwise almost regular space and fuzzy pairwise semi regular space in 2000. In this section we introduce fuzzy pairwise γ regular space, fuzzy pairwise γ normal space and study some basic properties of these spaces.

Definition 2.1:   A fuzzy bitpological space (X, τ_1, τ_2) is said to be (i, j)-fuzzy γ regular space if for each τ_i-γ closed set F and for each fuzzy point $x_p \notin F$ there exists τ_i-γ open set U and τ_j-γ open set V such that $x_p \in U$, $F \leq V$ with $U \cap V = 0$, $i, j = 1, 2$ and $i \neq j$.

A fuzzy bitpological space (X, τ_1, τ_2) is said to be fuzzy pairwise γ regular space if it is both (i, j)-fuzzy γ regular and (j, i)-fuzzy γ regular space.

Now, we give an example of (i, j) fuzzy γ regular space.

Example 2.2:   Let $X = \{a, b, c\}$ and μ_1, μ_2, μ_3 be fuzzy sets of X defined as follows

$$\mu_1(a) = 0.1 \qquad \mu_1(b) = 0.1 \qquad \mu_1(c) = 0.1$$

$$\mu_2(a) = 0.7 \qquad \mu_2(b) = 0.7 \qquad \mu_2(c) = 0.7$$

$$\mu_3(a) = 0.8 \qquad \mu_3(b) = 0.9 \qquad \mu_3(c) = 0.8$$

Consider fuzzy topologies $\tau_1 = \{0, \mu_1, \mu_2{}^c, 1\}$ and $\tau_2 = \{0, \mu_2, \mu_3, 1\}$.

Here, $F = \mu_3$ is τ_1-fuzzy γ closed set and $U = \mu_2$ is τ_1-fuzzy γ open set and $V = \mu_3$ is τ_2-fuzzy γ open set.

And also, $x_p = 0.7 \notin F$, $U \wedge V = \phi$ and $x_p = 0.7 \in U$, $F \leq V$.

Hence, (X, τ_1, τ_2) is (τ_1, τ_2) fuzzy γ regular space.

Theorem 2.3:   For a fuzzy bitopological space (X, τ_1, τ_2) these following are equalivalent:

(i) (X, τ_1, τ_2) is fuzzy pairwise γ regular space.

(ii) For each fuzzy point $x_p \in X$ and τ_i-fuzzy γ open set G containing x_p, $\exists$ a τ_i-fuzzy γ open set H such that $x_p \in H \leq (\tau_j$-γ clH$) \leq G$.

(iii) For each fuzzy point $x_p \in X$ and τ_i-γ closed set K and $x_p \notin K \exists$ a τ_i-fuzzy γ open set Γ such that $x_p \in \gamma$ and $((\tau_i$-γ int $\Gamma) \wedge K = O$.

Proof:

(i) $\rightarrow$ (ii) Let $x_p \in X$ and let G be a τ_i-fuzzy γ open set containing the fuzzy point x_p.

Then $(I_X - G)$ is τ_i-fuzzy γ-closed set in X and $x_p \notin I_X - G$

Since X is fuzzy pairwise γ regular space $\exists$ a τ_I–fuzzy γ open set H and τ_j-fuzzy γ-open set W such that $x_p \in H$; $I_X - G \leq W$ and $H \wedge W = O$. $I, j = 1, 2$ $(i \neq j)$

Then $H \subseteq I_X - W$ and τ_j-γcl$(H) \leq \tau_j$-γcl$(I_X - W) = I_X - W$

$\Rightarrow \qquad \tau_j$-γcl$(H) \leq I_X - W$

$\Rightarrow \qquad \tau_j$-γcl$(H) \wedge W = O$

But $\tau_j\text{-}\gamma cl(H) \wedge (I_X - G) \leq \tau_j\text{-}\gamma cl(H) \wedge W = O$

$\Rightarrow \qquad \tau_j\text{-}\gamma cl(H) \leq G.$

Hence $x_p \in H \leq \tau_j\text{-}\gamma cl(H) \leq G.$

(ii) $\rightarrow$ (iii)   Let $x_p \in X$ and K be τ_i-fuzzy γ closed set such that $x_p \notin K$. Then $I_X - K$ is τ_i-fuzzy γ open set and $x_p \in I_X - K$.

Therefore by (ii) $\exists$ a τ_i-fuzzy γ open set Γ such that $x_p \in \Gamma \leq \tau_j\text{-}\gamma cl(\Gamma) \leq I_X - K.$

Hence $\tau_j\text{-}\gamma cl(\Gamma) \wedge K = O.$

(iii) $\rightarrow$ (i)   Let K be a τ_i-fuzzy γ closed set such that $x_p \wedge K.$

Then by (iii) $\exists$ a τ_i-fuzzy γ open set U such that $x_p \in U$ and $\tau_j\text{-}\gamma cl(U) \wedge K = O.$

Then $\tau_j\text{-}\gamma\, cl(U) \leq I_X - K$

$\Rightarrow K \leq I_X - \tau_j\text{-}\gamma cl(U)$. Put $V = I_X - \tau_j\, \gamma cl(U)$

Then U is the τ_j-fuzzy γ open set such that $K \leq V$, $x_p \in V$ and $U \wedge V = O$, $i, j = 1, 2$ and $i \neq j.$

Hence (X, τ_1, τ_2) is fuzzy pairwise γ-regular space.

Definition 2.4:   A space (X, τ_1, τ_2) is said to be (i, j)-fuzzy γ normal space if for every τ_i-fuzzy γ closed set A and τ_j-fuzzy γ closed set B such that $A \wedge B = 0$ there exist τ_i-fuzzy γ open set U, τ_i-fuzzy γ open set V such that $U \wedge V = 0$ for $i, j = 1, 2; i \neq j.$

(X, τ_1, τ_2) is said to be fuzzy pairwise γ normal space if it is both (i, j)-fuzzy γ normal space and (j, i)-fuzzy γ normal space.

Theorem 2.5:   For a space (X, τ_1, τ_2) the following are equivalent:

   (i) (X, τ_1, τ_2) is a fuzzy pairwise γ normal space.

   (ii) For each τ_i-fuzzy γ closed set A and τ_j-fuzzy γ closed set H containing A, there is τ_i-fuzzy γ closed set U such that $A \leq U \leq (\tau_j\text{-fuzzy }\gamma clU) \leq H$

   (iii) For each τ_i-fuzzy γ closed set A and τ_j-fuzzy γ closed set B with $A \wedge B = 0$ there exist a τ_j-fuzzy γ open set U containing A such that $(\tau_i\text{- fuzzy }\gamma\, clU) \wedge B = 0.$

   (iv) For each τ_i-fuzzy γ closed set A and τ_j-fuzzy γ closed set B with $A \wedge B = 0$, there is a τ_j-fuzzy γ open set U containing A and τ_i-fuzzy γ open set V containing B such that $(\tau_i\text{-fuzzy }\gamma\, clU) \wedge (\tau_j\text{-fuzzy }\gamma\, clV) = 0.$

Proof:

   (i) $\rightarrow$ (ii)   Let (X, τ_1, τ_2) is a fuzzy pairwise γ normal space. Let A be a τ_i-fuzzy γ closed set and H be a τ_j-fuzzy γ open set such that $A \leq H.$

Then $I_X - H$ is a τ_j-fuzzy γ closed set in A and hence

$A \leq H$, $A \wedge (I_X - H) = 0$

Now, (x, τ_1, τ_2) is a fuzzy pairwise γ normal space

There is a τ_i-fuzzy γ open set U and τ_j-fuzzy γ open set W such that $A \leq H$, $I_X - H \leq W$ and $U \wedge W = 0$

$$\Rightarrow \qquad U \leq I_X - W.$$

Then $(\tau_j$-fuzzy γclU$) \leq (\tau_j$- fuzzy γcl $(I_X - W)) = I_X - W$

$\Rightarrow (\tau_j$- fuzzy γclU$) \wedge W = 0$

But, $(\tau_j$-fuzzy γclU$) \wedge (I_X - H) \leq (\tau_j$-fuzzy γclU$) \wedge W = 0$.

$\Rightarrow (\tau_j$-fuzzy γclU$) \wedge I_X - H = 0$.

Hence $(\tau_j$-fuzzy γclU$) \leq H$ and $A \leq U$

$\therefore A \leq U \leq (\tau_j$-fuzzy γclU$) \leq H$.

(ii) $\rightarrow$ **(iii)** Let A be a τ_i-fuzzy γ-closed set and B be a τ_j-fuzzy γ-closed set such that $A \wedge B = 0$. Then $A \leq I_X - B$ and $I_X - B$ is a τ_j-fuzzy γ-open set containing τ_i-fuzzy γ-closed set A. Now by (ii) there is a τ_j-fuzzy γ-open set U containing A such that τ_i-fuzzy γ cl$(U) \leq I_X - B$

Hence $(\tau_i$-fuzzy γ cl$(U)) \wedge B = 0$.

(iii) $\rightarrow$ **(iv)** Let A be a τ_i-fuzzy γ-closed set and B be a τ_j-fuzzy γ-closed set such that $A \wedge B = 0$. Now by (iii) there is a τ_i-fuzzy γ open set V containing B such that τ_j-fuzzy γcl $V \wedge A = 0$.

Now τ_j-fuzzy γ-cl$V$ and A are disjoint in X.

Hence again by (iii) there is a τ_j-fuzzy γ-open set U containing A such that $(\tau_j$-fuzzy γ-cl$U) \wedge (\tau_j$-fuzzy γ-cl$V) = 0$.

(iv) $\rightarrow$ **(i)** Let A be a τ_i-fuzzy γ-closed set and B be τ_j-fuzzy γ-closed set such that $A \wedge B = 0$. Then by (iv) there is a τ_i-fuzzy γ-open set V containing B and a τ_j-fuzzy γ-open set U containing A such that $(\tau_j$-fuzzy γ-cl$U) \wedge (\tau_j$-fuzzy γ-cl$V) = 0$.

But $0 \leq U \wedge V \leq (\tau_j$-fuzzy γ-cl$U) \wedge (\tau_j$-fuzzy γ-cl$V) = \phi$

$\therefore U \wedge V = \phi$

Hence (X, τ_1, τ_2) is a fuzzy pairwise γ-normal space.

Acknowledgement

I would like to thanks to UGC-NERO for giving me the financial support (MRP-2012-13).

References

1. Azad, K.K., 1981: On fuzzy continuity, *fuzzy almost continuity and fuzzy weakly continuity*, J. Math. Anal. Apply, **82**, 14-32.

2. Chang, C.L., 1968: *Fuzzy topological spaces*, J. Math. Anal. Appl., **24**, 182-190.

3. Concilio, A. Di., and Gerla, G., 1984: *Almost compactness in fuzzy topological spaces*, Fuzzy Sets and Systems. **13**, 187-192.

4. Hanafy, I.M., Mahmoud, F.S., and Khalaf, M.M., 2006: *Fuzzy Bitopology on fuzzy sets: Fuzzy pairwise γ continuity and fuzzy pairwise γ-retract*, Journal of Fuzzy Math. **14**(4), 893-905.

5. Hanafy, I.M., 1999: Fuzzy γ open sets and fuzzy γ continuity, *Journal of Fuzzy Math.* **7**(2).

6. Hyydar, A., 1987: Almost compactness and near compactness in fuzzy topological spaces. *Fuzzy Sets and Systems.* **22**, 289-295.

7. Imm, Y.B., 2006: Fuzzy γ-open sets and γ-rresolute open mappings, *Far East J. Math.Sci.* **21**, 259-267.

8. Imm, Y.B., 2007: Fuzzy pairwise γ-irresoluteness. International *J. Fuzzy and Intelligent Systems*, **7**, 188-192.

9. Imm, Y.B., Lee, E.P., and Park, S.W., 2006: Fuzzy γ-irresolute mappings on fuzzy topological space., *J. Fuzzy Math.* **14**, 605-612.

10. Imm, Y.B., Lee, E.P., and Park, S.W., 2002: Fuzzy pairwise γ-continuous mappings *J. Fuzzy Math.* **10**, 695-709.

11. Mahmoud, F.S., Fath, Alla, M.A., and Khalaf, M.M., 2004: Fuzzy γ open sets and fuzzy γ continuity in bifuzzy topological space, *Appl. Math. & Computation*, **153**, 117-126.

12. Mukherjee, M.N., and Sinha, S.P., 1990: On some near-fuzzy continuous functions between fuzzy topological spaces, *Fuzzy Sets and Systems.* **34**, 245-254.

13. Safiya, Abu A.S.M., and Fora, A.A., 1994: Compactness and weakly induced bifuzzy topological spaces, *Fuzzy Sets and Systems.* **62**, 89-96.

14. Sampath Kumar S., 1994: Semi open sets, semi-continuity and semi-open mappings infuzzy bitopological spaces. *Fuzzy Sets and Systems.* **64**, 421-426.

15. Kumar, Sampath, 2000: Some near fuzzy pairwise continuous mappings between fuzzy bitopological spaces, *Fuzzy Sets and Systems.* **109**, 251-262.

16. Zadeh, L.A., 1965: Fuzzy sets, *Inform. and Control*, **8**, 338-353.

On γ-Induced L-Fuzzy Supra Topological Spaces

Baby Bhattacharya[*] **and Sunny Biswas**[#]

Department of Mathematics, NIT Agartala
Barjala, Tripura
E-mail: babybhatt75@gmail.com[*]; sunnybiswas87@rediffmail.com[#]

ABSTRACT

In this paper we introduce and study the concept of Scott-γ-continuity and γ-induced L-fuzzy supra topological spaces. Scott-γ-continuous functions turn out to be the natural tool for studying the γ-induced L-fuzzy supra topological spaces. We discuss the connections between some separation and covering properties of an ordinary topological space and its corresponding γ-induced L-fuzzy supra topological space. Finally we study fuzzy generalized γ-continuous mapping and its properties by means of fuzzy generalized γ-closed set.

Keywords: Fuzzy lattice, Prime element, Molecule, γ-open set, Scott-γ-continuity, γ-induced L-fuzzy supra topological space, Fuzzy generalized γ-continuous mapping.

1. INTRODUCTION

In 1983 A.S. Mashhour et. al. [3] introduced the concept of supra topological spaces. In 1987 M.E. Abd. El-Monsef and A.E. Ramadan [1] introduced the concept of fuzzy supra topological spaces as follows: A family $F \subset I^X$ is called fuzzy supra topology on X if $0, 1 \in F$ and it is closed under arbitrary supremum. In 1997 Bhaumik and Mukherjee [8] introduced the concept of induced fuzzy supra topological space with the notion of s-lower semi continuous functions. And then after in 2003 Mukherjee and Bhattacharya [7] introduced the concept of induced L-fuzzy supra topological spaces. In Section 2, we introduce a new class of functions from a topological space (X, T) to a fuzzy lattice L with its Scott topology, called Scott-γ-continuous functions. After that, we investigate some of their properties and characterizations. We prove that the set $S_\gamma(T)$ of Scott-γ-continuous functions turn out to be the natural tool for studying γ-induced L-fuzzy supra topological space (γ-IL-FST space). In Section 3, we discuss the connections between several properties of an ordinary topological space (X, T) and its corresponding γ-IL-FST space $(X, S_\gamma(T))$. In Section 4, we study the concept of fuzzy generalized γ-closed set and fuzzy generalized γ-closed mapping.

Throughout this paper, X and Y denotes non-empty ordinary sets and $L = L(\leq, \vee, \wedge, ')$ always denotes a fuzzy lattice, and 0 and 1 are the smallest and the greatest element of L respectively. By a fuzzy lattice we mean a complete completely distributive lattice L, if L has an order reversing involution $a \to a'$ $(a \in L)$. A mapping from X into L is said to be an L-fuzzy set on X. The collection of all L-fuzzy sets on X denoted by L^X, can be naturally seen as a fuzzy lattice $(L^X, \leq, \vee, \wedge, ')$. We will denote by 1_A, the characteristic function of the ordinary subset A of X.

We state the following definitions and results for the ready references.

Definition 1.1 [13]: An element p of L is called a prime element iff $p \neq 1$ and if for arbitrary $a, b \in L$ we have $a \wedge b \leq p$ then $a \leq p$ or $b \leq p$. The set of all prime elements of L will be denoted by $pr(L)$.

In [20], Warner has determined the prime elements of the fuzzy lattice L^X. Here, $pr(L^X) = \{x_p : x \in X \text{ and } p \in pr(L)\}$ where for each $x \in X$ and each $p \in pr(L)$, $x_p : X \to L$ is the fuzzy set defined by

$$x_p(y) = \begin{cases} p, & \text{if } y = x \\ 1, & \text{otherwise} \end{cases}$$

These x_p are called the L-fuzzy points of X and we have x_p is a member of an L-fuzzy set g and we write $x_p \in g$ iff $g(x) \not\leq p$.

Definition 1.2 [15]: Let L be a fuzzy lattice. $\alpha \in L$ is called a union-irreducible (or a molecule or coprime) element of L, if for arbitrary $a, b \in L$, we have $\alpha \leq a \vee b \Rightarrow \alpha \leq a$ or $\alpha \leq b$. The set of all nonzero union-irreducible elements of L is denoted by $M(L)$.

Clearly, $p \in pr(L)$ iff $p' \in M(L)$.

Definition 1.3 [9]: Let (X, T) be a topological space and $a \in X$. A function $f : (X, T) \to I$ is called a Scott continuous (or lower semi continuous) at $a \in X$ iff for every $\alpha \in [0, 1]$ with $\alpha < f(a)$ there is a neighbourhood U of 'a' such that $\alpha < f(x)$ for every $x \in U$. f is called Scott continuous (or lower semi continuous) on X iff f is Scott continuous (or lower semi continuous) at every point of X.

Definition 1.4 [21]: The set $\omega(T)$ of Scott continuous function from a topological space (X, T) to L with its Scott topology is an L-fuzzy topology, called the induced L-fuzzy topology (IL-FT).

Result 1.5 [22]: The Scott topology of L is generated by the sets of the form $\{l \in L : l \not\leq p\}$ where $p \in pr(L)$.

Definition 1.6 [12]:   A subset A of a topological space X is said to be γ-open if $A \subset \mathrm{Cl}\,(\mathrm{Int}\,A) \cup \mathrm{Int}\,(\mathrm{Cl}\,A)$.

The complement of a γ-open set is said to be γ-closed.

2. SCOTT-γ-CONTINUOUS FUNCTION AND γ-INDUCED L-FUZZY SUPRA TOPOLOGICAL SPACES

Considering a fuzzy lattice L with its Scott topology we introduce the concept of Scott-γ-continuity. We obtain an L-fuzzy supra topological space from a given ordinary topological space. Let (X, T) be a topological space and $f : (X, T) \to L$ be a function where L has its Scott topology. By result 1.5, we have, f is Scott continuous iff for every $p \in pr(L), f^{-1}\,(\{l \in L : l \not\leq p\}) \in T$.

In ordinary topological space (X, T) we define the γ-open neighbourhood as follows:

Let p be a point in (X, T). A subset N of X is a γ-open neighbourhood of p iff N is a superset of a γ-open set S containing $p : p \in S \subset N$, where S is a γ-open set in (X, T).

Definition 2.1:   Let (X, T) be a topological space and $a \in X$. A function $f : (X, T) \to L$, where L has its Scott topology is said to be Scott-γ-continuous at $a \in X$ iff for every $p \in pr(L)$ with $f(a) \not\leq p$, there is a γ-open neighbourhood N of 'a' in (X, T) such that $f(x) \not\leq p$ for every $x \in N$, i.e. $N \subset f^{-1}\,(\{l \in L : l \not\leq p\})$. Then f is called Scott-γ-continuous on X iff f is Scott-γ-continuous at every point of X.

Proposition 2.2:   The characteristic function of every γ-open set is Scott-γ-continuous.

Proof: Let N be a γ-open set in a topological space (X, T) and $a \in X, p \in pr(L)$ with $1_N\,(a) \not\leq p$. Then $a \in N$ and N is a γ-open neighbourhood of 'a'. We also have $1_N(x) \not\leq p$ for every $x \in N$. Hence 1_N is Scott-γ-continuous at $a \in X$.

Proposition 2.3:   If $\{f_j : j \in \Lambda\}$ is an arbitrary family of Scott-γ-continuous functions from a topological space (X, T) to L, then $f = \bigvee\limits_{j \,\in\, \Lambda} f_j$ is also Scott-γ-continuous.

Proof: Let $p \in pr(L)$ and $a \in X$ with $f(a) = \bigvee\limits_{j \,\in\, \Lambda} f_j(a) \not\leq p$, then there is a $j \in \Lambda$ such that $f_j(a) \not\leq p$. Since f_j is Scott-γ-continuous at 'a', there is a γ-open neighbourhood N of 'a' such that $f_j(x) \not\leq p$ for all $x \in N$. Hence $f(x) = \bigvee\limits_{j \,\in\, \Lambda} f_j(x) \not\leq p$ for all $x \in N$. Thus f is Scott-γ-continuous at $a \in X$.

Since the intersection of two γ-open sets may not be γ-open, we have the following:

Proposition 2.4:   Let (X, T) be a topological space. If $f, g : (X, T) \to L$ are Scott-γ-continuous functions then $f \wedge g : (X, T) \to L$ is not Scott-γ-continuous.

Proposition 2.5:   Let (X, T) be a topological space. $f : (X, T) \to L$ is Scott-γ-continuous if and only if for every $p \in pr(L)$, $f^{-1}(\{l \in L : l \not\leq p\})$ can be expressed as a union of some γ-open sets in (X, T).

Proof: Let $p \in pr(L)$ and $x \in f^{-1}(\{l \in L : l \not\leq p\})$. Then $f(x) \not\leq p$. Here f is Scott-γ-continuous at x, thus there exists a γ-open set N_x in (X, T) such that $x \in N_x$ and $N_x \subset f^{-1}(\{l \in L : l \not\leq p\})$. Hence $f^{-1}(\{l \in L : l \not\leq p\}) = \cup N_x$ where N_x is γ-open.

On the other hand, let $a \in X$ and $p \in pr(L)$ with $f(a) \not\leq p$. Then $a \in f^{-1}(\{l \in L : l \not\leq p\})$. By the hypothesis there is a γ-open set N in (X, T) such that $a \in N$ and $N \subset f^{-1}(\{l \in L : l \not\leq p\})$ which implies that f is Scott-γ-continuous.

Theorem 2.6:   For topological space (X, T) the set $S_\gamma(T) = \{f \in L^X : f : (X, T) \to L$ is Scott-γ-continuous$\}$ is an L-fuzzy supra topology on X.

Proof: It follows immediately from propositions 2.2, 2.3 and 2.4.

Definition 2.7:   The L-fuzzy supra topology $S_\gamma(T) = \{f \in L^X : f : (X, T) \to L$ is Scott-γ-continuous$\}$ is called a γ-induced L-fuzzy supra topology (γ-IL-FST) and the space $(X, S_\gamma(T))$ is called γ-induced L-fuzzy supra topological space (γ-IL-FST space). The members of $S_\gamma(T)$ are called supra open L-fuzzy subsets.

Lemma 2.8:   If A is γ-open in a topological space (X, T) then $1_A \in S_\gamma(T)$.

Proof: By proposition 2.2 the lemma follows immediately.

Remark: Since every Scott continuous function from a topological space (X, T) to a fuzzy lattice L is Scott-γ-continuous, we have $\omega(T) \subset S_\gamma(T)$ where $\omega(T)$ is the L-fuzzy topology of Scott continuous functions from (X, T) to L.

Definition 2.9:   A function $f : (X, T) \to (Y, G)$ from a topological space (X, T) to another topological space (Y, G) is said to be γ-irresolute if the inverse image of each γ-open subset in Y is γ-open in X.

Definition 2.10:   A function $f : (X, S_\gamma(T_1)) \to (Y, S_\gamma(T_2))$ from a γ-IL-FST space to another γ-IL-FST space is said to be fuzzy supra γ-continuous if $f^{-1}(\lambda) \in S_\gamma(T_1)$ for every $\lambda \in S_\gamma(T_2)$.

Theorem 2.11:   A function $f : (X, S_\gamma(T_1)) \to (Y, S_\gamma(T_2))$ is fuzzy supra γ-continuous iff $f : (X, T_1) \to (Y, T_2)$ is γ-irresolute.

Proof: Suppose that $f : (X, S_\gamma(T_1)) \to (Y, S_\gamma(T_2))$ be fuzzy supra γ-continuous. Let A be a γ-open set in (Y, T_2) then $1_A \in S_\gamma(T_2)$. By fuzzy supra γ-continuity $f^{-1}(1_A) = 1_{f^{-1}(A)} \in S_\gamma(T_1)$. We shall show that $f^{-1}(A)$ is γ-open in (X, T_1). Let $p \in$

$pr(L)$ and $x \in f^{-1}(A)$. Then $1_{f^{-1}(A)}(x) \nleq p$. Since $1_{f^{-1}(A)}(x) \in S_\gamma(T_1)$, there exists a γ-open set N_x in (X, T_1) such that $x \in N_x$ and $N_x \subset 1_{f^{-1}(A)}^{-1}(\{l \in L : l \nleq p\}) = f^{-1}(A)$. Thus we have for each $x \in f^{-1}(A)$, there exists a γ-open set N_x in (X, T_1) such that $x \in N_x \subset f^{-1}(A)$. This shows that $f^{-1}(A)$ is γ-open set in (X, T_1). Consequently, $f : (X, T_1) \to (Y, T_2)$ is γ-irresolute.

Conversely, suppose $f : (X, T_1) \to (Y, T_2)$ is γ-irresolute. Take $\alpha \in S_\gamma(T_2)$, we shall show that $f^{-1}(\alpha) \in S_\gamma(T_1)$ i.e. $f^{-1}(\alpha) : (X, T) \to L$ is Scott-γ-continuous. Let $a \in X$ and $p \in pr(L)$ with $f^{-1}(\alpha)(a) \nleq p$. Then $\alpha(f(a)) \nleq p$. Since $\alpha : (Y, T_2) \to L$ is Scott-γ-continuous at $f(a) \in Y$, there exists a γ-open set N in (Y, T_2) such that $f(a) \in N$ and $\alpha(y) \nleq p$ for all $y \leq N$. Since N is γ-open in (Y, T_2), $f^{-1}(N)$ is also γ-open in (X, T_1). Now, we have, $a \in f^{-1}(N)$ which implies that there is a γ-open set B in (X, T_1) such that $a \in B \subset f^{-1}(N)$. Hence $f^{-1}(\alpha)(x) = \alpha(f(x)) \nleq p$ for every $x \in B$. This shows that $f^{-1}(\alpha)$ is Scott-γ-continuous. Consequently $f : (X, S_\gamma(T_1)) \to (Y, S_\gamma(T_2))$ is fuzzy supra γ-continuous.

3. COVERING AND SEPARATION PROPERTIES OF Γ-IL-FST SPACE $(X, S_\gamma(T))$

In this section, we study the connections between some separation and covering properties of an ordinary topological space and its corresponding γ-induced L-fuzzy supra topological space.

Definition 3.1: A topological space (X, T) is called γ-compact iff every γ-open cover of X admits a finite subcover of X.

Definition 3.2: A γ-IL-FT space $(X, S_\gamma(T))$ is said to be fuzzy supra γ-compact iff for every prime element p of L and every collection $\{f_j : j \in \Lambda\}$ of supra open L-fuzzy sets with $\left(\bigvee_{j \in \Lambda} f_j \right)(x) \leq p, \forall x \in X$, there is a finite subset Λ_0 of Λ such that $\left(\bigvee_{j \in \Lambda_0} f_j \right)(x) \nleq p$ for all $x \in X$.

Theorem 3.3: A γ-IL-FST space $(X, S_\gamma(T))$ is fuzzy supra γ-compact iff the corresponding topological space (X, T) is γ-compact.

Proof: Suppose that $(X, S_\gamma(T))$ is fuzzy supra γ-compact. Let $\{A_j : j \in \Lambda\}$ be an γ-open covering of X. Then $1_{A_j} \in S_\gamma(T)$ for every $j \in \Lambda$, as A_j is γ-open in (X, T). Thus $\{1_{A_j} : j \in \Lambda\}$ is a family of supra open L-fuzzy sets in $(X, S_\gamma(T))$ with $\left(\bigvee_{j \in \Lambda} 1_{A_j} \right)(x) \nleq p, \forall x \in X$. From the fuzzy supra γ-compactness of $(X, S_\gamma(T))$ there exists a finite subset Λ_0 of Λ such that $\left(\bigvee_{j \in \Lambda} 1_{A_j} \right)(x) \nleq p, \forall x \in X$. Hence

$X = \bigcup_{j \in \Lambda_0} A_j$. Thus (X, T) is γ-compact.

Conversely, let (X, T) be γ-compact and $p \in pr(L)$. Let $\mathcal{B} = \{f_j : j \in \Lambda\}$ be a family of supra open L-fuzzy sets in $(X, S_\gamma(T))$ with $\left(\underset{j \in \Lambda}{\vee} f_j \right) (x) \nleq p, \forall x \in X;$

where $f_j(x) = \alpha_j$ if $x \in A_j$ and $f_j(x) = 0$, otherwise , these A_j are γ-open in (X, T) and $\alpha_j \in L$ for every $j \in \Lambda$. Then for each $x \in X$, there is a $j \in \Lambda$ such that $f_j(x) \nleq p$ i.e. $\alpha_j \nleq p$. Let $\beta = \{A_j : $ there is a $j \in \Lambda$ such that $\alpha_j \nleq p$ and $f_j \in \mathcal{B}\}$. Then β is a family of γ-open cover of X. From the γ-compactness of (X, T) there exists a finite subfamily of β say β_0 where $\beta_0 = \{A_1, A_2, ..., A_n\}$ such that

$X = \overset{n}{\underset{j=1}{\cup}} A_j.$ Hence $\left(\overset{n}{\underset{j=1}{\vee}} f_j \right) (x) \nleq p$ for all $x \in X$ and thus $(X, S_\gamma(T))$ is fuzzy supra γ-compact.

In general topology, a topological space (X, T) is called completely Hausdorff iff for any distinct points x, y of X there are open sets U and V such that $x \in U$, $y \in V$ and $Cl(U) \cap Cl(V) = \phi$.

Definition 3.4:   A topological space (X, T) is called γ-completely Hausdorff iff for any distinct points x, y of X there are γ-open sets A and B such that $x \in A$, $y \in B$ and $Cl(U) \cap Cl(V) = \phi$.

Definition 3.5:   A γ-IL-FST space $(X, S_\gamma(T))$ is said to be fuzzy supra completely Hausdorff iff for every distinct points x, y of X and every $p, q \in pr(L)$ there exists supra open L-fuzzy sets f and g such that $x_p \in f, y_q \in g$ and for all $z \in X$, $Cl(f) \cap Cl(g) = 0$.

Theorem 3.6:   A topological space (X, T) is γ-completely Hausdorff iff the γ-IL-FST space $(X, S_\gamma(T))$ is fuzzy supra completely Hausdorff.

Proof: Let $x, y \in X (x \neq y)$ and $p, q \in pr(L)$. From the γ-complete Hausdorffness of (X, T) there exists two γ-open sets U and V in (X, T) such that $x \in U, y \in V$ and $Cl(U) \cap Cl(V) = \phi$. Now $1_U, 1_V \in S_\gamma(T)$, since U and V are γ-open sets in (X, T). Also $1_U(x) \nleq p$ and $1_V(y) \nleq q$ and for all $z \in X$, $Cl\ 1_U(z) = 1_{ClU}(z) = 0$ or, $Cl\ 1_V(z) = 1_{ClV}(z) = 0$, since $Cl(U) \cap Cl(V) = \phi$. Thus $Cl\ 1_U(z) \wedge Cl\ 1_V(z) = 0$ Hence $(X, S_\gamma(T))$ is fuzzy supra completely Hausdorff space.

Conversely, let $x, y \in X(x \neq y)$ and $p, q \in pr(L)$. From the fuzzy supra complete Hausdorffness of $(X, S_\gamma(T))$ there exists supra open L-fuzzy sets f, g which are defined by

$$f(z) = \begin{cases} \alpha \text{ if } z \in U \\ 0 \text{ otherwise} \end{cases}$$

and
$$g(z) = \begin{cases} \beta \text{ if } z \in V \\ 0 \text{ otherwise} \end{cases}$$

where U and V are γ-open sets in (X, T) and $\alpha, \beta \in L$ such that $x_p \in f, y_q \in g$ and for all $z \in X$, $Cl(f) \cap Cl(g) = 0$. Hence we have $x \in U, y \in V$ where U

and V are γ-open in (X, T) and $\mathrm{Cl}(U) \cap \mathrm{Cl}(V) = \phi$. Hence (X, T) is γ-completely Hausdorff.

4. FUZZY GENERALIZED γ-CLOSED SETS IN L-FUZZY TOPOLOGICAL SPACE AND FUZZY GENERALIZED γ-CONTINUOUS MAPPING

Definition 4.1: Let (X, τ) be a fuzzy topological space. An operator $\gamma : \tau \to I^X$ is a function from τ to the I^X such that $\beta \subseteq \beta^\gamma$ for each $\beta \in \tau$ where β^γ denotes the value of γ at β. The operation defined by $\gamma(\beta) = \beta$, $\gamma(\beta) = Cl(\beta)$ and $\gamma(\beta) = Int\, Cl(\beta)$ are examples of operator γ.

Definition 4.2: Let β be a subset of a space X. A fuzzy point $x_p \in \beta$ is said to be fuzzy γ-interior point of β iff there exist an fuzzy open neighbourhood N of x_p such that $N^\gamma \subseteq \beta$. The set of all such points is denoted by $F\, \mathrm{int}_\gamma (\beta)$.

Thus $F\, \mathrm{int}\gamma(\beta) = \{x_p \in \beta : x_p \in N \in \tau$ and $N^\gamma \subseteq \alpha\} \subseteq \beta$, β is fuzzy γ-open iff $\beta = F\, \mathrm{int}_\gamma(\beta)$.

In this section, by means of fuzzy γ-closed set and fuzzy generalized γ-closed set, we study fuzzy generalized γ-continuous mapping and its properties.

Definition 4.3: Let (L^X, δ) be an L-fuzzy topological space and $p \in pr(L)$ and $A \in L^X$. A is called an p-open set, if for any $x \in X$, $Int\, ClA(x) \leq p \Rightarrow A(x) \leq p$.

The set of all p-open set in (L^X, δ) is denoted by $O_p(\delta)$. Clearly $\forall p \in pr(\mathrm{L})$, $\delta \in O_p(\delta)$.

Definition 4.4: Let (L^X, δ) be an L-fuzzy topological space and $\gamma \in M(L)$ and $A \in L^X$, A is called an γ-closed set, if for any $x \in X$, $Cl\, Int\, A(x) \geq \gamma \Rightarrow A(x) \geq \gamma$.

The set of all γ-closed sets in (L^X, δ) is denoted by $C_\gamma(\delta)$. Clearly $\forall \gamma \in M(L)$, $\delta' \subset C_\gamma(\delta)$.

Definition 4.5: If λ is an L-fuzzy set in a L-fts L^X and $\gamma \in M(L)$, then $\mathrm{Cl}_\gamma(\lambda) = \cap\{\mu : \mu \cap \lambda\}$, μ "is fuzzy γ-closed set" is called a fuzzy γ-closure of λ.

An L-fuzzy set λ in a L-fts (L^X, δ) is fuzzy γ-closed iff $\lambda = Cl_\gamma(\lambda)$.

Theorem 4.6: Let (L^X, δ) be an L-fuzzy topological space and $\gamma \in M(L)$ and $\lambda \in L^X$. If λ is fuzzy γ-closed set then λ' is fuzzy γ'-open set.

Definition 4.7: Let (L^X, δ) be an L-fuzzy topological space and $\gamma \in M(L)$, and $\lambda \in L^X$. Then λ is called an fuzzy generalized γ-closed set (in short Fgγ-closed set), if $Cl_\gamma(\lambda) \leq \mu$ whenever $\lambda \leq \mu$ and μ is fuzzy γ'-open.

Theorem 4.8: If λ_1 and λ_2 are Fgγ-closed sets then $\lambda_1 \vee \lambda_2$ is a Fgγ-closed set.

Proof: It follows from the definition of Fgγ-closed and the fact that $Cl_\gamma(\lambda_1 \vee \lambda_2) = Cl_\gamma(\lambda_1) \vee Cl_\gamma(\lambda_2)$.

However, the intersection of two Fgγ-closed sets is not fuzzy generalized γ-closed set.

Definition 4.9: Let (L^X, δ) and (L^Y, σ) be two L-fuzzy topological spaces. A map $f : L^X \to L^Y$ is called fuzzy generalized γ-continuous (in short Fgγ-continuous) if the inverse image of every fuzzy γ-closed set in L^Y is Fgγ-closed in L^X.

Theorem 4.10: If $f : L^X \to L^Y$ is fuzzy γ-continuous then it is Fgγ-continuous. The converse is not true.

Theorem 4.11: Let $f : (L^X, \delta) \to (L^Y, \sigma)$ be a map. Then the following statements are equivalent:

(a) f is Fgγ-continuous.

(b) The inverse image of each fuzzy γ-open set in L^Y is Fgγ-open in L^X.

Definition 4.12: Let $\gamma \in M(L)$. A map $f : L^X \to L^Y$ is called Fgγ-c-irresolute, if the inverse image of every, Fgγ-closed set in L^Y is Fgγ-closed set in L^X.

The following are the properties of Fgγ-c-irresolute maps.

Theorem 4.13: $f : L^X \to L^Y$ is Fgγ-c-irresolute iff the inverse image of every Fgγ-open set in L^Y is Fgγ-open set in L^X.

Theorem 4.14: If $f : L^X \to L^Y$ is Fgγ-c-irresolute then it is Fgγ-continuous. But the converse of the theorem is not true.

Theorem 4.15: Suppose $f : L^X \to L^Y$, $g : L^Y \to L^Z$ be maps. Assume f is Fgγ-c-irresolute and g is Fgγ-continuous. Then gof is Fgγ-continuous.

Proof: Let λ be a γ-closed set in L^Z, since g is Fgγ-continuous, it follows that $g^{-1}(\lambda)$ is a Fgγ-closed set in L^Y. Now by assumption, $f^{-1}(g^{-1}(\lambda))$ is a Fgγ-closed in L^X. This shows that gof is Fgγ-continuous.

5. CONCLUSION

Connection between an ordinary topological space and its corresponding γ-IL-FST space is established. There is a scope to study the properties of fuzzy γ-closed set and fuzzy generalized γ-closed set in a L-fuzzy topological space.

Acknowledgements

The authors express their sincere thanks to the referees for their valuable comments and suggestions towards the improvement of the paper.

References

1. Abd El-Monsef, M.E., and Ramadan, A.E., 1987: *On fuzzy supratopological spaces*, Indian J. pure appl.Math., **18**(4), 322-329.

2. Mashhour, A.S., M.E. Abd. El. Monsef, and Deeb, S.N.El., 1982: *On pre-continuous and week pre-continuous mapping*, Pre-Math Phys. Soc. *Egypt* **53**, 47-53.

3. Mashhour, A.S., Allam, A.A., Mahmoud, F.S., and Khedr, F.H. 1983: On supratopological spaces, *Indian J. pure appl. Math.*, **14**(4), 502-510.

4. Andrrijevic, D., 1996: *On b-open sets*, Mat. Vesnik. **48**, 59-64.

5. Gizem Gunel and Gulhan Aslim, 2011: *Notes on L-fuzzy γ-open sets*, Filomat **25**(2), 173-185.

6. Azad, K.K., 1981: *On fuzzy semicontinuty, fuzzy almost continuity and fuzzy weakly continuity*, J. Math. Anal. Appl. **82**, 14-32.

7. Mukherjee, A. and Bhattacharya, B., 2003: *Pre-Induced L-Supra Topological Spaces*, Indian J. pure appl. Math., **34**(10), 1487-1493.

8. Bhaumik, R.N., and Mukherjee, A., 1997: *Induced fuzzy supra topological spaces*, Fuzzy Sets and Systems **91**: 123-126.

9. Bourbaki, N., 1996: *Elements of Mathematics, General Topology Part-I*, Addssion-Wesely, Reading, M.A.

10. Chang, C.L., 1968: *Fuzzy topological spaces*, J. Math. Anal. Appl. **24**, 182-190.

11. Daraby, B., and Nimse, S.B., 2007: *On Fuzzy Generalized -closed set and its application*, Filomat **25**(2), 99-108.

12. El-Atik, A.A., 1997: *A study on some types of mappings on topological spaces.* M. Sci. thesis, Tanta Univ., Egypt,.

13. Gierz, G., et al., 1980: *A compendium of continuous lattices*, Springer, Berlin.

14. Goguen, J.A., 1967: *L-fuzzy sets*, J. Math. Anal. Appl. **18**, 145-174.

15. Guojun, Wang, 1992: *Theory of topological molecular lattices*, Fuzzy Sets and Systems **47**, 351-376.

16. Hanafy, I.M., 1999: *Fuzzy γ-open sets and fuzzy γ-continuity*, J. Fuzzy Math. **7**, 419-430.

17. Lowen, R., 1976: *Fuzzy topological space and fuzzy compactness*, J. Math. Anal. Appl. **56**, 621-633.

18. Mirmiran, Majid, 2011: *Weak Insertion of a γ-continuous Function*, Applied Mathematics, **1**(1), 1-3.

19. Steen, L.A., Seebach, J.A., 1970: *Counter examples in topology*, New York.

20. Warner, M.W., 1991: *Frame-Fuzzy points and membership*, Fuzzy Sets and Systems **42**, 335-344.

21. Warner, M.W., 1990: *Fuzzy topology with respect to continuous lattices*, Fuzzy Sets and Systems **35**, 85-91.

22. Warner, M.W., Mc Lean, R.G., 1993: *On compact Hausdorff L-fuzzy spaces*, Fuzzy Sets and Systems **56**, 103-110.

L-Fuzzy Topological Spaces Induced by Scott-σ-Continuous Function

Baby Bhattacharya[*] **and Jayasree Chakraborty**[#]
Department of Mathematics, NIT Agartala
Barjala, Jirania, Tripura
E-mail: babybhatt75@gmail.com[*]; chakrabortyjayasree1@gmail.com[#]

ABSTRACT

The aim of this paper is to introduce and to study the concept of Scott-σ-continuity and σ-induced L-fuzzy topological space. Some properties of newly defined functions (namely Scott-σ-continuous function) are studied. The Scott-σ-continuous function turns out to be the natural tool for studying the σ-induced L-fuzzy topological space. We also discuss the connection between some separation and covering properties of an ordinary topological space and its corresponding σ-induced L-fuzzy topological space. Lastly the concept of fuzzy σ-generalized closed sets in L-fuzzy topological space are given.

Keywords: Fuzzy lattice, prime element, regular F_σ-subset, Scott-σ-continuous function, σ-induced L-fuzzy topological space, fuzzy σ-generalized closed sets.

1. INTRODUCTION

The concept of induced fuzzy topological space was first introduced by Weiss in 1975 [12]. In 1999, Aygun, Warner, Kudri [2] introduced the concept of completely induced L-fuzzy topological space. The concept of regular G_δ-subset was introduced by Mack in 1970 [7]. The complement of a regular G_δ-subset is called a regular F_σ-subset. In this paper we introduce a new class of functions from a topological space (X, T) to a fuzzy lattice L with its Scott topology called Scott-continuous functions. Then we study some of their properties and characterizations. We also prove that the set $F_\sigma(T)$ of Scott-σ-continuous functions from a topological space (X, T) to L with its Scott topology, is an L-fuzzy topology. Scott-σ-continuous functions turn out to be the natural tool for studying σ-induced L-fuzzy topological space. Then we discuss the connection between several properties of an ordinary topological space (X, T) and its corresponding σ-induced *L*-fuzzy topological space $(X, F_\sigma(T))$. In Section 4, we study the concept of fuzzy σ-generalized closed set.

We state the following definitions and results for the ready references.

Definition 1.1 [7]: A subset H of a topological space X is called a regular G_δ-subset if H is an intersection of a sequence of closed sets whose interiors contain H. Equivalently, if $H = \cap_i G_i = \cap_i \mathrm{cl}_X G_i$ for $I \in N$, where each G_i is open in X, then H is regular G_δ-subset of X. The complement of a regular G_δ-subset is called a regular F_σ-subset.

Definition 1.2 [3]: Let (X, T) be a topological space and $a \in X$. A function $f : (X, T) \to I$ is called a Scott continuous (or lower semi continuous) at $a \in X$ iff for every $\alpha \in [0, 1]$ with $\alpha < f(a)$ there is a neighbourhood U of 'a' such that $\alpha < f(x)$ for every $x \in U$. f is called Scott continuous (or lower semi continuous) on X iff f is Scott continuous (or lower semi continuous) at every point of X.

Definition 1.3 [10]: The set $\omega(T)$ of Scott-continuous functions from a topological space (X, T) to L with its Scott-topology is an *L*-fuzzy topology, called the induced *L*-fuzzy topology(IL-FT).

Definition 1.4 [4]: An element p of L is called prime iff $p \neq 1$ and whenever $a, b \in L$ with $a \wedge b \leq p$ then $a \leq p$ or $b \leq p$. The set of all prime elements of L will be denoted by $pr(L)$.

In [11] has determined the prime elements of the fuzzy lattice L^X. Here

$$pr(L^X) = \{x_p : x \in X \text{ and } p \in pr(L)\}$$

where for each $x \in X$ and each $p \in pr(L)$, $x_p : X \to L$ is the *L*-fuzzy set defined by

$$x_p(y) = \begin{cases} p, & \text{if } y = x \\ 1, & \text{otherwise} \end{cases}$$

These x_p are called the *L*-fuzzy points of X and we have x_p is member of an *L*-fuzzy set g and we write $x_p \in g$ iff $g(x) \not\leq p$.

Definition 1.5 [5]: Let L be a fuzzy lattice. $\alpha \in L$ is called a union-irreducible (or a molecule or coprime) element of L, if for arbitrary $a, b \in L$, we have $\alpha \leq a \vee b \Rightarrow \alpha \leq a$ or $\alpha \leq b$. The set of all nonzero union-irreducible elements of L is denoted by $M(L)$. Clearly, $p \in pr(L)$ iff $p' \in M(L)$.

Proposition 1.6 [11]: The set of the form $\{l \in L : \delta << l\}$ are Scott-open.

Result 1.7 [8]: The set of the form $\{l \in L : 1 \not\leq p\}$ where $p \in pr(L)$ generates the Scott topology of L.

Definition 1.8 [1]: A topological space (X, T) is said to be almost regular iff for each non empty regular closed subset F of X and for each point $x \in F'$, there exist disjoint open sets U and V such that $x \in U$ and $F \subset V$.

Definition 1.9 [9]:   A topological space (X, T) is completely Hausdroff iff every distinct points x, y of X, there are open sets U and V such that $x \in U$, $y \in V$ and $\mathrm{Cl}(U) \cap \mathrm{Cl}(V) = \phi$.

Definition 1.10 [11]:   Let (X, τ) be an L-fuzzy topological space. A non empty sub family β of τ is a basis for τ iff for every $x_p \in pr(L^X)$ and every $f \in \tau$ with $x_p \in f$, there is a $g \in \beta$ such that $x_p \in g \leq f$.

2. SCOTT-σ-CONTINUOUS FUNCTIONS AND THEIR PROPERTIES

In this section we introduce a new class of functions called Scott-σ-continuous function and study their properties.

In an ordinary topological space (X, T) we define a regular F_σ-neighborhood of a point $p \in X$ in the following way:

A subset N of a topological space (X, T) is said to be a regular F_σ-neighbourhood of a point $p \in X$ if there exists a regular F_σ-subset U containing p such that $p \in U \subset N$. Where U is a regular F_σ-subset in (X, T).

Definition 2.1:   Let (X, T) be a topological space and $a \in X$. A function $f : (X, T) \to L$ where L has its Scott topology is said to be Scott-σ-continuous at $a \in X$ iff for every $p \in pr(L)$ with $f(a) \nleq p$, there is a regular F_σ-neighbourhood N of 'a' in (X, T) such that $f(x) \nleq p$, $\forall x \in N$ i.e. $N \subset f^{-1}(\{l \in L : l \nleq p\})$. Then f is called Scott-σ-continuous on X iff f is Scott-σ-continuous at every point of X.

Proposition 2.2:   The characteristic function of every regular F_σ-subset is Scott-σ-continuous.

Proof: Let A be a regular F_σ-subset in a topological space (X, T) and the characteristic function 1_A of A is defined as

$$1_A(x) = \begin{cases} 1 \text{ if } x \in A \\ 0 \text{ if } x \in X - A \end{cases}$$

We have to show that 1_A is Scott-σ-continuous i.e. $1_A(x) \nleq p$, for every $x \in A$ and $p \in pr(L)$. Let $a \in X$, $p \in pr(L)$ with $1_A(a) \nleq p$. Then $a \in A$ and A is a regular F_σ-neighbourhood of a. We also have $1_A(x) \nleq p$, for all $x \in A$. Hence 1_A is Scott-σ-continuous at $a \in X$.

Proposition 2.3:   If $\{f_j : j \in \Lambda\}$ is an arbitrary family of Scott-σ-continuous functions from a topological space $(X, T) \to L$, then $f = \bigvee_{j \in \Lambda} f_j$ is also Scott-σ-continuous.

Proof: Let $p \in pr(L)$ and $a \in X$ with $f(a) = \bigvee_{j \in \Lambda} f_j(a) \nleq p$, then there is a $j \in \Lambda$ such that $f_j(a) \nleq p$, since f_j is Scott-σ-continuous at 'a', there is regular

F_σ-neighbourhood N of 'a' such that $f_j(x) \not\leq p$, $\forall x \in N$. Hence $f(x) = \bigvee_{j \in \Lambda} f_j(x) \not\leq p$, $\forall x \in N$. Thus f is Scott-σ-continuous at $a \in X$.

Proposition 2.4: Let (X, T) be a topological space. If f, g: $(X, T) \to L$ are Scott-σ-continuous functions then $f \wedge g : (X, T) \to L$ is Scott-σ-continuous as well.

Proof: Let $a \in X$ and $p \in pr(L)$ with $(f \wedge g)(a) \not\leq p$. Then $f(a) \not\leq p$ and $g(a) \not\leq p$. Since f and g are Scott-σ-continuous at 'a', there are regular F_σ-neighbourhoods U and V of 'a' such that $f(x) \not\leq p$ for all $x \in U$ and $g(x) \not\leq p$ for all $x \in V$. Let $W = U \cap V$. Then W is a regular F_σ-neighbourhood of 'a', since the intersection of two regular F_σ-subsets is regular F_σ-subset and p is prime. We have $(f \wedge g)(x) \not\leq p$ for all $x \in W$. Hence $(f \wedge g)$ is Scott-σ-continuous at $a \in X$.

Proposition 2.5: Let (X, T) be a topological space. The function $f : (X, T) \to L$ is Scott-σ-continuous iff for every $p \in pr(L)$, $f^{-1}(\{l \in L : l \not\leq p\})$ can be expressed as a union of some regular F_σ-subsets in (X, T).

Proof: Necessity: Let $p \in pr(L)$ and $x \in f^{-1}(\{l \in L : l \not\leq p\})$. Then $f(x) \not\leq p$. Since f is Scott-σ-continuous at x, then there exist a regular F_σ-subset N_x in (X, T) such that $x \in N_x$ and $N_x \subset f^{-1}\{l \in L : l \not\leq p\}$. Hence $f^{-1}\{l \in L : l \not\leq p\} = \cup N_x$ where N_x is regular F_σ-subset.

Sufficiency: Let $a \in X$ and $p \in pr(L)$ with $f(a) \not\leq p$. Then $a \in f^{-1}(\{l \in L : l \not\leq p\})$. By the hypothesis, there is a regular F_σ-subset N in (X, T) such that $a \in N$ and $N \subset f^{-1}(\{l \in L : l \not\leq p\})$, which implies that f is Scott-σ-continuous.

Theorem 2.6: For a topological space (X, T) the set $\mathcal{F}_\sigma(T) = \{f \in L^X : f:(X, T) \to L$ is Scott-σ-continuous$\}$ is an L-fuzzy topology on X.

Proof: From the proposition 2.2, 2.3 and 2.4 proof is obvious.

Remark: Since every Scott-σ-continuous function from a topological space (X, T) to a fuzzy lattice L is Scott continuous, we have $\mathcal{F}_\sigma(T) \subset \omega(T)$, where $\omega(T)$ is the L-fuzzy topology of Scott continuous functions from (X, T) to L.

Definition 2.7: The L-fuzzy topology defined above is called a σ-induced L-fuzzy topology (σ-IL-FT) and the space $(X, \mathcal{F}_\sigma(T))$ is called σ-induced L-fuzzy topological space and the members of $(X, \mathcal{F}_\sigma(T))$ are called fuzzy open sets in $(X, \mathcal{F}_\sigma(T))$.

Proposition 2.8: For a topological space (X, T), the following family forms a base for $\mathcal{F}_\sigma(T)$

$$\mathcal{F}_\sigma(\beta) = \{Z^\beta : Z \text{ is regular } F_\sigma\text{-subset in } (X, T), \beta \in L\}$$

Where $Z^\beta : X \to L$ is defined by $Z^\beta(x) = \beta$ if $x \in Z$ and $Z^\beta(x) = 0$ otherwise.

Proof: Let $f \in F_\sigma(T)$ and $x_p \in pr(L^X)$ with $f(x) \not\leq p$. Then f is Scott-σ-continuous and $f(x) \not\leq p$. By the continuity of **L[4]** there exists $\eta \in L$ and $\eta << f(x)$ and $\eta \not\leq p$. We take $\beta \in L$ with $\eta << \beta << f(x)$. Hence $f(x) \in \{l \in L : \beta << l\}$. Since $\{l \in L : \beta << l\}$ is Scott open [**by proposition 1.6**], there is a $q \in pr(L)$ such that $f(x) \in \{l \in L : l \not\leq q\} \subset \{l \lfloor L : \beta << l\}$ **by proposition 1.7.** Then $f(x) \not\leq q$. From the Scott-σ-continuity of f, there exist a regular F_σ-subset Z in (X, T) such that $x \in Z$ and $f(z) \not\leq q$, $\forall z \in Z$. Thus $\beta << f(z)$ for every $z \in Z$ and hence $\beta \leq f(z)$ for every $z \in Z$. Moreover, $\beta \not\leq p$ because $\beta >> \eta \not\leq p$. Consequently $x_p \in Z^\beta$. $Z^\beta \in \mathcal{F}_\sigma(\beta)$ and $Z^\beta \leq f$. Thus by **proposition 1.10** $\mathcal{F}_\sigma(\beta)$ is a base for $\mathcal{F}_\sigma(T)$.

Lemma 2.9: If A is a regular F_σ-subset in a topological space (X, T) then $1_A \in \mathcal{F}_\sigma(T)$.

Proof: By proposition 2.2, the lemma follows immediately.

Definition 2.10: A function $f : (X, \mathcal{F}_\sigma(T_1)) \to (Y, \mathcal{F}_\sigma(T_2))$ from a σ-IL-FT space to another σ-IL-FT space is said to be fuzzy continuous if $f^{-1}(\lambda) \in \mathcal{F}_\sigma(T_1)$ for every $\lambda \in \mathcal{F}_\sigma(T_2)$.

3. CONNECTION BETWEEN PROPERTIES OF A TOPOLOGICAL SPACE AND ITS CORRESPONDING σ-INDUCED L-FUZZY TOPOLOGICAL SPACE

In this section, we study the connections between separation and covering properties of an ordinary topological space (X, T) and its corresponding σ-induced L-fuzzy topological space $(X, \mathcal{F}_\sigma(T))$.

Definition 3.1: A topological space (X, T) is called σ-completely Hausdroff iff for any distinct points x, y of X there are regular F_σ-subsets A and B such that $x \in A$, $y \in B$ and $Cl(A) \cap Cl(B) = \phi$.

Definition 3.2: A σ-ILFT space $(X, \mathcal{F}_\sigma(T))$ is said to be fuzzy completely Hausdroff iff for every distinct points x, y of X and every $p, q \in pr(L)$ there exists L-fuzzy sets f and g such that $x_p \in f$, $y_q \in g$ and $\forall z \in X$, $Cl(f)(z) = 0$ or $Cl(g)(z) = 0$.

Theorem 3.3: The topological space (X, T) is σ-completely Hausdroff iff the σ-IL-FT space $(X, \mathcal{F}_\sigma(T))$ is fuzzy completely Hausdroff.

Proof: Let $x, y \in X (x \neq y)$ and $p, q \in pr(L)$. By σ-complete Hausdroffness of (X, T), there exists two regular F_σ-subsets U and V in (X, T), such that $x \in U$ and $y \in V$ and $Cl(U) \cap Cl(V) = \phi$. Then $1_U, 1_V \in \mathcal{F}_\sigma(T)$ because U and V are regular F_σ-subsets in (X, T). We also have $1_U(x) \not\leq p$, $1_V(y) \leq q$ and $\forall z \in X$. $Cl(1_U)(z) = 1_{Cl(U)}(z) = 0$ or $Cl(1_V)(z) = 1_{Cl(V)}(z) = 0$ because $Cl(U) \cap Cl(V) = \phi$. Consequently $(X, \mathcal{F}_\sigma(T))$ is a fuzzy completely Hausdroff.

Conversely, let $x, y \in X (x \neq y)$ and $p, q \in pr(L)$. From the fuzzy complete Hausdrofness of $(X, \mathcal{F}_\sigma(T))$, there exists basic open L-fuzzy sets f, g which are defined by respectively $f(z) = \gamma$ if $z \in U, f(z) = 0$ otherwise and $g(z) = \beta$ if $z \in V, g(z) = 0$ otherwise, where U and V are regular F_σ-subsets in (X, T) and $\gamma, \beta \in L$ such that $x_p \in f, y_q \in g$ and $(\forall z \in X)$, $Cl(f)(z) = 0$ or $Cl(g)(z) = 0$. Hence we have $x \in U$ and $y \in V$ where U and V are regular F_σ-subsets in (X, T) and $Cl(U) \cap Cl(V) = \phi$. Hence (X, T) is σ-completely Hausdroff.

Since the intersection of two regular F_σ-subsets is regular F_σ-subset, then the family of all regular F_σ-subsets in (X, T) forms a base for a smaller topology T_σ on X, called the σ-semi-regularization of T. A topological space (X, T) is sad to be σ-semi-regular iff $T = T_\sigma$ .i.e. (X, T_σ) space is the σ-semi-regularization topological space. Obviously $T_\sigma \subset T$.

Analogous to the almost regular [1] space we define the following:

Definition 3.4:   A topological space (X, T) is said to be σ-almost regular iff for each non empty regular G_δ-subset F of X and each point $x \in F'$, there exists disjoint open sets U and V such that $x \in U$ and $F \subset V$.

Obviously, every σ-almost regular topological space is almost regular.

Proposition 3.5:   A topological space (X, T) is said to be σ-almost regular if for each non empty regular G_δ-subset F of X and each point $x \in F'$, there exists disjoint regular F_σ-subset U and V such that $x \in U$ and $F \subset V$.

Proof: It is obvious from the definition of σ-almost regular. Since regular F_σ-subset implies open set.

Lemma 3.6:   Let (X, T) be a topological space and (X, T_σ) be its σ-semi-regularization topological space. Then (X, T) is σ-almost regular iff (X, T_σ) is regular.

Theorem 3.7:   A topological space (X, T) is σ-almost regular iff the σ-induced L-fts $(X, \mathcal{F}_\sigma(T))$ is fuzzy regular.

Proof: Let $p \in pr(L), x \in X$ and let f be a closed L-fuzzy set in $(X, \mathcal{F}_\sigma(T))$ such that there is $y \in X$ with $y_p \notin f'$ and $f(x) = 0$. Then $x_p \in f'$ and $f' \in \mathcal{F}_\sigma(T)$. So **by the proposition 2.8** there is a basic open L-fuzzy set $g \in L^X$ defined by $g(z) = \gamma$ if $z \in A$ and $g(z) = 0$ otherwise, where A is regular F_σ-subset in (X, T) and $\gamma \in L$ such that $x_p \in g \leq f'$. Then, we have $g(x) \nleq p$ which implies that $g(x) = \gamma \nleq p$ and hence $x \in A$. Since A' is regular G_δ-subset in (X, T) and (X, T) is σ-almost regular, then there are regular F_σ-subsets U, V in (X, T) such that $x \in U, A' \subset V$ and $U \cap V = \phi$. Let $\eta = 1_U$ and $\theta = 1_V$. Then $\eta, \theta \in \mathcal{F}_\sigma(T), \eta(x) = 1 \nleq p$ and for every $y_p \notin f'$. $y_p \in \theta$ because $y_p \in f'$ implies that $g(y) \leq p$ and hence $g(y) = 0$ because $\gamma \nleq p$. This means that $y \in A'$. Thus $y \in V$ and $\theta(y) = 1 \leq p$. In addition, $(\forall z \in X) \eta(z) = 0$ or $\theta(z) = 0$. Consequently $(X, \mathcal{F}_\sigma(T))$ is fuzzy regular.

Conversely, let K be a non-empty regular G_δ-subset in (X, T) and let $x \in K$. Then 1_K is closed in $(X, \mathcal{F}_\sigma(T))$. $1_K(x) = 0$ and there exist a $y \in X$ with $y_p \notin 1'_K$ because K is non-empty set. From the fuzzy regularity of $(X, \mathcal{F}_\sigma(T))$, there exist basic open L-fuzzy sets g, $h \in L^X$ defined by $g(z) = \gamma$ if $z \in A$ and $g(z) = 0$ otherwise and $h(z) = b$ if $z \in B$ and $h(z) = 0$ otherwise, where A and B are regular F_σ-subsets in (X, T) and γ, $\beta \in L$ such that $x_p \in g$, $y_p \in h$ for every $y_p \notin 1'_K$ and (for all $z \in X$) $g(z) = 0$ or $h(z) = 0$. Hence $x \in A$, $K \subset B$ and $A \cap B = \phi$. Thus **by the proposition 3.5** (X, T) is σ-almost regular.

Definition 3.8:   A topological space (X, T) is called σ-nearly compact iff every regular F_σ-cover of X has a finite subcover.

Definition 3.9 [8]:   An L-fts (X, T) is said to be fuzzy compact iff for every prime element p of L and every collection $(f_i)_{i \in J}$ of open L–fuzzy sets with $\left(\underset{i \in J}{\vee} f_i \right)(x) \nleq p$ for all $x \in X$, there is a finite subset F of J such that $\left(\left(\underset{i \in F}{\vee} f_i \right)(x) \nleq p \right)$ for all

$x \in X$.

Lemma 3.10:   A topological space (X, T) is σ-nearly compact if (X, T_σ) is compact. Using the definition **3.9[8]** we prove the following theorem.

Theorem 3.11:   A σ-ILFT space is fuzzy compact iff the corresponding topological space (X, T) is σ-nearly compact.

Proof: Let us assume that (X, T) is σ-nearly compact. Let $p \in pr\ (L)$ and $\mathcal{B} = \{f_j : j \in \Lambda\}$ be a family of basic open L-fuzzy sets in $(X, \mathcal{F}_\sigma(T))$ with $(\vee f_j)_{j \in \Lambda}(x) \nleq p$ for all $x \in X$; where $f_j(x) = \lambda_j$ if $x \in A_j$ and $f_j(x) = 0$ otherwise, then A_j are regular F_σ-subset in (X, T) and $\lambda_j \in L$ for every $j \in \Lambda$. Then for each $x \in X$ there is $j \in \Lambda$ such that $f_j(x) \nleq p$ i.e. $\lambda_j \nleq p$ and let $\gamma = \{A_j : $ there is an $j \in \Lambda$ such that $\lambda_j \nleq p$ and $f_j \in \mathcal{B}\}$. Then γ is a family of regular F_σ-subsets in (X, T) covering X. In other words, γ is a regular F_σ-cover of X. From the σ-compactness of (X, T), there exists a finite subfamily of γ say γ_0

where $\gamma_0 = \{A_1, A_2, \ldots, A_n\}$ such that $X = \overset{n}{\underset{j=1}{\cup}} A_n$. Hence $\left(\overset{n}{\underset{j=1}{\vee}} f_j(x) \right) \leq p$ for all

$x \in X$ and thus $(X, \mathcal{F}_\sigma(T))$ is fuzzy compact.

Conversely, suppose $(X, \mathcal{F}_\sigma(T))$ is fuzzy compact. Let $\{A_j : j \in \Lambda\}$ be a regular F_σ-cover of X. Then $1_A \in \mathcal{F}_\sigma(T)$ for every $j \in \Lambda$ and A_j is regular F_σ-subset in (X, T). Thus $\left\{ 1_{A_j} : j \in \Lambda \right\}$ is a family of open L-fuzzy sets in

$(X, \mathcal{F}_\sigma(T))$ with $\left(\underset{j \in \Lambda}{\vee} 1_{A_j} \right)(x) \nleq p$ for all $x \in X$. From the fuzzy compactness of

$(X, \mathcal{F}_\sigma(T))$ there exist a finite subset Λ_0 of Λ such that $\left(\underset{j \in \Lambda_0}{\vee} 1_{A_j} \right)(x) \nleq p$ for all

$x \in X$. Hence $X = \underset{j \in \Lambda_0}{\cup} A_j$. Thus (X, T) is σ-nearly compact.

4. FUZZY σ-GENERALIZED CLOSED SETS IN L-FUZZY TOPOLOGICAL SPACE

Definition 4.1: Let (L^X, δ) be an L-fuzzy topological space and $\sigma \in M(L)$ and $A \in L^X$, A is called a σ-closed set, if for any $x \in X$, Cl Int ClA$(x) \geq \sigma \Rightarrow$ $A(x) \geq \sigma$.

The set of all σ-closed set in (L^X, δ) is denoted by $\text{Cl}_\sigma(\delta)$.

Definition 4.2: If λ is an L-fuzzy set in a L-fts L^X and $\sigma \in M(L)$, then $\text{Cl}_\sigma(\lambda)$ $= \cap\{\mu : \mu \geq \lambda\}$, μ is fuzzy σ- closed set is called a fuzzy σ-closure of λ.

An L-fuzzy set λ in a L-fts (L^X, δ) is fuzzy σ-closed iff $\lambda = C_\sigma(\lambda)$.

Definition 4.3: Let (L^X, δ) be an L-fuzzy topological space and $\sigma \in M(L)$ and $\lambda \in L^X$. Then λ is called an fuzzy σ- generalized closed set (in short F_σ-gclosed set), if $\text{Cl}_\sigma(\lambda) \leq \mu$ whenever $\lambda \leq \mu$ and μ is fuzzy σ′-open.

Remark: Let (L^X, δ) be an L-fuzzy topological space and $\sigma \in M(L)$ and $\lambda \in L^X$. If λ is fuzzy σ-closed set then λ' is fuzzy σ′-open set.

Theorem 4.4: If λ_1 and λ_2 are F_σ-gclosed then $\lambda_1 \vee \lambda_2$ is a F_σ-gclosed.

Proof: It is obvious.

Theorem 4.5: Intersection of two F_σ-gclosed closed sets is not generally F_σ-gclosed set.

Example 4.6: The intersection of two F_σ-gclosed sets is not F_σ-gclosed set.

Let $X = \{x_1, x_2, x_3\}$. Define $f_1, f_2, f_3, f_4, f_5 : X \rightarrow [0, 1]$ as follows

$f_1 = 0_X, f_2 = 1_X, f_3 = \{(x_1, 0.5), (x_2, 0.5), (x_3, 0.5)\}$,

$f_4 = \{(x_1, 0.4), (x_2, 0.3), (x_3, 0.2)\}$,

$f_5 = \{(x_1, 0.7), (x_2, 0.6), (x_3, 0.8)\}$

Clearly $\delta = \{f_1, f_2, f_3, f_4, f_5\}$ is an L-fts on X.

Define $\lambda_1, \lambda_2 : X \rightarrow [0, 1]$ as follows:

$\lambda_1 = \{(x_1, 0.9), (x_2, 0.4), (x_3, 0.3)\}$

and $\lambda_2 = \{(x_1, 0.5), (x_2, 0.4), (x_3, 0.9)\}$.

Here λ_1 and λ_2 are F_σ-gclosed sets but

$\lambda_1 \wedge \lambda_2 = \{(x_1, 0.5), (x_2, 0.4), (x_3, 0.3)\}$

is not F_σ-gclosed set where $\sigma = \dfrac{1}{2}$.

5. CONCLUSION

Connection between an ordinary topological space and its corresponding σ-IL-FT space is established. There is a scope to study the properties of fuzzy σ-closed set and fuzzy σ-generalized closed set in a *L*-fuzzy topological space.

Acknowledgement

The authors express their sincere thanks to the referees for their valuable comments and suggestions towards improvement of the paper.

References

1. Singhal, M.K., and Arya, S.P, 1969: *On almost regular spaces*, Glasnik Mat24 (4), 89-99.

2. Aygun, H, Warmer, M.W., 1999: Kudri, S.R.T, *Completely induced L-fuzzy topological Spaces*, Fuzzy Sets and Systems 103: 513-523.

3. Bourbaki, N., 1996: *Elements of Mathematics, General Topology Part-I*, Addssion-Wesely, Reading, M.A.

4. Gierz, G., et al., 1980: *A compendium of Continuous Lattice*, Springer, Berlin.

5. Wang, Gujon, 1992: *Theory of topological molecular lattices*, Fuzzy Sets and Systems 47, 351-376.

6. Lane E.P., 1979: *Weak C insertion of a continuous function*, Notices, Amer. Math. Soc. 26, A-231.

7. Mack, J.E., *Countably paracompactness,Weak normality properties*, Trans. Amer. Math. Soc. 148, 256-272.

8. Warner, M.W., Mc Lean, R.G., 1993: *On compact Hausdroff L-fuzzy spaces*, Fuzzy Sets and Systems 56, 103-110.

9. Steen, L.A., and Seebach, J.A., 1970: *Counter example in topology*, New York.

10. Warner, M.W, 1990: *Fuzzy topology with respect to continuous lattices*, Fuzzy Sets and Systems 35, 85-91.

11. Warner, M.W., 1991: *Frame-Fuzzy points and membership*, Fuzzy Sets and Systems 42, 335-344.

12. Weiss, M.D., 1975: *Fixed Points, separation and induced topologies for fuzzy sets*, J. Math. Anal. Appl. 50: 142-150.

Rough Sets, Fuzzy Sets and Soft Computing
Editor: S. Bhattacharya Halder

Characterizations of ij-Fuzzy pre-paracompact Space in Fuzzy Bitopological Spaces

Anjan Mukherjee[1*] and Bishnupada Debnath[2#]

[1]*Department of Mathematics, Tripura University, Suryamaninagar Agartala, Tripura*
[2]*Lalsingmura H.S. School, Bishalgarh, Sipahijala, Tripura*
E-mail: anjan2002_m@yahoo.co.in[*];
debnath_bishnupada@rediffmail.com[#]

ABSTRACT

Dieudonne J. [4] introduced the notion of paracompactness in general topological spaces. In 2001, Martin M. K.[6] initiated the phenomenon of paracompactness in bitopological spaces and thereafter AL-Zoubi K. & AL-Ghour S. introduced the notion of P3-paracompactness of topological space in terms of preopen sets in 2007. But very recently AL-Swidi L.A.H. [On Pre-paracompactness in Bitopological Spaces; European Journal of Scientific Research; Vol. 46, No.2 (2010), 165-171] gave some characterizations of pre-paracompactness by utilizing ijpreopen sets in bitopological spaces. In this paper, we generalize the concepts of pre-paracompact spaces in fuzzy bitopological space with the help of ij-fuzzy preopen sets. We also show various characterizations, properties of pre-paracompactness and its relationships with other types of spaces.

Ams Subject Classification: 54 A 40; 03 E 72

Keywords: ij-fuzzy pre-locally finite collection, ij-fuzzy pre-paracompact, fuzzy pairwise preirresolute function, some new separation axioms etc.

1. INTRODUCTION

The concepts of fuzzy regular open, fuzzy regular closed, fuzzy semi-open, fuzzy semi-closed and fuzzy pre-open sets have been introduced by S. Sampath Kumar [13, 14] in a fuzzy bitopological space. Throughout the paper (X, τ_1, τ_2) and (Y, μ_1, μ_2) (or simple X and Y) denote fuzzy bitopological spaces (briefly, fbts). If A is a fuzzy subset of X, we shall denote the closure of A and the interior of A in (X, τ_1, τ_2) by $\tau_i\text{-cl}(A)$ and $\tau_i\text{-int}(A)$, respectively, where $i = 1, 2$, and $i, j = 1, 2; i \neq j$.

A fuzzy subset A of X is said to be *ij*-fuzzy pre-open [resp. *ij*-fuzzy semi-open, *ij*-fuzzy regular open, *ij*-fuzzy regular closed] if $A \leq \tau_i\text{-int}(\tau_j\text{-cl}(A))$ [resp. $A \leq \tau_i\text{-cl}(\tau_j\text{-int}(A))$, $A = \tau_i\text{-int}(\tau_j \text{ cl}(A))$, $A = \tau_i\text{-cl}(\tau_j\text{-int}(A))$]. The family of all *ij*-fuzzy semi-open (resp. *ij*-fuzzy regular open and *ij*-fuzzy preopen) sets of X is denoted by *ij*-FSO(X) (resp. *ij*-FRO(X) and *ij*-FPO(X)). The intersection of all *ij*-fuzzy preclosed sets which contain A is called the *ij*-fuzzy preclosure of A and is denoted by *ij*-FPcl(A). Obviously, *ij*-FPcl$A$ is the smallest *ij*-fuzzy preclosed set which contains A. In this paper, we introduce *ij*-fuzzy pre-paracompactness in fuzzy bitopological spaces. Moreover, various characterizations, properties of *ij*-fuzzy pre-paracompactness and its relationships with other types of spaces are also cited.

Definition 1.1:   A fuzzy bitopological space (X, τ_1, τ_2) is called *ij*-fuzzy locally indiscrete if every τ_i-fuzzy open subset of X is τ_j-fuzzy closed.

Definition 1.2: A collection $C = \{\lambda_k : k \in \Gamma\}$ of fuzzy subsets of X is called,

1. locally finite with respect to the topology τi (respectively, *ij*-fuzzy strongly locally finite), if for each fuzzy point $x_p \in X$, there exists $\mu \in \tau_i$ (respectively, $\mu \in$ *ij*-FRO(X)) containing x_p and which intersects at most finitely many members of C;

2. *ij*-fuzzy P-locally finite if for each fuzzy point $x_p \in X$, there exists a *ij*-fuzzy preopen set μ in X containing x_p and which intersects at most finitely many members of C.

From definitions the following implications are obvious:

ij-fuzzy Strongly locally finite $\Rightarrow$ τ_i-fuzzy locally finite $\Rightarrow$ ij-fuzzy P-locally finite.

Definition 1.3:   A fuzzy bitopological space X is called $(\tau_i\text{-}\tau_j)$-paracompact with respect to the topology τ_i, if every τ_i-fuzzy open cover of X has τ_j -fuzzy open refinement which is locally finite with respect to the topology τ_i.

We give the following lemmas without proof:

Lemma 1.4: For a fuzzy bitopological space X, the followings are equivalent:
   (a) X is *ij*-fuzzy locally indiscrete
   (b) Every fuzzy subset of X is *ij*-fuzzy preopen
   (c) Every τ_j-fuzzy closed subset of X is *ij*-fuzzy preopen.

Lemma 1.5: If λ is a *ij*-fuzzy preopen subset of X, then τ_i -Cl(λ) is (τ_i, τ_j)-fuzzy regular closed.

Lemma 1.6: Let λ and β be subsets of a space X. Then
   (a) If $\lambda \in$ *ij*-FPO(X) and $\beta \in \tau_i$, then $(\lambda \wedge \beta) \in$ *ij*-FPO(X).
   (b) If $\lambda \in$ *ij*-FPO(X) and $\beta \in$ *ij*-FSO(X), then $(\lambda \wedge \beta) \llcorner$ *ij*-FPO(β, $\tau_{1\beta}$, $\tau_{2\beta}$).

(c) $\lambda \in ij\text{-FPO }(\beta, \tau_{1\beta}, \tau_{2\beta})$ and $\beta \in ij\text{-FPO}(X)$, then $\lambda \in ij\text{-FPO}(X)$.

Theorem 1.7: Let $\mathcal{C} = \{\lambda_k : k \in \Lambda\}$ be a collection of ij-fuzzy semi-open subsets of a fuzzy bitopological space X, then $\mathcal{C}$ is ij-fuzzy P-locally finite if and only if it is ij-fuzzy strongly locally finite.

Proof:

Necessary Part: Since every ij-fuzzy regular open subset of X is ij-fuzzy preopen, so $\mathcal{C}$ is ij-fuzzy P-locally finite.

Sufficiency Part: Let $x_p \in X$ and η_x be a ij-fuzzy pre-open set of X containing x_p and $\eta_x \wedge \lambda_k \neq 0_X$, for each $k = 1, 2, ..., n$. Choose $\gamma_x = \tau_i\text{-Int}(\tau_j\text{-Cl}(\eta_x))$, then γ_x is a ij-fuzzy regular open which contains x_p. We show for every $k \in \Lambda/\{k_1, k_2, ..., k_n\}$, $\gamma_x \wedge \lambda_k = 0_X$. For each $k \in \lambda$, choose $\mu_k \in \tau_j$ such that $\mu_k \leq \alpha_k \leq (\tau_j\text{-Cl}(\mu_k)$. Now, if $\gamma_x \wedge \lambda_k \neq 0_X$, then $\tau_i\text{-Cl}(\mu_k) \wedge \gamma_x \neq 0_X$, and so $\tau_i\text{-Cl}(\mu_k) \wedge \tau_i\text{-Int}(\tau_j\text{-Cl}(\eta_x) \neq 0_X$, which implies that there exist $z_p \in \tau_i\text{-Cl}(\mu_k)$ and $z_p \in \tau_i\text{-Int}(\tau_j\text{-Cl}(\eta_x)$. So for each τ_i-open set v_z containing z_p such that $v_z \wedge \mu_k \neq 0_X$, but $\tau_i\text{-Int}(\tau_j\text{-Cl}(\eta_x)$ is a τ_i-open set containing z_p which implies that $\mu_k \wedge \tau_i\text{-Int}(\tau_j\text{-Cl}(\eta_x)) \neq 0_X$. Since, $\tau_i\text{-Int}(\tau_j\text{-Cl}(\eta_x) \leq \tau_j\text{-Cl}(\eta_x)$, so we get that $\mu_k \wedge \tau_j\text{-Cl}(\eta_x)) \neq 0_X$, then there exists $z_q \in \mu_k$ and $z_q \in \tau_j\text{-Cl}(\eta_x)$. Therefore for each τ_j-open set β_q containing z_q, $\eta_x \wedge \beta_q \neq 0_X$, but μ_k is τ_j-open set containing z_q, So $\eta_x \wedge \mu_k \neq 0_X$, which implies $0_X \neq \eta_x \wedge \mu_k \leq \eta_x \wedge \alpha_k$, which contradicts the hypothesis. Thus $k \in \Lambda/\{k_1, k_2, ..., k_n\}$.

Now using the above theorem and the fact that every ij-fuzzy regular open set is τ_j-fuzzy open set and every τ_j-fuzzy open, ij-fuzzy semiopen set, we obtain the following corollaries:

Corollary 1.8: Let $\mathcal{C} = \{\lambda_k : k \in \Gamma\}$ be a collection of τ_j-fuzzy open (ij-fuzzy regular open) subsets of a fuzzy bitopological space X. If $\mathcal{C}$ is ij-fuzzy P-locally finite, then $\mathcal{C}$ is locally finite with respect to the topology τ_j.

Now by using Theorem 1.7, Corollary 1.8 and the Definition 1.3, we obtain the following corollary:

Corollary 1.9: A fuzzy bitopological space X is $(\tau_j\text{-}\tau_j)$-fuzzy paracompact with respect to the topology τ_j if and only if every τ_j-fuzzy open cover of X has a open τ_j-fuzzy open refinement which is ij-fuzzy p-locally finite.

Theorem 1.10: Let $\mathcal{C} = \{\lambda_k : k \in \Gamma\}$ be a collection of fuzzy subsets of a fbts X. Then the following statements are true:

(a) $\mathcal{C}$ is ij-fuzzy P-locally finite if and only if the collection $\{ij\text{-FPCl}(\lambda_k) : k \in \Gamma\}$ is ij-fuzzy P-locally finite.

(b) If $\mathcal{C} = \{\lambda_k : k \in \Gamma\}$ is ij-fuzzy P-locally finite, then $\vee\, ij\text{-FPCl}(\lambda_k) = ij\text{-FPCl}(\vee \lambda_k : k \in \Gamma)$

(c) $\mathcal{C}$ is locally finite with respect to the topology τ_i if and only if the collection $\{ij\text{-FPCl}\,(\lambda_k): k \in \Gamma\}$ is locally finite with respect to the topology τ_i.

Proof:

(a) Suppose that, $\mathcal{C}$ is ij-fuzzy P-locally finite. For each fuzzy point $x_p \in X$, $\exists$ a ij-fuzzy pre-open set μ_x containing x_p which meets finitely many of the sets λ_k, say $\lambda_{k_1}, \lambda_{k_2}, \lambda_{k_3}, \ldots, \lambda_{k_n}$. Since $\lambda_{k_m} \leq ij\text{-FPCl}(\lambda_{k_m})$, for each $m = 1, 2, 3, \ldots, n$. Thus μ_x meets $ij\text{-FPCl}(\lambda_{k_1})$, $ij\text{-FPCl}(\lambda_{k_2})$, $ij\text{-FPCl}(\lambda_{k_3})$, $\ldots, ij\text{-FPCl}(\lambda_{k_n})$. Therefore, $\{ij\text{-FPCl}(\lambda_k): k \in \Gamma\}$ is ij-fuzzy P-locally finite.

Conversely, let $x_p \in X$. Then $\exists$ a ij-fuzzy preopen set μ_x containing x_p which meets finitely many of the sets $ij\text{-FPCl}(\lambda_k)$, say $ij\text{-FPCl}(\lambda_{k_1})$, $ij\text{-FPCl}(\lambda_{k_2})$, $ij\text{-FPCl}(\lambda_{k_3}), \ldots, ij\text{-FPCl}(\lambda_{k_n})$, we get that $\mu_x \wedge ij\text{-FPCl}(\lambda_{k_m}) \neq 0_X$ for each $m = 1, 2, 3, \ldots, n$. Let $z_t \in \mu_x$ and $z_t \in ij\text{-FPCl}(\lambda_{k_m})$, which implies that for each ij-fuzzy pre-open ν_z such that $\nu_z \wedge \lambda_{k_m} \neq 0_X$ but μ_x is ij-fuzzy pre-open containing z_t, so $\mu_x \wedge \lambda_{k_m} \neq 0_X$ for each $m = 1, 2, 3, \ldots, n$. Thus $\mathcal{C}$ is ij-fuzzy P-locally finite.

(b) Suppose that, $\mathcal{C}$ is ij-fuzzy P-locally finite, which can be easily got $\{\vee\, ij\text{-FPCl}(\lambda_k)\} \leq ij\text{-FPCl}(\vee\, \lambda_k)$. On the other hand, let $z_t \in ij\text{-FPCl}(\vee\, \lambda_k)$. Then every ij-fuzzy pre-open set ν_z such that $\nu_z \wedge (\vee\, \lambda_k) \neq 0_X$. By hypothesis, $\exists$ a ij-fuzzy pre-open set μ_z containing z_t which needs only finitely many of the sets λ_k, say $\lambda_{k_1}, \lambda_{k_2}, \lambda_{k_3}, \ldots, \lambda_{k_n}$. Thus for each ij-fuzzy pre-open set ν_z containing z_t, $\nu_z \vee (\wedge\, \lambda_{k_m}) \neq 0_X$, which implies that $z_t \in ij\text{-FPCl}(\vee\, \lambda_{k_m})$ $= \vee\, ij\text{-FPCl}(\lambda_{k_m})$, $\exists$ a 'm' such that $z_t \in ij\text{-FPCl}(\lambda_{k_m})$. Therefore we get that $z_t \in (ij\text{-FPCl}(\lambda_k)$ and hence $\vee\, ij\text{-FPCl}(\lambda_k) = ij\text{-FPCl}(\vee\, \lambda_k)$.

(c) Suppose that, $\mathcal{C}$ is fuzzy locally finite with respect to the topology τ_i, which implies that for each fuzzy point $x_p \in X$, $\exists$ a τ_i-fuzzy open set μ_x meets only finitely many of the sets λ_k, say $\lambda_{k_1}, \lambda_{k_2}, \lambda_{k_3}, \ldots, \lambda_{k_n}$ but $\lambda_{k_m} \leq ij\text{-FPCl}(\lambda_{k_m})$, we get that μ_x which meets $ij\text{-FPCl}(\lambda_{k_1})$, $ij\text{-FPCl}(\lambda_{k_2})$, $ij\text{-FPCl}(\lambda_{k_3}), \ldots, ij\text{-FPCl}\,(\lambda_{k_n})$. Thus $\{ij\text{-FPCl}(\lambda_k): k \in \Gamma\}$ is fuzzy locally finite with respect to the topology τ_i.

Conversely, let $x_p \in X$. Then $\exists$ a τ_i-fuzzy open set μ_x which meets only finitely many of the sets $ij\text{-FPCl}(\lambda_k)$, say $ij\text{-FPCl}(\lambda_{k_1})$, $ij\text{-FPCl}(\lambda_{k_2})$, $ij\text{-FPCl}(\lambda_{k_3})$, $\ldots, ij\text{-FPCl}(\lambda_{k_n})$. Now choose a fuzzy point $z_t \in \mu_x$ and $z_t \in ij\text{-FPCl}(\lambda_{k_m})$, for each $m = 1, 2, 3, \ldots, n$, therefore for each ij-fuzzy pre-open ν_z containing z_t such that $\nu_z \wedge \lambda_{k_m} \neq 0_X$, but $z_t \in \mu_x$, then we get μ_x which meets only finitely many of sets λ_k's. Hence, C is fuzzy locally finite with respect to the topology τ_i.

Definition 1.11:   A mapping $f: (X, \tau_1, \tau_2) \to (Y, \rho_1, \rho_2)$ is called

1. fuzzy paiwise pre-irresolute **[13]** if for each $V \in ij\text{-FPO}(Y), f^{-1}(V) \in ij\text{-FPO}(X)$;

2. ij-fuzzy strongly pre-continuous if for each $\lambda \in ij\text{-FPO}(Y)$, $f^{-1}(\lambda) \in \tau_i$.

3. ij-fuzzy pre-closed if for each $\lambda \in ij\text{-FPC}(X), f^{-1}(\lambda) \in ij\text{-FPC}(Y)$.

4. ij-fuzzy strongly pre-closed if for each $\lambda \in \tau_i\text{-FC}(X), f(\lambda) \in ij\text{-FPC}(Y)$.

5. ij-fuzzy pre-open if for each $\lambda \in ij\text{-FPO}(X), f(\lambda) \in ij\text{-FPO}(Y)$.

6. $(\tau_i\text{-}\rho_i)$-continuous $[(\tau_i\text{-}\rho_i)\text{-open and }(\tau_i\text{-}\rho_i)\text{-closed}]$ if $f: (X, \tau_i) \to (Y, \rho_i)$ for $i = 1, 2$ are $(\tau_1\text{-}\rho_1)$-continuous and $(\tau_2\text{-}\rho_2)$-continuous $[(\tau_1\text{-}\rho_1)\text{-open,}$ $(\tau_2\text{-}\rho_2)\text{-open and }(\tau_1\text{-}\rho_1)\text{-closed, }(\tau_2\text{-}\rho_2)\text{-closed}]$.

Lemma 1.12:   Let $f: (X, \tau_1, \tau_2) \to (Y, \rho_1, \rho_2)$ be a function defined by $f(x) = y, \exists\, x \in X$. Then the followings are true:

1. f is ij-fuzzy preclosed if and only if for every fuzzy point y_p of Y and $\mu \in ij\text{-FPO}(X)$ such that $(f^{-1}(y))_p \leq \mu$, there is $\nu \in ij\text{-FPO}(Y)$ such that $p \leq \nu(y)$ and $f^{-1}(\nu) \leq \mu$.

2. f is ij-fuzzy strongly preclosed if and only if for every fuzzy point y_p of Y and $\mu \in \tau_i$ containing $(f^{-1}(y))_p$, there is $\nu \in ij\text{-FPO}(Y)$ such that $p \leq \nu(y)$ and $f^{-1}(\nu) \leq \mu$.

3. If f be $(\tau_i\text{-}\rho_i)$-continuous, $(\tau_i\text{-}\rho_i)$-open function, then f is ij-fuzzy pre-open function.

Lemma 1.13: Let $f: (X, \tau_1, \tau_2) \to (Y, \rho_1, \rho_2)$ be a fuzzy pairwise preirresolute function. If $\mathcal{C} = \{\lambda_k: k \in \Gamma,$ an index set$\}$ is a ij-fuzzy P-locally finite collection in Y, then $f^{-1}(\mathcal{C}) = \{f^{-1}(\lambda_k): k \in \Gamma\}$ is a ij-fuzzy P-locally finite collection in X.

Theorem 1.14: Let $f: (X, \tau_1, \tau_2) \to (Y, \rho_1, \rho_2)$ be a ij-fuzzy strongly pre-continuous function. If $\mathcal{C} = \{\lambda_k: k \in \Gamma\}$ is a ij-fuzzy P-locally finite collection in Y, then $f^{-1}(\mathcal{C}) = \{f^{-1}(\lambda_k): k \in \Gamma\}$ is a locally finite collection with respect to the topology τ_i.

Definition 1.15:   A fuzzy subset λ of a fuzzy bitopological space X is called ij-fuzzy strongly compact if for every cover of λ by ij-fuzzy pre-open subsets of X has a finite subcover.

Theorem 1.16:   Let $f: (X, \tau_1, \tau_2) \to (Y, \rho_1, \rho_2)$ be a ij-fuzzy strongly closed function such that $[f^{-1}(y)]_p$ is compact relative to the topology τ_i for each fuzzy point $y_p \in Y$. If $\mathcal{C} = \{\lambda_k: k \in \Gamma\}$ is a locally finite collection with respect to the topology τ_i, then $f(\mathcal{C}) = \{f(\lambda_k): k \in \Gamma\}$ is a ij-fuzzy P-locally finite collection in Y.

2. *ij*-FUZZY PRE-PARACOMPACT SPACES

In this section, we introduce the generalized paracompactness and separation axioms using the notions of *ij*-fuzzy preopen sets in bitopoogical spaces, and give some characterization of these types of spaces and study the relationships between them and other well known spaces.

Definition 2.1: A fuzzy bitopological space X is called *ij*-fuzzy pre-paracompact if every τ_i-fuzzy open cover of X has a *ij*-fuzzy P-locally finite *ij*-fuzzy pre-open refinement.

Definition 2.2: Let X be a fuzzy bitopological space. The space X is said to be:

1. fuzzy pairwise Hausdorff space [10] if and only if for all distinct fuzzy points x_r, y_s in X, there exists τ_i-fuzzy open U and τ_j-fuzzy open V such that $x_r \in U$, $y_s \in V$ and $U \wedge V = \Phi$;

2. fuzzy pairwise regular [1] if and only if for each τ_i-fuzzy closed set F and $x_r \in X$, $x_r \notin F$, there are two disjoint fuzzy open sets $U \in \tau_i$, $V \in \tau_j$ such that $F \leq U$, $x_r \in V$;

3. fuzzy pairwise P-regular [11] if and only if for each τ_j-fuzzy closed set F and $x_r \in X$, $x_r \notin F$, there are two disjoint *ij*-fuzzy preopen sets U and V, such that $x_r \in U$ and $F \leq V$;

4. fuzzy pair wise P-normal [11] if and only if whenever A and B are disjoint τ_i-fuzzy closed sets in X, there are disjoint *ij*-fuzzy preopen sets U and V with $A \leq U$ and $B \leq V$;

5. fuzzy pairwise P-T_1-space if for each distincts fuzzy points x_r and y_s in X, there are *ij*-fuzzy pre-open sets U and V in X such that $x_r \in U$ and $y_s \in V$;

6. fuzzy pairwise-T_1-space if for each distincts fuzzy points x_r and y_s in X, there are τ_i-fuzzy open set U and τ_j-fuzzy open set V in X such that $x_r \in U$ and $y_s \in V$;

7. fuzzy pairwise P-T_3-space if it is both fuzzy pairwsise P-regular space and P-T_1-space;

8. fuzzy pairwise P-T_4-space if it is both fuzzy pairwsise P-normal space and P-T_1-space.

Lemma 2.3:   A fuzzy bitopological space X is fuzzy pairwise pre-regular if and only if for each τ_i-fuzzy open set μ and $x_r \in \mu$, then there exists $\nu \in$ *ij*-FPO(X) such that $x_r \in \nu \leq$ *ij*-FPCl(ν) $\leq \mu$.

Proof: Easy and so omitted.

Theorem 2.4: Every *ij*-fuzzy pre-paracompact pairwise Hausdorff fuzzy bitopological space X is fuzzy pairwise p-regular.

Proof: Let λ be a τ_i-fuzzy closed set and let $x_p \notin \lambda$. For each $y_p \in \lambda$ choose an τ_i-fuzzy open set μ_y and τ_i-fuzzy open set β_x such that $y_p \in \mu_y$, $x_p \in \beta_x$ and $\mu_y \wedge \beta_x = 0_X$, so we get that $x_p \notin ij$-FPCl(μ_y). Therefore, the family $\mu = \{\mu_y : y_p \in \lambda\}$ $\vee \{X - \lambda\}$ is an τ_i-fuzzy open cover of X and so it has a *ij*-fuzzy P-locally finite *ij*-fuzzy pre-open refinement Π. Put $\nu = \{\beta \in \Pi \colon \beta \wedge \lambda \neq 0_X\}$, then ν is a *ij*-fuzzy pre-open set containing λ and ij-PCl$(\nu) = \{ij$-FPCl$(\beta) \colon \beta \in \Pi$ and $\beta \wedge \lambda \neq 0_X\}$ [by Theorem 1.10(b)]. Therefore $\mu = X - ij$-PCl(ν) is a *ij*-fuzzy pre-open set containing x_p such that μ and ν are disjoint subsets of X. Thus X is fuzzy pairwise pre-regular. From the above theorem, the following corollaries can easily be obtained:

Corollary 2.5: Every *ij*-fuzzy pre-paracompact pairwise Hausdorff fbts X is fuzzy pairwise p-normal.

Theorem 2.6: Let (X, τ_1) and (X, τ_2) be fuzzy regular spaces. Then (X, τ_1, τ_2) is *ij*-fuzzy preparacompact if and only if every τ_i-fuzzy open cover $\mathcal{C} = \{\lambda_k \colon k \in \Gamma$, an index set$\}$ of X has a *ij*-fuzzy P-locally finite *ij*-fuzzy pre-closed refinement Σ.

Proof: Let $\mathcal{C} = \{\lambda_k \colon k \in \Gamma$, an index set$\}$ be τ_i-fuzzy open cover of X. For each $x_p \in X$, we choose a member $\mu_x \in C$ and by regularity of (X, τ_1) and (X, τ_2), $\exists \tau_i$-fuzzy open set ν_x containing x_p such that τ_i-Cl$(\nu_x) \leq \mu_x$. Therefore $\Psi = \{\nu_x \colon x_p \in X\}$ is a τ_i-fuzzy open cover of X and so by hypothesis it has a *ij*-fuzzy P-locally finite *ij*-fuzzy pre-closed refinement, say $\wedge = \{\lambda_k \colon k \in \Gamma$, an index set$\}$. Now consider the collection ij-FPCl$(\Omega) = \{ij$-FPCl$(\eta_k) \colon k \in \Gamma\}$ as a *ij*-fuzzy P-locally finite [by Theorem 1.10(a)] *ij*-fuzzy preclosed subsets (X, τ_1, τ_2). Since for every $k \in \Gamma$, ij-FPCl$(\eta_k) \leq ij$-FPCl$(\nu_x) \leq \tau_i$-Cl$(\nu_x) \leq \mu_x$ for some $\mu_x \leq \mathcal{C}$. Therefore, ij-FPCl(Ω) is a refinement of $\mathcal{C}$.

Conversely, let $\mathcal{C}$ be τ_i -fuzzy open cover of X and Ψ be a *ij*-fuzzy P-locally finite *ij*-fuzzy pre-closed refinement of $\mathcal{C}$. For each $x_p \in X$, we choose a member $\eta_x \in ij$-FPO(X) such that $x_p \in \eta_x$ and η_x intersects at most finitely many members of $\mathcal{C}$. Let Σ be a *ij*-fuzzy preclosed *ij*-fuzzy P-locally finite refinement of $\Omega = \{\eta_x \colon x_p \in X\}$. For each $\nu \in \Psi$, let $\nu' = X - \beta$, where $\beta \in \Sigma$ and $\beta \wedge \nu = 0_X$. Then $\{\nu' \colon \nu \in \Psi$, an index set$\}$ is a *ij*-fuzzy pre-open cover of X. Now for each $\nu \in \Psi$, choose $\nu_\nu \in \mathcal{C}$ such that $\nu \leq \mu_\nu$, therefore the collection $\{\mu_\nu \wedge \nu' \colon \nu \in \Psi\}$ is a *ij*-fuzzy P-locally finite *ij*-fuzzy pre-open refinement of $\mathcal{C}$ and thus (X, τ_1, τ_2) is *ij*-fuzzy pre-paracompact.

Theorem 2.7: Let λ be a *ij*-fuzzy regular closed subset of a *ij*-fuzzy pre-paracompact fuzzy bitopological space (X, τ_1, τ_2). Then $(\lambda, \tau_1, \tau_2)$ is *ij*-fuzzy pre-paracompact.

Proof: Let $\Psi = \{\nu_k : k \in \Gamma,$ an index set$\}$ be τ_i-fuzzy open cover of λ in the subspace $(\lambda, \tau_{1\lambda}, \tau_{2\lambda})$. So for each $k \in \Gamma$, $\exists$ a $\mu_k \in \tau_i$ such that $\nu_k = \lambda \wedge \mu_k$, then the collection $\Xi = \{\mu_k : k \in \Gamma\} \wedge \{X - \lambda\}$ is a τ_i-fuzzy open cover of ij-fuzzy pre-paracompact space (X, τ_1, τ_2) and so it has a ij-fuzzy P-locally finite ij-fuzzy preopen refinemen, say $\Xi' = \{W_m : m \in \Delta,$ an index set$\}$, then by Lemma 1.6(b) and by the fact that ij-FRO$(X) \leq ij$-FSO(X), the collection $\Psi' = \{\lambda \wedge W_m : m \in \Delta\}$ is a ij-fuzzy P-locally finite ij-fuzzy preopen refinement of Ψ in $(\lambda, \tau_{1\lambda}, \tau_{2\lambda})$. This completes the proof.

Theorem 2.8: Let λ and β be fuzzy subsets of a fbts (X, τ_1, τ_2) such that $\lambda \leq \beta$.

1. If $\beta \in ij$-FPO(X) and λ is ij-fuzzy pre-paracompact relative to $(\beta, \tau_{1\beta}, \tau_{2\beta})$, then λ is ij-fuzzy preparacompact relative to (X, τ_1, τ_2).

2. If $\beta \in ij$-FSO(X) and λ is ij-fuzzy pre-paracompact relative to (X, τ_1, τ_2), then λ is ij-fuzzy preparacompact relative to $(\beta, \tau_{1\beta}, \tau_{2\beta})$.

Proof:

1. Let $\Xi = \{\mu_k : k \in \Gamma,$ an index set$\}$ be τ_i-fuzzy open cover of λ in X. Then the collection $\Xi_\beta = \{\beta \wedge \mu_k : k \in \Gamma\}$ is a τ_{i_β}-fuzzy open cover of λ in $(\beta, \tau_{1\beta}, \tau_{2\beta})$. Since λ is ij-fuzzy pre-paracompact relative to $(\beta, \tau_{1\beta}, \tau_{2\beta})$, hence Ξ_β has ij-fuzzy P-locally finite ij-fuzzy pre-open refinement Ω_β in $(\beta, \tau_{1\beta}, \tau_{2\beta})$. Since $\beta \in ij$-FPO(X) and $\mu_k \in \tau_i$, $\forall\, k \in \Gamma$, $i = 1, 2$ so by Lemma 1.6(a), $\beta \wedge \mu_k \in ij$-FPO$(X)$, $\forall\, k \in \Gamma$. Then the collection Ω_β is ij-fuzzy P-locally finite ij-fuzzy pre-open refinement of Ξ in (X, τ_1, τ_2). Therefore, λ is ij-fuzzy pre-paracompact relative to (X, τ_1, τ_2).

2. Let $\Psi = \{\nu_k : k \in \Gamma,$ an index set$\}$ be an τ_{i_β}-fuzzy open cover of λ in $(\beta, \tau_{1\beta}, \tau_{2\beta})$. For every $k \in \Gamma$, we choose $\mu_k \in \tau_i$ such that $\nu_k = \beta \wedge \mu_k$. Then the collection $\Xi = \{\mu_k : k \in \Gamma,$ an index set$\}$ is a τ_i-fuzzy open cover of λ relative to (X, τ_1, τ_2) and so it has a ij-fuzzy P-locally finite ij-fuzzy pre-open refinement of Ξ in (X, τ_1, τ_2), say, Σ. Thus by Lemma 1.6 (b), the collection $\Sigma_\beta = \{\eta \wedge \beta : \eta \in \Sigma\}$ is a ij-fuzzy P-locally finite ij-fuzzy pre-open refinement of Ξ in $(\beta, \tau_{1\beta}, \tau_{2\beta})$. Hence, λ is ij-fuzzy preparacompact relative to $(\beta, \tau_{1\beta}, \tau_{2\beta})$.

Corollary 2.9: Let λ be a fuzzy subset of a fbts (X, τ_1, τ_2).

1. If $\lambda \in ij$-FPO(X) and the subspace $(\lambda, \tau_{1\lambda}, \tau_{2\lambda})$ is ij-fuzzy pre-paracompact, then λ is ij-fuzzy pre-paracompactrelativeto (X, τ_1, τ_2).

2. If $\lambda \in ij$-FSO(X) and λ is ij-fuzzy pre-paracompact relative to (X, τ_1, τ_2), then the subspace $(\lambda, \tau_{1\lambda}, \tau_{2\lambda})$ is ij-fuzzy pre-paracompact.

Proof: Follows from the above theorem.

Lemma 2.10: Let $f: (X, \tau_1, \tau_2) \to (Y, \rho_1, \rho_2)$ be a *ij*-fuzzy pre-closed function defined by $f(x) = y$, $\forall\, x \in X$ such that $f^{-1}(y_p)$ is *ij*-fuzzy strongly compact relative to X for every fuzzy point $y_p \in Y$. If $\mathcal{C} = \{\lambda_k : k \in \Gamma$, an index set$\}$ is a *ij*-fuzzy P-locally finite collection in X, then $f(\mathcal{C}) = \{f(\lambda_k): k \in \Gamma\}$ is a *ij*-fuzzy P-locally finite collection in Y.

Proof: Let y_p be a fuzzy point of Y. Then for each fuzzy point $x_p \in [f^{-1}(y)]_p$ choose $\lambda_x \in ij\text{-PO}(X)$ such that $x_p \in \lambda_x$ and λ_X intersects at most finitely many members of $\mathcal{C}$. The collection $\{\lambda_X: x_p \in [f^{-1}(y)]_p\}$ is a *ij*-fuzzy pre-open cover of $[f^{-1}(y)]_p$ and so there exists finite number of points $x_1, x_2, ..., x_n$ of $[f^{-1}(y)]_p$ such that $[f^{-1}(y)]_p \leq \vee\, \lambda_{x_i} = \lambda$. Since f is *ij*-fuzzy pre-closed function, $\exists\, ij$-fuzzy pre-open set in Y such that $y_p \in \vee$ and $f^{-1}(\vee) \leq \lambda$. Therefore, $\vee$ intersects at most finitely many members of $f(\mathcal{C}) = \{f(\lambda_k): k \in \Gamma\}$ and thus $f(\mathcal{C})$ is a *ij*-fuzzy P-locally finite collection in Y.

Theorem 2.11: Let $f: (X, \tau_1, \tau_2) \to (Y, \rho_1, \rho_2)$ be $(\tau_i\text{-}\rho_i)$-fuzzy continuous, $(\tau_i\text{-}\rho_i)$-fuzzy open, *ij*-fuzzy strongly *ij*-fuzzy pre-open and surjective function such that for each fuzzy point $y_p \in Y$, $(y)]_p$ is *ij*-fuzzy strongly compact relative to (X, τ_1, τ_2). If X is *ij*-fuzzy pre-para compact, then so is Y.

Proof: Let $\Psi = \{v_k: k \in \Gamma$, an index set$\}$ be ρ_i-fuzzy open cover of (Y, ρ_1, ρ_2). Since f is $(\tau_i\text{-}\rho_i)$fuzzy continuous, the collection $\Xi = f^{-1}(\Psi) = \{f^{-1}(v_k): k \in \Gamma\}$ is an τ_i-fuzzy open cover of *ij*-fuzzy pre-paracompact space (X, τ_1, τ_2) and so it has a *ij*-fuzzy P-locally finite *ij*-fuzzy preopen refinement, say Ω. Again, since, f is $(\tau_i\text{-}\rho_i)$-fuzzy continuous and $(\tau_i\text{-}\rho_i)$-fuzzy open, then by Lemma 1.12(3), f is *ij*-fuzzy pre-open function and so by Lemma 2.10, the collection $f(\Omega)$ is *ij*-fuzzy P-locally finite *ij*-fuzzy preopen refinement of Ψ. Thus Y is *ij*-fuzzy pre-paracompact space.

Theorem 2.12: Let $f: (X, \tau_1, \tau_2) \to (Y, \rho_1, \rho_2)$ be $(\tau_i\text{-}\rho_i)$-fuzzy continuous, $(\tau_i\text{-}\rho_i)$-fuzzy closed, fuzzy pairwise pre-irresulate and surjective map such that for each fuzzy point $y_p \in Y$, $[f^{-1}(y)]_p$ is τ_i-fuzzy compact in (X, τ_i). If Y is *ij*-fuzzy pre-paracompact, then so is X.

Proof: Let $\Xi = \{\mu_k: k \in \Gamma$, an index set$\}$ be τ_i-fuzzy open cover of fbts (X, τ_1, τ_2). For each fuzzy point $y_t \in Y$, Ξ is a τ_i-fuzzy open cover of τ_i-compact subspace $f^{-1}(y)$ and so $\exists$ a finite subcover Γ^* of Γ such that $f^{-1}(y) \leq \vee\, \mu_k$. Put $\mu_y = \vee\, \mu_k$ is τ_i-open in (X, τ_i). As f is a $(\tau_i\text{-}\rho_i)$-fuzzy closed function, for each $y_t \in Y$, $\exists\, \rho_i$-fuzzy open subset v_y of Y such that $y_t \in v_y$ and $f^{-1}(v_y) \leq \mu_y$. Then the collection $\Psi = \{v_y: y_t \in Y\}$ is ρ_i-fuzzy open cover of *ij*-fuzzy pre-paracompact space (Y, ρ_1, ρ_2) and so it has a *ij*-fuzzy P-locally finite *ij*-fuzzy pre-open refinement, say $\Omega = \{\eta_m: m \in \Delta$, an index set$\}$. Since, f is a fuzzy pairwise pre-irresulate, then the collection $f^{-1}(\Omega) = \{f^{-1}(\eta_m): m \in \Delta\}$ is *ij*fuzzy pre-open *ij*-fuzzy P-locally finite cover of (X, τ_1, τ_2) such that $\exists\, m \in \Delta, \eta_m \leq \mu_y$ for some $y_t \in Y$. Finally, the collection $\{f^{-1}(\eta_m) \wedge \mu_k: m \in \Delta, k \in \Gamma\}$ is a *ij*-fuzzy P-locally finite *ij*-fuzzy pre-open refinement of Ξ. Thus X is *ij*-fuzzy pre-paracompact space.

References

1. Kandil, A., Nouth A.A., and El-Sheikh, S.A. 1995: On fuzzy bitopological Spaces, Fuzzy Sets and Systems 74, 353-363.

2. Shahana, B., 1991: On fuzzy strongly continuity and fuzzy pre-continuity, Fuzzy Sets and Systems, 44, 303-308.

3. Kelly, B., 1963: "Bitopological Spaces", Proc. London Math. Soc., 3, PP. 17-89.

4. Dieudonne, J., 1944: "Une generalization des spaces compacts", J. Math. Pures. Appl. 23: 65-76.

5. Azad, K.K., 1981: On fuzzy continuity, fuzzy almost continuity and fuzzy semi continuity; *Jarn. Math. Anal. Appl*; 82, 14-32.

6. Martin, M.K., 2001: A note on Raghavan-Reilly's Pairwise paracompactness", Int. J. Math. And Math. Sci. Vol. 24, No. 2, 139-143.

7. Singal, M.K., and Rajvanshi, N., 1992: Regularly open sets in fuzzy topological spaces; Fuzzy Sets and Systems; 50, 343-353.

8. Stone, M.H., 1937: "Applications of the theory of Boolen rings to general topology", Trans. Amer. Math. Soc., 41; 375-381.

9. Pao-Ming, P., andYing-Ming, L., 1980: Fuzzy Topology I: Nbd structure of a fuzzy point and Moore Smith convergence; *J. Math. Anal. and Appl.*; 76(2), 571-599.

10. Srivastava R., and Srivastava, M., 1997: On Pairwise Hausdorff fuzzy bitopological spaces, J. Fuzzy Math. 5, 553-564.

11. Srivastava, R., and Srivastava, M., 2007: On certain separation axioms in fuzzy bitopological spaces, Far East J. Math. Sci.; Vol.-27, 3, 579-587.

12. Srivastava R., and Srivastava, M., 2001: On fuzzy pairwise T0-and fuzzy parwise T1bitopological spaces, Indian J. of Pure Appl. Math. 32, 387-396.

13. Sampath, S. Kumar, 1994: "On fuzzy pairwise -continuity and fuzzy pairwise pre-continuity; Fuzzy Sets and Systems; 62, 231-238.

14. Sampath. S.. Kumar. 1994: Semi-open sets, semi continuity and semi-open mappings in fuzzy bitopological space; Fuzzy sets and systems; 64, 421-426.

15. Zadeh, L.A., 1981: Fuzzy sets; *Information and Control*; 8(1965); 338-353. Azad K. K.; On fuzzy continuity, fuzzy almost continuity and fuzzy semi continuity; *Jarn. Math. Anal. Appl*; 82, 14-32.

An Inventory Model for Deteriorating Items with Two Credit Period in Fuzzy Environment

P. Majumder[*] and U.K. Bera[#]

*Department of Mathematics, National Institute of Technology
Agartala, Barjala, Jirania
E-mail: pinki.mjmdr@rediffmail.com[*]; bera_uttam@yahoo.co.in[#]*

ABSTRACT

The present study investigates the economic order quantity (EOQ) — based inventory model for a retailer under two levels of trade credit to reflect the supply chain management situation in the fuzzy sense. It is assumed that the retailer can obtain full trade credit offered by the supplier and the retailer also offers a trade credit to customers. Here demand rate, ordering cost and unit selling price of the item are imprecise in nature. Under these conditions, the retailer can obtain the most benefits. Study also investigates the retailer's inventory policy for deteriorating items in a supply chain management situation as a cost minimization problem in the fuzzy sense. Due to this impreciseness a fuzzy model is converted into a crisp model by graded mean integration method. The goal of this proposed work is to find minimum average cost & optimal cycle time. A numerical example is then used to illustrate the model. Finally the model is solved by Generalized Reduced Gradient (GRG) method and using LINGO software.

Keywords: Fuzzy demand, Delay Payments, Graded mean integration method.

1. INTRODUCTION

The basic EOQ model is based on the implicit assumption that the payment must be made for the items as soon as the retailer receives them from a supplier. However, in practice, the supplier will allow a certain fixed period (credit period) for settling the amount that the supplier owes to retailer for the items supplied. Before the end of the trade credit period, the retailer can sell the goods and accumulate revenue and earn interest. Usually a higher interest is charged if the payment is not settled by the end of the trade credit period. In a real world, the supplier often makes use of this policy to promote his commodities. In this regard, a number of research papers appeared which deal with the EOQ problem under fixed credit period. A number of researchers have already worked on the integrated Economic order quantity (EOQ) or Economic production quantity model (EPQ) on supply chain system such as Goyal [6], Dava [5], Aggarwal and Jaggi [1], Chang et al., [2]; Chung [3,4]; Huang [8]; Jamal et al., [9], Mahata and Goswami [11], Huang [7], Mahata and Mahata [10]. The present study investigates the economic order quantity (EOQ)-based inventory model for a retailer under two levels of trade credit to reflect the supply chain management situation in the fuzzy sense. It is assumed that the retailer maintains a powerful position and can obtain the full trade credit offered by supplier and the retailer just offers partial trade credit to customers. Furthermore, the demand rate and the inventory costs namely holding cost, ordering cost, purchase cost and selling price may be flexible with some vagueness as to their values. In real life situations, all these parameters in an inventory model are uncertain, imprecise and the determination of the optimum cycle time becomes a non-stochastic vague decision making process. Again, for this type of models, statistical estimations proved to be inefficient because of the lack of statistical observations. In this situation, a suitable way to model these imprecise data is to use fuzzy sets. The ill-formed vagueness in the above parameters are introduced making them fuzzy in nature and then the model is formulated in a fuzzy environment. We use the Graded Mean Integration Representation method for defuzzifying the fuzzy annual total variable cost.

In this paper, the fuzzy inventory model is discussed considering demand and unit selling price as Trapezoidal fuzzy number. Accordingly the total fuzzy inventory cost was found and graded mean integration representation method is used in subsequence to defuzzify the trapezoidal fuzzy total inventory cost.

2. ASSUMPTION AND NOTATION

To develop this model we use the following assumption and notation.

Notation

$\tilde{D}$ = Fuzzy annual demand,

$\tilde{A}$ = Fuzzy cost of placing an order.

c = Unit purchasing price,

σ = Fuzzy unit selling price

h = Unit holding cost per year excluding interest charges.

I_e = Interest which can be earned per dollar per year.

I_c = Interest payable per dollar in investment in inventory per year.

$TRC_1(T)$ = Total relevant cost per unit time when $N \le M \le T$.

$TRC_2(T)$ = Total relevant cost per unit time when $N \le T \le M$.

$TRC_3(T)$ = Total relevant cost per unit time When $0 \le T \le N$.

$\widetilde{TRC}_1(T)$ = Frelevant cost per unit time in $N \le M \le T$.

$\widetilde{TRC}_2(T)$ = total relevant cost per unit time in $N \le T \le M$.

$\widetilde{TRC}_3(T)$ = total relevant cost per unit time in $0 \le T \le N$.

$TR_1C_1(T)$ = Defuzzified value of fuzzy total cost $\widetilde{TRC}_1(T)$.

$TR_2C_2(T)$ = Defuzzified value of fuzzy total cost $\widetilde{TRC}_2(T)$.

$TR_3C_3(T)$ = Defuzzified value of fuzzy total cost $\widetilde{TRC}_3(T)$.

T^*_1 = Optimal value of T_1,

T_2^* = Optimal value of T_2

T_3^* = Optimal value of T_3

Assumption

  (i) Inventory deals with single item only.

  (ii) Replenishment rate is infinite and is instantaneous.

  (iii) Shortages are not allowed.

  (iv) Lead time is zero.

  (v) The deterioration rate θ is constant and it is assumed to be small.

  (vi) There is no repair or replenishment of deteriorated unit in the given cycle.

  (vii) Fuzzy annual demand $\tilde{D}$ is taken a trapezoidal fuzzy number.

  (viii) When $T \ge M$, the account is settled at time $T = M$ and retailer starts paying for the interest charges on the items in stock with rate I_c. When $T \le M$, the account is settled at $T = M$ and the retailer does not need to pay interest charge.

  (ix) The retailer can accumulate revenue and earn interest after his/her customer pays for the amount of purchasing cost to the retailer until the end of the trade credit period offered by the supplier. That is, the retailer can accumulate revenue and earn interest during the period N to M with rate I_e under the condition of trade credit.

  (x) $s \ge c$; $I_c \ge I_e$, $M \ge N$

3. MATHEMATICAL FORMULATION

The inventory level decreases owing to demand as well as deterioration. Thus, the change of inventory level can be represented by the following differential equation

$$\frac{dI(T)}{dt} + \theta I(T) = -D \tag{1}$$

With the initial condition

$$I(0) = Q, \qquad I(T) = 0 \tag{2}$$

Now solution of (1) is

$$I(t) = \frac{D}{\theta} [e^{\theta(T-t)} - 1], \quad 0 \le t \le T$$

and the order quantity is

$$Q = I(0) = \frac{D}{\theta} [e^{\theta T} - 1]$$

1. Annual ordering cost $= \dfrac{A}{T}$

2. Annual procurement cost $= \dfrac{cQ}{T} = \dfrac{cD}{\theta T} (e^{\theta T} - 1)$

3. Annual inventory holding cost(excluding interest charges)

$$= \frac{h}{T} \int_0^T I(t)\, dt = \frac{hD}{\theta^2 T} (e^{\theta T} - 1 - \theta T)$$

Depending upon M, N and T following three cases may arise:

Case 1: $N < M \le T$,   Case 2: $N \le T \le M$,   Case 3: $T \le N < M$

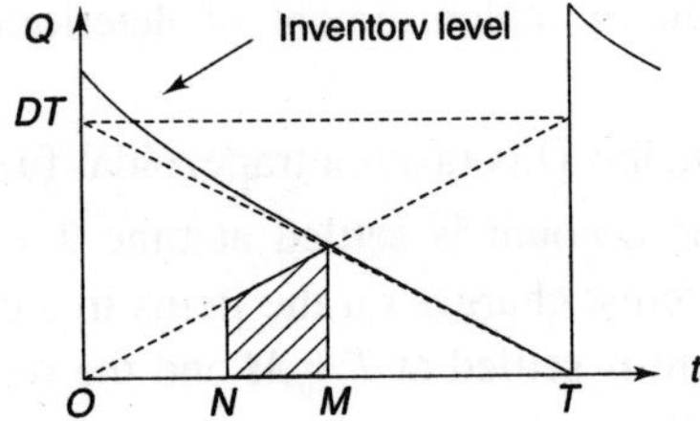

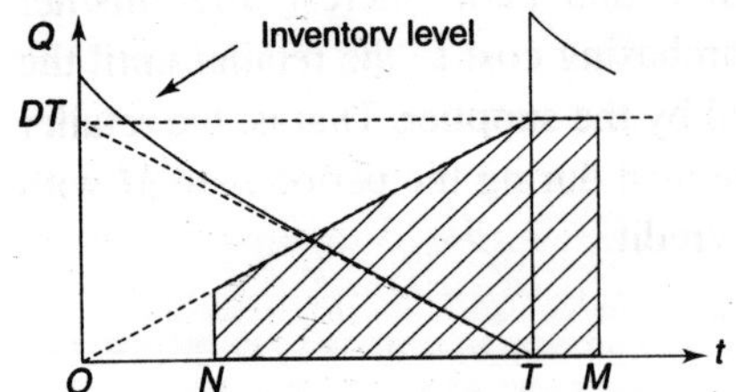

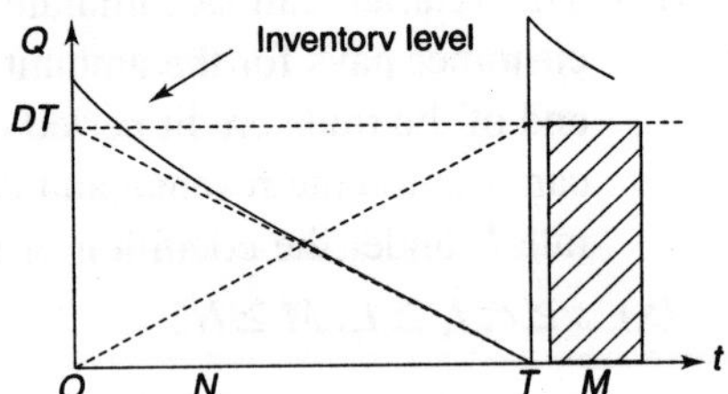

Case 1: $(M \le T)$:   Annual interest payable

$$= \frac{cI_c}{T} \int_M^T I(t)\, dt$$

$$= \frac{cI_c}{T} \cdot \frac{D}{\theta^2} [e^{\theta(T-M)} - 1 - \theta(T-M)]$$

$$\text{Annual interest earned} = \frac{SI_e}{T} \int_N^M D(t)\, dt = \frac{S}{2T} DI_e (M^2 - N^2)$$

Case 2: $(N \leq T \leq M)$: In this case annual interest payable = 0

$$\text{Annual interest earned} = \frac{SI_e}{T} \int_N^T D(t)\, dt + DT(M-T)$$

$$= \frac{S}{2T} DI_e (2MT - T^2 - N^2)$$

Case 3: $(T \leq N)$: In this case, annual interest payable = 0

Annual interest earned = $SI_e\, T(M-N)D$.

4. FUZZY MATHEMATICAL MODEL

Here $\tilde{A} = (a_1, a_2, a_3, a_4)$, $\tilde{D} = (d_1, d_2, d_3, d_4)$, and $\tilde{S} = (s_1, s_2, s_3, s_4)$ be the trapezoidal number based on the above assumptions, fuzzy model can be expressed as in different cases.

Case 1:

Total cost per unit time when $N < M \leq T$ is given by,

$$TRC_1(T) = \frac{A}{T} + \frac{C}{\theta T}(e^{\theta T} - 1)D + \frac{h}{\theta^2 T}(e^{\theta T} - 1 - \theta T)D$$

$$+ \frac{cI_c}{\theta^2 T}(e^{\theta(T-M)} - 1 - \theta(T-M))D - D\frac{I_e}{2T}(M^2 - N^2)S$$

$\therefore$ The total fuzzy cost per unit time is given by,

$$T\tilde{R}C_1(T) = \frac{\tilde{A}}{T} + \frac{C}{\theta T}(e^{\theta T} - 1)\tilde{D} + (e^{\theta T} - 1 - \theta T)\tilde{D}$$

$$+ \frac{cI_c}{\theta^2 T}(e^{\theta(T-M)} - 1 - \theta(T-M))\tilde{D} - \frac{I_e}{2T}(M^2 - N^2)\tilde{S} * \tilde{D}$$

$$= \left(\left\{ \frac{a_1}{T} + \frac{c}{\theta T}(e^{\theta T} - 1)d_1 + \frac{h}{\theta^2 T}(e^{\theta T} - 1 - \theta T)d_1 \right.\right.$$

$$\left. + \frac{cI_c}{\theta^2 T}(e^{\theta(T-M)} - 1 - \theta(T-M))d_1 - \frac{I_e}{2T}(M^2 - N^2)s_4 d_4 \right\},$$

$$\left\{ \frac{a_2}{T} + \frac{c}{\theta T}(e^{\theta T} - 1)d_2 + \frac{h}{\theta^2 T}(e^{\theta T} - 1 - \theta T)d_2 \right.$$

$$\left. + \frac{cI_c}{\theta^2 T}(e^{\theta(T-M)} - 1 - \theta(T-M))d_2 - \frac{I_e}{2T}(M^2 - N^2)s_3 d_3 \right\},$$

$$\left\{\frac{a_3}{T} + \frac{c}{\theta T}\,(e^{\theta T} - 1)\,d_3 + \frac{h}{\theta^2 T}\,(e^{\theta T} - 1 - \theta T)\,d_3\right.$$

$$\left. + \frac{cI_c}{\theta^2 T}\,(e^{\theta(T-M)} - 1 - \theta(T-M))\,d_3 - \frac{I_e}{2T}\,(M^2 - N^2)\,s_2 d_2\right\},$$

$$\left.\left\{\frac{a_4}{T} + \frac{c}{\theta T}\,(e^{\theta T} - 1)\,d_4 + \frac{h}{\theta^2 T}\,(e^{\theta T} - 1 - \theta T)\,d_4\right.\right.$$

$$\left.\left.\left. + \frac{cI_c}{\theta^2 T}\,(e^{\theta(T-M)} - 1 - \theta(T-M))\,d_4 - \frac{I_e}{2T}\,(M^2 - N^2)\,s_1 d_1\right\}\right)\right.$$

$$TR_1C_1\,(T) = \frac{1}{6}\left[\left\{\frac{a_1}{T} + \frac{c}{\theta T}\,(e^{\theta T} - 1)\,d_1 + \frac{h}{\theta^2 T}\,(e^{\theta T} - 1 - \theta T)\,d_1\right.\right.$$

$$\left. + \frac{cI_c}{\theta^2 T}\,(e^{\theta(T-M)} - 1 - \theta(T-M))\,d_1 - \frac{I_e}{2T}\,(M^2 - N^2)\,s_4 d_4\right\}$$

$$+ 2\left[\left\{\frac{a_2}{T} + \frac{c}{\theta T}\,(e^{\theta T} - 1)\,d_2 + \frac{h}{\theta^2 T}\,(e^{\theta T} - 1 - \theta T)\,d_2\right.\right.$$

$$\left. + \frac{cI_c}{\theta^2 T}\,(e^{\theta(T-M)} - 1 - \theta(T-M))\,d_2 - \frac{I_e}{2T}\,(M^2 - N^2)\,s_3 d_3\right\}$$

$$+ 2\left[\left\{\frac{a_3}{T} + \frac{c}{\theta T}\,(e^{\theta T} - 1)\,d_3 + \frac{h}{\theta^2 T}\,(e^{\theta T} - 1 - \theta T)\,d_3\right.\right.$$

$$\left. + \frac{cI_c}{\theta^2 T}\,(e^{\theta(T-M)} - 1 - \theta(T-M))\,d_3 - \frac{I_e}{2T}\,(M^2 - N^2)\,s_2 d_2\right\}$$

$$+ \left\{\frac{a_4}{T} + \frac{c}{\theta T}\,(e^{\theta T} - 1)\,d_4 + \frac{h}{\theta^2 T}\,(e^{\theta T} - 1 - \theta T)\,d_4\right.$$

$$\left.\left.\left. + \frac{cI_c}{\theta^2 T}\,(e^{\theta(T-M)} - 1 - \theta(T-M))\,d_4 - \frac{I_e}{2T}\,(M^2 - N^2)\,s_1 d_1\right\}\right]\right]$$

Therefore $\dfrac{d(TR_1C_1\,(T))}{dT} = 0$, gives

$$-2\theta^2\,(a_1 + 2a_2 + 2a_3 + a_4) + 2(c\theta + h)\,((\theta T - 1)e^{\theta T} + 1)\,(d_1 + 2d_2$$
$$+ 2d_3 + d_4) + 2cI_c\,(d_1 + 2d_2 + 2d_3 + d_4)\,((\theta T - 1)e^{\theta(T-M)} + 1 - M\theta)$$
$$+ I_e\theta^2\,(M^2 - N^2)\,(s_4 d_4 + 2s_3 d_3 + 2s_2 d_2 + s_1 d_1) = 0 \qquad (3)$$

Case 2:

Total cost per unit time when $N \le T \le M$ is given by,

$$TRC_2\,(T) = \frac{A}{T} + \frac{C}{\theta T}\,(e^{\theta T} - 1)\,D + \frac{h}{\theta^2 T}\,(e^{\theta T} - 1 - \theta T)\,D$$

$$- D\,\frac{I_e}{2T}\,(2MT - T^2 - N^2)\,S$$

Similarly,

$$TR_2C_2\,(T) = \frac{1}{6}\left[\left\{\left\{\frac{a_1}{T} + \frac{c}{\theta T}\,(e^{\theta T} - 1)\,d_1 + \frac{h}{\theta^2 T}\,(e^{\theta T} - 1 - \theta T)\,d_1\right.\right.\right.$$

$$\left. - \frac{I_e}{2T}\,(2MT - T^2 - N^2)\,s_4 d_4\right\} + 2\left\{\frac{a_2}{T} + \frac{c}{\theta T}\,(e^{\theta T} - 1)\,d_2\right.$$

$$\left. + \frac{h}{\theta^2 T}\,(e^{\theta T} - 1 - \theta T)\,d_2 - \frac{I_e}{2T}\,(2MT - T^2 - N^2)\,s_3 d_3\right\}$$

$$+ 2\left\{\frac{a_3}{T} + \frac{c}{\theta T}\,(e^{\theta T} - 1)\,d_3 + \frac{h}{\theta^2 T}\,(e^{\theta T} - 1 - \theta T)\,d_3\right.$$

$$\left. - \frac{I_e}{2T}\,(2MT - T^2 - N^2)\,s_2 d_2\right\} + \left\{\frac{a_4}{T} + \frac{c}{\theta T}\,(e^{\theta T} - 1)\,d_4\right.$$

$$\left.\left.\left. + \frac{h}{\theta^2 T}\,(e^{\theta T} - 1 - \theta T)\,d_4 - \frac{I_e}{2T}\,(2MT - T^2 - N^2)\,s_1 d_1\right\}\right]\right]$$

$$\frac{d(TR_2C_2\,(T))}{dT} = 0, \text{ gives}$$

$$-2\theta^2\,(a_1 + 2a_2 + 2a_3 + a_4) + 2(c\theta + h)\,((\theta T - 1)e^{\theta T} + 1)\,(d_1 + 2d_2$$
$$+ 2d_3 + d_4) + I_e\theta^2\,(T^2 - N^2)\,(s_4 d_4 + 2s_3 d_3 + 2s_2 d_2 + s_1 d_1) = 0 \qquad (4)$$

Case 3:

Total cost per unit time when $T \leq N \leq M$ is given by,

$$TRC_3\,(T) = \frac{A}{T} + \frac{C}{\theta T}\,(e^{\theta T} - 1)\,D + \frac{h}{\theta^2 T}\,(e^{\theta T} - 1 - \theta T)\,D$$

$$- D\,I_e\,(M - N)\,S$$

Similarly,

$$TR_3C_3\,(T) = \frac{1}{6}\left[\left\{\left\{\frac{a_1}{T} + \frac{c}{\theta T}\,(e^{\theta T} - 1)\,d_1 + \frac{h}{\theta^2 T}\,(e^{\theta T} - 1 - \theta T)\,d_1\right.\right.\right.$$

$$\left. - I_e\,(M - N)\,s_4 d_4\right\} + 2\left\{\frac{a_2}{T} + \frac{c}{\theta T}\,(e^{\theta T} - 1)\,d_2\right.$$

$$\left. + \frac{h}{\theta^2 T}\,(e^{\theta T} - 1 - \theta T)\,d_2 - I_e\,(M - N)\,s_3 d_3\right\}$$

$$+ 2\left\{\frac{a_3}{T} + \frac{c}{\theta T}\,(e^{\theta T} - 1)\,d_3 + \frac{h}{\theta^2 T}\,(e^{\theta T} - 1 - \theta T)\,d_3\right.$$

$$\left. - I_e\,(M - N)\,s_2 d_2\right\} + \left\{\frac{a_4}{T} + \frac{c}{\theta T}\,(e^{\theta T} - 1)\,d_4\right.$$

$$\left.\left.\left. + \frac{h}{\theta^2 T}\,(e^{\theta T} - 1 - \theta T)\,d_4 - I_e\,(M - N)\,s_1 d_1\right\}\right]\right]$$

$$\frac{d(TR_3C_3\,(T))}{dT} = 0, \text{ gives}$$

$$-\theta^2\,(a_1 + 2a_2 + 2a_3 + a_4) + (c\theta + h)\,((\theta T - 1)e^{\theta T} + 1)\,(d_1 + 2d_2 + 2d_3 + d_4) = 0 \tag{5}$$

The annual total relevant cost for the retailer can be expressed as

TRC(*T*) = Ordering cost + procurement cost + inventory holding cost + interest payable-interest earned

$$TRC(T) = \begin{cases} TR_1C_1\,(T); & \text{if } T \geq M \\ TR_2C_2\,(T); & \text{if } N \leq T \leq M \\ TR_3C_3\,(T); & \text{if } 0 < T \leq N < M \end{cases} \tag{6}$$

where,

$$TR_1C_1\,(T) = \frac{1}{6}\Bigg[\frac{1}{T}\,(a_1 + 2a_2 + 2a_3 + a_4) + \frac{C}{\theta T}\,(e^{\theta T} - 1)\,(d_1 + 2d_2 + 2d_3 + d_4) + \frac{h}{\theta^2 T}\,(e^{\theta T} - 1 - \theta T)\,(d_1 + 2d_2 + 2d_3 + d_4) + \frac{cI_c}{\theta^2 T}\,(e^{\theta(T - M)} - 1 - \theta(T - M))\,(d_1 + 2d_2 + 2d_3 + d_4) - \frac{I_e}{2T}\,(M^2 - N^2)\,(s_4d_4 + 2s_3d_3 + 2s_2d_2 + s_1d_1)\Bigg] \tag{7}$$

$$TR_2C_2\,(T) = \frac{1}{6}\Bigg[\frac{1}{T}\,(a_1 + 2a_2 + 2a_3 + a_4) + \frac{C}{\theta T}\,(e^{\theta T} - 1)\,(d_1 + 2d_2 + 2d_3 + d_4) + \frac{h}{\theta^2 T}\,(e^{\theta T} - 1 - \theta T)\,(d_1 + 2d_2 + 2d_3 + d_4) - \frac{I_e}{2T}\,(2MT - T^2 - N^2)\,(s_4d_4 + 2s_3d_3 + 2s_2d_2 + s_1d_1)\Bigg] \tag{8}$$

$$TR_3C_3\,(T) = \frac{1}{6}\Bigg[\frac{1}{T}\,(a_1 + 2a_2 + 2a_3 + a_4) + \frac{C}{\theta T}\,(e^{\theta T} - 1)\,(d_1 + 2d_2 + 2d_3 + d_4) + \frac{h}{\theta^2 T}\,(e^{\theta T} - 1 - \theta T)\,(d_1 + 2d_2 + 2d_3 + d_4) - I_e\,(M - N)\,(s_4d_4 + 2s_3d_3 + 2s_2d_2 + s_1d_1)\Bigg] \tag{9}$$

Since $TR_1C_1\,(M) = TR_2C_2\,(M)$ and $TR_2C_2\,(N) = TR_3C_3\,(N)$. Therefore TRC(*T*) is continuous and well defined on $T > 0$

Equation (7)..(9) yield

$$T\tilde{R}_1C_1\,(T) = \frac{1}{6}\Bigg[-\frac{1}{T^2}\,(a_1 + 2a_2 + 2a_3 + a_4) + \frac{C}{\theta T}\,(\theta e^{\theta T})\,(d_1 + 2d_2$$

$$+ 2d_3 + d_4) - \frac{c}{\theta T^2} (e^{\theta T} - 1) (d_1 + 2d_2 + 2d_3 + d_4)$$

$$+ \frac{h}{\theta^2 T} (\theta e^{\theta T} - \theta) (d_1 + 2d_2 + 2d_3 + d_4)$$

$$- \frac{h}{\theta^2 T^2} (e^{\theta T} - 1 - \theta T) (d_1 + 2d_2 + 2d_3 + d_4)$$

$$+ \frac{cI_c}{\theta^2 T} (\theta e^{\theta(T-M)} - \theta) (d_1 + 2d_2 + 2d_3 + d_4)$$

$$- \frac{cI_c}{\theta^2 T^2} (e^{\theta(T-M)} - 1 - \theta(T-M)) (d_1 + 2d_2 + 2d_3 + d_4)$$

$$\left. + \frac{I_e}{2T^2} (M^2 - N^2) (s_4 d_4 + 2s_3 d_3 + 2s_2 d_2 + s_1 d_1) \right] \tag{10}$$

$$T\tilde{R}_2 C_2 (T) = \frac{1}{6} \left[-\frac{1}{T^2} (a_1 + 2a_2 + 2a_3 + a_4) + \frac{C}{\theta T} (\theta e^{\theta T}) (d_1 + 2d_2 \right.$$

$$+ 2d_3 + d_4) - \frac{c}{\theta T^2} (e^{\theta T} - 1) (d_1 + 2d_2 + 2d_3 + d_4)$$

$$+ \frac{h}{\theta^2 T} (\theta e^{\theta T} - \theta) (d_1 + 2d_2 + 2d_3 + d_4)$$

$$- \frac{h}{\theta^2 T^2} (e^{\theta T} - 1 - \theta T) (d_1 + 2d_2 + 2d_3 + d_4)$$

$$- \frac{I_e}{2T} (2M - 2T) (s_4 d_4 + 2s_3 d_3 + 2s_2 d_2 + s_1 d_1)$$

$$\left. + \frac{I_e}{2T^2} (2M - T^2 - N^2) (s_4 d_4 + 2s_3 d_3 + 2s_2 d_2 + s_1 d_1) \right] \tag{11}$$

$$T\tilde{R}_3 C_3 (T) = \frac{1}{6} \left[-\frac{1}{T^2} (a_1 + 2a_2 + 2a_3 + a_4) + \frac{C}{\theta T} (\theta e^{\theta T}) (d_1 + 2d_2 \right.$$

$$+ 2d_3 + d_4) - \frac{c}{\theta T^2} (e^{\theta T} - 1) (d_1 + 2d_2 + 2d_3 + d_4)$$

$$+ \frac{h}{\theta^2 T} (\theta e^{\theta T} - \theta) (d_1 + 2d_2 + 2d_3 + d_4)$$

$$\left. - \frac{h}{\theta^2 T^2} (e^{\theta T} - 1 - \theta T) (d_1 + 2d_2 + 2d_3 + d_4) \right] \tag{12}$$

The objective in this paper is to find an optimal cycle time to minimize the annual total relevant cost for the retailer. For this the optimal cycle time $T_1{}^*$ by setting equation (10) equal to 0; is the root of the following equation:

$$- 2\theta^2 (a_1 + 2a_2 + 2a_3 + a_4) + 2(c\theta + h) ((\theta T - 1)e^{\theta T} + 1) (d_1 + 2d_2$$
$$+ 2d_3 + d_4) + 2cI_c (d_1 + 2d_2 + 2d_3 + d_4) ((\theta T - 1)e^{\theta(T - M)} + 1 - M\theta)$$
$$+ I_e\theta^2 (M^2 - N^2) (s_4d_4 + 2s_3d_3 + 2s_2d_2 + s_1d_1) = 0 \qquad (13)$$

Equation (13) is the optimality condition of (7).

Let
$$f_1(t) = - 2\theta^2 (a_1 + 2a_2 + 2a_3 + a_4) + 2(c\theta + h) ((\theta T - 1)e^{\theta T} + 1)$$
$$(d_1 + 2d_2 + 2d_3 + d_4) + 2cI_c (d_1 + 2d_2 + 2d_3 + d_4)$$
$$((\theta T - 1)e^{\theta(T - M)} + 1 - M\theta) + I_e\theta^2 (M^2 - N^2) (s_4d_4 + 2s_3d_3$$
$$+ 2s_2d_2 + s_1d_1)$$

Then both $f_1(T)$ and $T\tilde{R}_1 C_1(T)$ have the same sign and domain.
We also have,

$$f_1(T) = 2(c\theta + h) (\theta^2 Te^{\theta T} + 1) (d_1 + 2d_2 + 2d_3 + d_4) + 2cI_c (d_1$$
$$+ 2d_2 + 2d_3 + d_4) (\theta^2 Te^{\theta(T - M)}) > 0, \quad \text{if } T > 0$$

Hence $f_1(T)$ is increasing on $(0, \infty)$.

Also,
$$f_1(0) = - \theta^2 [2(a_1 + 2a_2 + 2a_3 + a_4) + I_e (N^2 - M^2) (s_4d_4 + 2s_3d_3$$
$$+ 2s_2d_2 + s_1d_1)] + 2cI_c (d_1 + 2d_2 + 2d_3 + d_4)$$
$$(1 - M\theta - e^{- \theta M})$$

and $\lim\limits_{T \to \infty} f_1(T) = \infty > 0$

Since $1 - \theta M < e^{-\theta M}$, so we restrict to the condition of

$$[2(a_1 + 2a_2 + 2a_3 + a_4) + I_e (N^2 - M^2) (s_4d_4 + 2s_3d_3 + 2s_2d_2 + s_1d_1)] > 0$$

In the rest of our mathematical analysis. Hence we see that

$$T\tilde{R}_1 C_1(T) = \begin{cases} < 0; & \text{if } T \in (0, T_1^*) \\ = 0; & \text{if } T = T_1^* \\ > 0; & \text{if } T \in (T_1^*, \infty) \end{cases} \qquad (14)$$

Based upon the above arguments, we sure that the solution of (13) not only exist but also is unique.

Therefore, T_1^* is the unique non negative solution of (13). Similarly, the optimal cycle time T_2^*, obtained by setting the equation (11) equal to zero.

Similarly, we have

$$T\tilde{R}_2 C_2(T) = \begin{cases} < 0; & \text{if } T \in (0, T_2^*) \\ = 0; & \text{if } T = T_2^* \\ > 0; & \text{if } T \in (T_2^*, \infty) \end{cases} \qquad (15)$$

Therefore, T_2^* is the unique non negative solution.

Similarly, the optimal cycle time T_3^*, obtained by setting the equation (12) equal to zero.

Similarly we have

$$T\tilde{R}_3 C_3\ (T) = \begin{cases} < 0; \text{ if } T \in (0,\ T_3{}^*) \\ \ = 0; \text{ if } T = T_3{}^* \\ \ > 0; \text{ if } T \in (T_3{}^*,\ \infty) \end{cases} \qquad (16)$$

Therefore, $T_3{}^*$ is the unique non negative solution.

5. DECISION RULES OF THE OPTIMAL CYCLE TIME T*

At present, we will establish an easy procedure to determine the optimal cycle time to simplify the solution procedure. Equation (10), (11) , (12) yield that

$$T\tilde{R}_1 C_1\ (M) = T\tilde{R}_2 C_2\ (M)$$

$$= \frac{1}{6}\left[-\frac{1}{M^2}\ (a_1 + 2a_2 + 2a_3 + a_4) + \frac{C}{\theta M}\ (\theta e^{\theta M})\ (d_1 + 2d_2 \right.$$

$$+ 2d_3 + d_4) - \frac{C}{\theta M^2}\ (e^{\theta M} - 1)\ (d_1 + 2d_2 + 2d_3 + d_4)$$

$$- \frac{h}{\theta^2 M^2}\ (\theta e^{\theta M} - 1 - \theta M)\ (d_1 + 2d_2 + 2d_3 + d_4)$$

$$+ \frac{I_e}{2M^2}\ (M^2 - N^2)\ (s_4 d_4 + 2s_3 d_3 + 2s_2 d_2 + s_1 d_1)$$

$$\left. + \frac{h}{\theta^2 M}\ (\theta e^{\theta M} - \theta)\ (d_1 + 2d_2 + 2d_3 + d_4) \right]$$

and

$$T\tilde{R}_2 C_2\ (N) = T\tilde{R}_3 C_3\ (N)$$

$$= \frac{1}{6}\left[-\frac{1}{N^2}\ (a_1 + 2a_2 + 2a_3 + a_4) + \frac{C}{\theta N}\ (\theta e^{\theta N})\ (d_1 + 2d_2 \right.$$

$$+ 2d_3 + d_4) - \frac{C}{\theta N^2}\ (e^{\theta N} - 1)\ (d_1 + 2d_2 + 2d_3 + d_4)$$

$$- \frac{h}{\theta^2 N}\ (\theta e^{\theta N} - \theta)\ (d_1 + 2d_2 + 2d_3 + d_4)$$

$$\left. - \frac{h}{\theta^2 N^2}\ (\theta e^{\theta N} - 1 - \theta N)\ (d_1 + 2d_2 + 2d_3 + d_4) \right]$$

Furthermore, we let

$$\Delta_1 = -2\theta^2\ (a_1 + 2a_2 + 2a_3 + a_4) + 2\ (c\theta + h)\ ((\theta M - 1)e^{\theta M} + 1)$$
$$(d_1 + 2d_2 + 2d_3 + d_4) + I_e \theta^2\ (M^2 - N^2)\ (s_4 d_4 + 2s_3 d_3$$
$$+ 2s_2 d_2 + s_1 d_1)$$

$$\Delta_2 = -2\theta^2 (a_1 + 2a_2 + 2a_3 + a_4) + 2(c\theta + h)((\theta N - 1)e^{\theta N} + 1)(d_1 + 2d_2 + 2d_3 + d_4)$$

We have

$$\Delta_1 - \Delta_2 = 2(c\theta + h)[(\theta M e^{\theta M} - e^{\theta M} + 1) - (\theta N e^{\theta N} - e^{\theta N} + 1)(d_1 + 2d_2 + 2d_3 + d_4) + I_e\theta^2(M^2 - N^2)(s_4 d_4 + 2s_3 d_3 + 2s_2 d_2 + s_1 d_1) \tag{17}$$

Let, $K(x) = xe^x - e^x + 1$, then $K(x) = xe^x > 0$. Consequently, $K(x) \geq K(y)$ if $x > y > 0$. So that $K(\theta M) \geq K(\theta N)$ as $\theta M > \theta N > 0$

This gives $(\theta M e^{\theta M} - e^{\theta M} + 1) \geq (\theta N e^{\theta N} - e^{\theta N} + 1)$

From (17), we have $\Delta_1 \geq \Delta_2$. Now we have the following theorem

Theorem 1:

(a) If $\Delta_1 > 0$ and $\Delta_2 \geq 0$, then $TRC(T^*) = TRC(T_3^*)$ and $T^* = T_3^*$

(b) If $\Delta_1 > 0$ and $\Delta_2 < 0$, then $TRC(T^*) = TRC(T_2^*)$ and $T^* = T_2^*$

(c) If $\Delta_1 \leq 0$ and $\Delta_2 < 0$, then $TRC(T^*) = TRC(T_1^*)$ and $T^* = T_1^*$

Based on the above theorem a numerical example is set.

6. NUMERICAL ILLUSTRATION

To solve the following model we use the following data

Numerical Example: Let $\tilde{A} = (40, 42, 44, 46)$, $\tilde{D} = (4400, 4700, 5000, 5300)$, $\tilde{S} = (100, 102, 104, 106)$ be trapezoidal fuzzy numbers, $h = \$ 6$ unit/year, $c = \$ 100$/unit/year, $\theta = 0.1(0.1)\ 0.3$, $I_C = \$0.17/\$$/year, $I_e = \$0.14/\$$/year

Table Optimal result of the proposed model

θ	$M = 0.015,\ N = 0.01$			
	Δ_1	Δ_2	T^*	$TRC(T^*)$
0.1	<0	<0	0.02355316	487535.4
0.2	<0	<0	0.02062446	488070.1
0.3	<0	<0	0.01856960	488545.8

θ	$M = 0.30,\ N = 0.25$			
	Δ_1	Δ_2	T^*	$TRC(T^*)$
0.1	>0	>0	0.03325346	484084.1
0.2	>0	>0	0.02606974	484795.3
0.3	>0	>0	0.0214446	485378.6

Contd..

Contd..

θ	$M = 0.015, N = 0.01$			
	Δ_1	Δ_2	T^*	$TRC(T^*)$
0.1	<0	<0	0.02355316	487535.4
0.2	<0	<0	0.02062446	488070.1
0.3	<0	<0	0.01856960	488545.8

7. DISCUSSION

In this paper, an inventory model of deteriorating item is proposed. From the above tables we have the following observation. When deterioration rate is increasing then the optimal cycle time T^* is decreasing and the optimal annual total cost $TRC(T^*)$ is increasing. The proposed model can assist the retailer in accurately determining the optimal cycle time, the optimal total relevant cost in fuzzy environment. Moreover, the proposed model can be used in inventory control of certain deteriorating items such as food items, photographic film, electronic components, fashionable commodities, glass products and others.

8. CONCLUSION

In this paper, we have developed an EOQ-based inventory model for deteriorating items to determine the optimal ordering policies of a retailer under two levels of trade credit to reflect the supply chain management situation in the fuzzy sense. It is assumed that the supplier offers a full trade credit period to the retailer and the retailer also offers a full trade credit to customers. Furthermore, the demand rate, ordering cost and unit selling price of the items are taken as fuzzy numbers. Based on above situations, we investigate the retailer's inventory system as a cost minimization problem to determine the retailer's optimal inventory policy under the supply chain management in the fuzzy sense. The annual total variable cost for the retailer in the fuzzy sense is defuzzified using the Graded Mean Integration Representation method. From the viewpoint of the costs, decision rules to find the optimal cycle time contains three cases: (i) $T \leq N$ (ii) $N \leq T \leq M$ and (iii) $T \geq M$. In order to obtain the optimal ordering policy, we propose one theorem. Numerical examples are given to illustrate this model.

The proposed model can be extended in several ways. For instance, we may extend the constant deterioration rate to a two-parameter Weibull distribution. In addition, we could consider the replenishment rate as finite, demand as a function of time, selling price, and others. We should also extend this model by taking demand, selling price, holding price as a random number, hybrid number etc. Finally, we could generalize the model to allow for shortages, quantity discounts, and others.

References

1. Aggarwal, S.P., and Jaggi, C.K., 1995: *Ordering policies of deteriorating items under permissible delay in payments*, Journal of the Operational Research Society, 46, 658-662.

2. Chang, H.J., Hung, C.H., Dye C.Y., 2001: *An inventory model for deteriorating items with linear trend demand the Condition of permissible delay in payments*, Production Planning and Control 12, 274-282.

3. Chang, H.J., Hung, C.H., Dye, C.Y., 2001: *An inventory model for deteriorating items with linear trend demand the Condition of permissible delay in payments*, Production Planning and Control 12, 274-282.

4. Chung, K.J., 1998b: "Economic order quantity model when delay in payments is permissible", Journal of Information & Optimization Sciences 19, 411-416.

5. Dava, U., Dava, U., 1985: *On Economic order quantity under conditions of permissible delay in payments by Goyal*, Journal of Operation Research Society, Vol 36, pp.1069.

6. Goyal S.K., 1985: *Economic order quantity under conditions of permissible delay in payments*, Journal of Operation Research Society 36, 335-338.

7. Huang, Y.F., 2007: "Optimal retailers replenishment decision in the EPQ model under two levels of trade credit policy", European Journal of Operation research 176.1577-1591.

8. Huang, Y.F. 2003: *Optimal retailer's ordering policies in the EOQ model under trade credit financing*, Journal of the Operational Research Society, vol. 54, pp. 1011-1015.

9. Jamal, A.M.M., Sarker, B.R., and Wang, S., 2000: *Optimal payment time for a retailer under permitted delay in payment by the wholesaler*, International Journal of Production Economics 66: 59-66.

10. Mahata, G.C. and Mahata, P., 2009: *Optimal retailer's ordering policies in the EOQ Model for deteriorating items under trade credit financing in supply chain*, International Journal of Engineering and Applied Sciences 5:5.

11. Mahata, G.C., and Goswami, A, 2006: *Production Lot-size Model with Fuzzy Production Rate and Fuzzy Demand Rate for deteriorating item under permissible delay in payments*, OPSEARCH, vol. 43, No. 4, pp 358-375.

Multi-item Solid Transportation Problem with Safety Measure

A. Baidya[1*], U.K. Bera[1#] and M. Maiti[2†]
[1]*Department of Mathematics, National Institute of Technology, Agartala, Jirania, West Tripura*
[2]*Department of Applied Mathematics, Vidyasagar University, Midnapore, West Bengal*
E-mail: abhijitnita@yahoo.in[*], uttam_bera@yahoo.co.in[#];
mmaiti2005@yahoo.co.in.[†]

ABSTRACT

The goal of this work is to solve an interval valued multi-item solid transportation problem (MIIVSTP) with safety measure. In this paper, we introduce a new concept, 'safety factor', in transportation problem. While transporting items from origins to destinations through different modes, common difficulties/risks involved are poor road condition, insurgency etc in some routes. The present study thus introduced the concept of desired total safety function taking all the probable risks into consideration. However, depending upon the nature of safety factor, we formulate five models without and with safety factor; this factor may be crisp, fuzzy, interval, stochastic in nature. The models were developed based on Generalized Reduced Gradient (GRG) method using LINGO 13.0 software. Numerical examples are used to illustrate the model and methodologies.

Keywords: Solid Transportation Problem, Safety Constraint, Possibility, Necessity, Credibility, Chance Constraint Programming.

1. INTRODUCTION

Hitchcock [5] defined the transportation problems (TPs) as one of the most common special type linear programming problems involving constraints. The Solid Transportation Problem (STP) is a generalization of the well-known TP in which conveyances are taken into account in the constraint set and it is of great use in public distribution systems. In some cases transportation parameters i.e., costs, sources, etc are expressed in the form of intervals. Pal and Sengupta [9], [10], [11], F. Jimenez and Verdegay [6], Das and Maiti [3] developed the Interval-valued Transportation Problem. Possibility theory was proposed by Zadeh [12] and developed by Dubois and Prade [4] and other researchers. Possibility theory is a new form of information theory which is related to, but independent of both fuzzy sets and probability theory, which provides an alternative and convenient framework for modeling of real-world fuzzy decision systems mathematically. As the average of possibility and necessity measures, credibility measure was introduced by Liu and Liu [7]. Here we use Chance Constraint Programming Technique to solve problems involving chance constraints, i.e., constraints having finite probability of being violated. This technique was originally developed by Charnes and Cooper [1], [2]. Recently L. Sahoo et al. [8] solve Genetic algorithm based multi-objective reliability optimization in interval environment.

`The safety factor along the different routes via different modes are normally estimated by the expects/ administrator. These values may be deterministic, fuzzy, random or interval numbers. Here decision maker (DM) likes to minimize the transportation cost choosing the particular routes and modes of transportation for particular so that total safety for the system is greater or equal to a predefined safety value.

In this paper, we formulate a MIIVSTP with safety factors for different routes, modes and items. Here, DM desired to have total safety factors for the transportation system more than or equal to a pre-fixed safety measure. Different modes are formulated taking safety factors as crisp, fuzzy, random and interval numbers and solved using the appropriate methods. The model and methodologies are illustrated with numerical examples.

2. SCOPE OF AN INTERVAL OBJECTIVE FUNCTION IN THE LIGHT OF MAXIMIZATION/MINIMIZATION PROBLEM

The objective of a conventional linear programming problem (LPP) is to maximize or minimize the value of its (one only, single-valued) objective function satisfying a given set of restriction. However, a single-objective interval linear programming problem (ILPP) contains an interval-valued objective function (IOF).Let us consider the following problem:

Maximize/Minimize

$$Z = \sum_{j=1}^{N} [C_{Lj}, C_{Rj}] x_j \tag{1}$$

Subject to, {Set of feasibility constraints}

As an interval can be represented by any two of its four attributes(viz., left limit, right limit, mid-value and width), then by using attributes mid-value and width(say), the ILPP(1) can be reduced into a bi-objective LPP as follows:

Max/Min{mid-value of the IOF} $\tag{2}$

Min{Width of the IOF} $\qquad$ (3)

Subject to {Set of feasible Constraints} $\qquad$ (4)

Henceforth we develop a composite goal to define the IOF as follows:
For the maximization problem:

$$\text{Maximize } \lambda \times m(Z) - (1 - \lambda) \times w(Z) \qquad (5)$$

w.r.t (1) and $\qquad 0 \leq \lambda \leq 1 \qquad$ (6)

For minimization problem:

$$\text{Minimize} \qquad ZZ = \lambda \times m(Z) + (1 - \lambda) \times w(Z) \qquad (7)$$

w.r.t (1) and $\qquad 0 \leq \lambda \leq 1 \qquad$ (8)

The lemda (λ) factor defines the DM's pessimistic or optimistic bias. If $\lambda = 1$, (5) and (7) show DM's absolute optimistic bias and if $\lambda = 0$, (7) and (7) indicate, on the contrary, the pessimistic DM's attitude [11]. With $\lambda = 0.5$ or with similar other value, a similar proportional balance between DM's optimistic and pessimistic preference may be thought of.

4. FORMULATION OF THE SOLID TRANSPORTATION PROBLEM

4.1 Notation

(i) M : number of origins/sources of the transportation problem.

(ii) N : number of destinations/demands of the transportation problem.

(iii) K : number of conveyances i.e. different modes of transporting units from sources to destinations.

(iv) E_k : k-th conveyance capacity of the transportation problem.

(v) O_i^q : amount of homogeneous product available at the i-th origin.

(vi) D_j^q : demand at the j-th destination.

(vii) C_{ijk}^q : per unit transportation cost from i-th origin to j-th destination by k-th conveyance of q-th item.

(viii) x_{ijk}^q : the amount transported from i-th origin to j-th destination by k-th conveyance.

(ix) s_{ijk}^q : the safety factor when an item is transformed from i-th origin to j-th destination by k-th conveyance of q-th item. If q-th item is transported from source i to destination j by conveyance k, then the safety factor s_{ijk}^q is considered. This implies that if $x_{ijk}^q > 0$, then we consider the safety factor for this route as a part of the safety constraint. Thus for the convenience of modeling, the following notation is introduced:

$$y_{ijk}^q = \begin{cases} 1 \text{ for } x_{ijk}^q > 0 \\ 0 \text{ otherwise} \end{cases}$$

Model 1

Formulation of MIIVSTP without Safety Factor:

$$\text{Minimize} \qquad Z = \sum_{q=1}^{Q} \sum_{i=1}^{M} \sum_{j=1}^{N} \sum_{k=1}^{K} \left[C_{ijkL}^q, C_{ijkU}^q \right] x_{ijk}^q. \tag{9a}$$

Subject to

$$\sum_{j=1}^{N} \sum_{k=1}^{K} x_{ijk}^q = O_i^q, \quad i = 1, 2, \ldots, M \tag{9b}$$

$$\sum_{i=1}^{M} \sum_{k=1}^{K} x_{ijk}^q = D_i^q, \quad j = 1, 2, \ldots, M \tag{9c}$$

$$\sum_{q=1}^{Q} \sum_{i=1}^{M} \sum_{j=1}^{N} x_{ijk}^q = E_k, \quad k = 1, 2, \ldots, K \tag{9d}$$

$$x_{ijk}^q \leq 0 \quad \text{for all } i, j, k, q \tag{9e}$$

The problem is feasible if and only if $O \cap D \cap E \neq \phi$, where

$$O = \sum_{q=1}^{Q} \sum_{i=1}^{M} O_i^q = \left[\sum_{q=1}^{Q} \sum_{i=1}^{M} o_{i1}^q, \sum_{q=1}^{Q} \sum_{i=1}^{M} o_{i2}^q \right],$$

$$D = \sum_{q=1}^{Q} \sum_{j=1}^{N} D_i^q = \left[\sum_{q=1}^{Q} \sum_{j=1}^{N} d_{j1}^q, \sum_{q=1}^{Q} \sum_{j=1}^{N} d_{j2}^q \right],$$

$$E = \sum_{k=1}^{K} E_k = \left[\sum_{k=1}^{K} e_{k1}, \sum_{k=1}^{K} e_{k2} \right].$$

Model 2   Formulation of MIIVSTP with Safety Factor:

Model 2a   Formulation of MIIVSTP with Safety Factor as a crisp number:

$$\text{Minimize} \qquad Z = \sum_{q=1}^{Q} \sum_{i=1}^{M} \sum_{j=1}^{N} \sum_{k=1}^{K} \left[C_{ijkL}^q, C_{ijkU}^q \right] x_{ijk}^q \tag{10}$$

Subject to the constraints (9b) – (9e)

$$S = \sum_{q=1}^{Q} \sum_{i=1}^{M} \sum_{j=1}^{N} \sum_{k=1}^{K} s_{ijk}^q y_{ijk}^q > B \tag{10a}$$

where s_{ijk}^q are crisp numbers, B is the desired safety measure for the whole transportation system and S is the total safety measure.

Model 2b   Formulation of MIIVSTP with safety Factor (under Possibility, Necessity and Credibility) as uncertain number:

It is the same as problem (8) where S is replaced as

$$\tilde{S} = \sum_{q=1}^{Q} \sum_{i=1}^{M} \sum_{j=1}^{N} \sum_{k=1}^{K} \tilde{s}_{ijk}^q \, y_{ijk}^q = \Sigma \left(s_{ijk1}^q, s_{ijk2}^q, s_{ijk3}^q \right) y_{ijk}^q$$

$$= \left(\Sigma s_{ijk1}^q \, y_{ijk}^q, \ \Sigma s_{ijk2}^q \, y_{ijk}^q, \ \Sigma s_{ijk3}^q \, y_{ijk}^q \right)$$

$$= (a_1, a_2, a_3), \text{ say} = \tilde{a}, \text{ say} \tag{10b}$$

where $\Sigma = \sum\limits_{q=1}^{Q} \sum\limits_{i=1}^{M} \sum\limits_{j=1}^{N} \sum\limits_{k=1}^{K}$.

It is same as problem (8), where the fuzzy safety constraints represented by (8b) is replaced by

[Model 2b(i)] Pos $(\tilde{S} \geq B) > \varepsilon$ (under possibility) $\tag{11}$

[Model 2b(ii)] Nec $(\tilde{S} \geq B) > \varepsilon$ (under necessity) $\tag{12}$

[Model 2b(iii)] Cr $(\tilde{S} \geq B) > \varepsilon$ (under credibility) $\tag{13}$

where $\tilde{S}$ be the total fuzzy safety measure.

Model 2c and 2d: Formulations of MIIVSTP with safety factor as interval and stochastic are

It is same as problem (10), where the fuzzy safety cost represented by (10b) is replaced by

[Model 2c] $\sum\limits_{q=1}^{Q} \sum\limits_{i=1}^{M} \sum\limits_{j=1}^{N} \sum\limits_{k=1}^{K} \left[s_{ijkL}^q, s_{ijkR}^q \right] y_{ijk}^q \geq [B_L, B_R]$ $\tag{14}$

[Model 2d] $\sum\limits_{q=1}^{Q} \sum\limits_{k=1}^{K} \sum\limits_{j=1}^{N} \sum\limits_{i=1}^{M} \tilde{s}_{ijk}^q \, y_{ijk}^q \geq B,$ $\tag{15}$

respectively.

5.   NUMERICAL EXAMPLE AND DISCUSSION

5.1   Input Data

A marketing company procures two types of items as rice and wheat from the three production sources and supply to two different destinations through the two different types of conveyances where availabilities, demands and conveyances capacity are interval in nature. i.e., we consider the following (3 × 2 × 2) MIIVSTP:

Supplies:

$$O_1^{\ 1} = [6, 16],\ O_2^{\ 1} = [8, 23],\ O_3^{\ 1} = [16, 30],\ O_1^{\ 2} = [21, 30],$$
$$O_2^{\ 2} = [8, 23],\ O_3^{\ 2} = [2, 12].$$

Demands:

$$D_1^{\ 1} = [8, 17],\ D_2^{\ 1} = [14, 29],\ D_1^{\ 2} = [23, 32],\ D_2^{\ 2} = [14, 19].$$

Conveyances Capacities:

$$E_1 = [26, 41],\ E_2 = [7, 42].$$

Desired minimum total safety measure for the system ($= B$) = 4.3.

Interval Transportation Costs:

$$C_{111}^1 = [38, 42],\ C_{211}^1 = [83, 86],\ C_{311}^1 = [7, 8],\ C_{112}^1 = [71, 75],$$
$$C_{212}^1 = [41, 43],\ C_{312}^1 = [10, 13],\ C_{121}^1 = [72, 76],\ C_{221}^1 = [71, 73],$$
$$C_{321}^1 = [49, 51],\ C_{122}^1 = [96, 100],\ C_{222}^1 = [52, 54],\ C_{322}^1 = [70, 73],$$
$$C_{111}^2 = [82, 86],\ C_{211}^2 = [80, 84],\ C_{311}^2 = [9, 11],\ C_{112}^2 = [16, 20],$$
$$C_{212}^2 = [37, 40],\ C_{312}^2 = [8, 12],\ C_{121}^2 = [85, 87],\ C_{221}^2 = [68, 72],$$
$$C_{321}^2 = [50, 54],\ C_{122}^2 = [95, 98],\ C_{222}^2 = [50, 51],\ C_{322}^2 = [71, 74].$$

Crisp Safety Factors:

$$S_{111}^1 = .5,\ S_{211}^1 = .85,\ S_{311}^1 = .75,\ S_{212}^1 = .75,\ S_{312}^1 = .55,\ S_{121}^1 = .75,$$
$$S_{221}^1 = .5,\ S_{321}^1 = .35,\ S_{122}^1 = .95,\ S_{222}^1 = .55,\ S_{322}^1 = .75,\ S_{111}^2 = .85,$$
$$S_{211}^2 = .85,\ S_{311}^2 = .3,\ S_{112}^2 = .85,\ S_{212}^2 = .6,\ S_{312}^2 = .5,\ S_{121}^2 = .35,$$
$$S_{221}^2 = .45,\ S_{321}^2 = .3,\ S_{122}^2 = .95,\ S_{222}^2 = .55,\ S_{322}^2 = .75.$$

Fuzzy Safety Factors:

$$\tilde{S}_{111}^1 = (.45, .5, .55),\ \tilde{S}_{211}^1 = (.8, .85, .9),\ \tilde{S}_{311}^1 = (.25, .3, .35),$$
$$\tilde{S}_{112}^1 = (.7, .75, .8),\ \tilde{S}_{212}^1 = (.75, .8, .85),\ \tilde{S}_{312}^1 = (.5, .55, .6),$$
$$\tilde{S}_{121}^1 = (.7, .75, .8),\ \tilde{S}_{221}^1 = (.45, .5, .55),\ \tilde{S}_{321}^1 = (.3, .35, .4),$$
$$\tilde{S}_{122}^1 = (.9, .95, .1),\ \tilde{S}_{222}^1 = (.5, .55, .6),\ \tilde{S}_{322}^1 = (.7, .75, .8),$$
$$\tilde{S}_{111}^2 = (.8, .85, .9),\ \tilde{S}_{211}^2 = (.8, .85, .9),\ \tilde{S}_{311}^2 = (.25, .3, .35),$$
$$\tilde{S}_{112}^2 = (.8, .85, .9),\ \tilde{S}_{212}^2 = (.55, .6, .65),\ \tilde{S}_{312}^2 = (.45, .5, .55),$$
$$\tilde{S}_{121}^2 = (.3, .35, .4),\ \tilde{S}_{221}^2 = (.4, .45, .5),\ \tilde{S}_{321}^2 = (.25, .3, .35),$$
$$\tilde{S}_{122}^2 = (.9, .95, .1),\ \tilde{S}_{222}^2 = (.5, .55, .6),\ \tilde{S}_{322}^2 = (.7, .75, .8).$$

Interval Safety Factors:

$$S^1_{111} = [.4, .6], \ S^1_{211} = [.8, .9], \ S^1_{311} = [.2, .4], \ S^1_{112} = [.7, .8],$$

$$S^1_{212} = [.7, .8], \ S^1_{312} = [.5, .6], \ S^1_{121} = [.7, .8], \ S^1_{221} = [.4, .6],$$

$$S^1_{321} = [.3, .4], \ S^1_{122} = [.9, .1], \ S^1_{222} = [.5, 68], \ S^2_{322} = [.7, .8],$$

$$S^2_{111} = [.8, .9], \ S^2_{211} = [.8, .9], \ S^2_{311} = [.2, .4], \ S^2_{112} = [.8, .9],$$

$$S^2_{212} = [.5, .7], \ S^2_{312} = [.4, .6], \ S^2_{121} = [.3, .4], \ S^2_{221} = [.4, .5],$$

$$S^2_{321} = [.3, .4], \ S^2_{122} = [.9, 1], \ S^2_{222} = [.5, .6], \ S^2_{322} = [.7, .8].$$

Table 1 Other assumed parametric values

		Model		
	2a	2b(i), 2b(ii), 2b(ii)	2c	2d
Data	$B > 4.3$	$\varepsilon = .85$ and $B = 9.5$	$B = [B_L, B_R] = [6, 6.5]$	$k_\alpha = 3.2, \ B = 17$

Input for Safety Constraints with Stochastic Safety Factors:

$$\tilde{S}^1_{111} = 13, \ \tilde{S}^1_{211} = 19, \ \tilde{S}^1_{311} = 5, \ \tilde{S}^1_{112} = 15, \ \tilde{S}^1_{212} = 2, \ \tilde{S}^1_{312} = 6,$$

$$\tilde{S}^1_{121} = 14, \ \tilde{S}^1_{221} = 2, \ \tilde{S}^1_{321} = 4, \ \tilde{S}^1_{122} = 17, \ \tilde{S}^1_{222} = 2, \ \tilde{S}^2_{322} = 2,$$

$$\tilde{S}^2_{111} = 4, \ \tilde{S}^2_{211} = 16, \ \tilde{S}^2_{311} = 7, \ \tilde{S}^2_{112} = 10, \ \tilde{S}^2_{212} = 13, \ \tilde{S}^2_{312} = 8,$$

$$\tilde{S}^2_{121} = 18, \ \tilde{S}^2_{221} = 12, \ \tilde{S}^2_{321} = 7, \ \tilde{S}^2_{122} = 17, \ \tilde{S}^2_{222} = 13, \ \tilde{S}^2_{322} = 6.$$

$$\left(\sigma^2_{111}\right)^1 = 3, \ \left(\sigma^2_{211}\right)^1 = 7, \ \left(\sigma^2_{311}\right)^1 = 3, \ \left(\sigma^2_{112}\right)^1 = 2, \ \left(\sigma^2_{212}\right)^1 = 5,$$

$$\left(\sigma^2_{312}\right)^1 = 2, \ \left(\sigma^2_{121}\right)^1 = 4, \ \left(\sigma^2_{221}\right)^1 = 2, \ \left(\sigma^2_{321}\right)^1 = 4, \ \left(\sigma^2_{122}\right)^1 = 3,$$

$$\left(\sigma^2_{222}\right)^1 = 2, \ \left(\sigma^2_{322}\right)^1 = 2, \ \left(\sigma^2_{111}\right)^2 = 1, \ \left(\sigma^2_{211}\right)^2 = 4, \ \left(\sigma^2_{311}\right)^2 = 2,$$

$$\left(\sigma^2_{112}\right)^2 = 2, \ \left(\sigma^2_{212}\right)^2 = 6, \ \left(\sigma^2_{312}\right)^2 = 3, \ \left(\sigma^2_{121}\right)^2 = 6, \ \left(\sigma^2_{221}\right)^2 = 4,$$

$$\left(\sigma^2_{321}\right)^2 = 2, \ \left(\sigma^2_{112}\right)^2 = 5, \ \left(\sigma^2_{222}\right)^2 = 3, \ \left(\sigma^2_{322}\right)^2 = 1,$$

This problem is feasible since $A \cap B \cap E = [61, 154] \cap [59, 97] \cap [33, 83] = [61, 83] \neq \phi$ (is non empty).

Model 1:

This interval valued solid transportation problem without safety factors with the above data is reduced to

Minimize

$$Z = \sum_{q=1}^{2} \sum_{i=1}^{3} \sum_{j=1}^{2} \sum_{k=1}^{2} C^q_{ijk} \ x^q_{ijk}$$

Subject to

$$\sum_{j=1}^{2}\sum_{k=1}^{2} x^1_{1jk} = [6,\ 16],\ \sum_{j=1}^{2}\sum_{k=1}^{2} x^1_{2jk} = [8,\ 23],$$

$$\sum_{j=1}^{2}\sum_{k=1}^{2} x^1_{3jk} = [16,\ 50],\ \sum_{j=1}^{2}\sum_{k=1}^{2} x^2_{1jk} = [21,\ 30],$$

$$\sum_{j=1}^{2}\sum_{k=1}^{2} x^2_{2jk} = [8,\ 23],\ \sum_{j=1}^{2}\sum_{k=1}^{2} x^2_{3jk} = [2,\ 12],$$

$$\sum_{i=1}^{3}\sum_{k=1}^{2} x^1_{i1k} = [8,\ 17],\ \sum_{i=1}^{3}\sum_{k=1}^{2} x^1_{i2k} = [14,\ 29],$$

$$\sum_{i=1}^{3}\sum_{k=1}^{2} x^2_{i1k} = [23,\ 32],\ \sum_{i=1}^{3}\sum_{k=1}^{2} x^2_{i2k} = [14,\ 19],$$

$$\sum_{q=1}^{2}\sum_{i=1}^{3}\sum_{j=1}^{2} x^q_{ij1} = [26,\ 41],\ \sum_{q=1}^{2}\sum_{i=1}^{3}\sum_{j=1}^{2} x^q_{ij2} = [7,\ 42] \tag{16}$$

$$x^q_{ijk} \geq 0 \text{ or all } i,\ j,\ k,\ q.$$

In a Transportation problem, the feasibility constraints are always equality constraints. So if demand, supply and conveyances capacities are all interval numbers, an equality constraint with decision variables in the left side can be written as a deterministic set of constraints as follows:

The composite Objective Function is in its final form,

$$\text{Min } ZZ = \lambda \text{ Min } (m\ (Z)) + (1 - \lambda) \text{ Min } (w\ (Z)),\ 0 \leq \lambda \leq 1.$$

Subject to the above constraints [14] and $x^q_{ijk} \geq 0$ for all $i,\ j,\ k,\ q$.

Now, with $\lambda = 0.5$, we have

$$\text{Min } ZZ = 0.5[\text{Min } (m\ (Z)) + \text{Min } (w\ (Z))]$$
$$= 21\ x^1_{111} + 37.5\ x^1_{112} + 38\ x^1_{121} + 50\ x^1_{122} + 43\ x^1_{211}$$
$$+ 21.5\ x^1_{212} + 36.5\ x^1_{221} + 27\ x^1_{222} + x^1_{222} + 4\ x^1_{311}$$
$$+ 6.5\ x^1_{312} + 25.5\ x^1_{321} + 36.5\ x^1_{322} + 43\ x^2_{111} + 10\ x^2_{112}$$
$$+ 43.5\ x^2_{121} + 49\ x^2_{122} + 42\ x^2_{211} + 20\ x^2_{212} + 36\ x^2_{221}$$
$$+ 25.5\ x^2_{222} + 5.5\ x^2_{311} + 6\ x^2_{312} + 27\ x^2_{321} + 37\ x^2_{322}.$$

Subject to above constraint (14) and $x^q_{ijk} \geq 0$ for all $i,\ j,\ k,\ q$.

Model 2a, 2b(i), 2b(ii), 2b(iii), 2c and 2d:

Formulations of these models are same as (10a) - (10m) along with additional constraint(s) due to safety constraint which differs for different models.

For Model – 2a
Crisp Safety constraint:

$$S = \sum_{q=1}^{2} \sum_{i=1}^{3} \sum_{j=1}^{2} \sum_{k=1}^{2} s_{ijk}^{q} \, y_{ijk}^{q}$$

is reduced to

$$S = 0.5\, y_{111}^{1} + 0.75\, y_{112}^{1} + 0.75\, y_{121}^{1} + 0.95\, y_{122}^{1} + 0.85\, y_{211}^{1}$$
$$+ 0.75\, y_{212}^{1} + 0.5\, y_{221}^{1} + 0.55\, y_{222}^{1} + 0.3\, y_{311}^{1} + 0.55\, y_{312}^{1}$$
$$+ 0.35\, y_{321}^{1} + 0.75\, y_{322}^{1} + 0.85\, y_{111}^{2} + 0.85\, y_{112}^{2} + 0.35\, y_{121}^{2}$$
$$+ 0.95\, y_{122}^{2} + 0.85\, y_{211}^{2} + 0.6\, y_{212}^{2} + 0.45\, y_{221}^{2} + 0.55\, y_{222}^{2}$$
$$+ 0.3\, y_{311}^{2} + 0.5\, y_{312}^{2} + 0.3\, y_{321}^{2} + 0.75\, y_{322}^{2}$$

For Model – 2b
Fuzzy Safety constraint:

$$\tilde{S} = \sum_{q=1}^{2} \sum_{i=1}^{3} \sum_{j=1}^{2} \sum_{k=1}^{2} \tilde{s}_{ijk}^{q} \, y_{ijk}^{q}$$

$\Rightarrow$
$$\tilde{S} = (.45, .5, .55)\, y_{111}^{1} + (.7, .75, .8)\, y_{112}^{1} + (.7, .75, .8)\, y_{121}^{1}$$
$$+ (.9, .95, .1)\, y_{122}^{2} + (.8, .85, .9)\, y_{211}^{1} + (.7, .75, .8)\, y_{212}^{1}$$
$$+ (.45, .5, .55)\, y_{221}^{1} + (.5, .55, .6)\, y_{222}^{1} + (.25, 0.3, .35)\, y_{311}^{1}$$
$$+ (.5, .55, .6)\, y_{312}^{1} + (.3, .35, .4)\, y_{321}^{1} + (.7, .75, .8)\, y_{322}^{1}$$
$$+ (.8, .85, .9)\, y_{111}^{2} + (.8, .85, .9)\, y_{112}^{2} + (.3, .35, .4)\, y_{121}^{2}$$
$$+ (.9, .95, 1)\, y_{122}^{2} + (.8, .85, .9)\, y_{211}^{2} + (.55, .6, .65)\, y_{212}^{2}$$
$$+ (.4, .45, .5)\, y_{221}^{2} + (.5, .55, .6)\, y_{222}^{2} + (.25, .3, .35)\, y_{311}^{2}$$
$$+ (.45, .5, .55)\, y_{312}^{2} + (.25, .3, .35)\, y_{321}^{2} + (.7, .75, .8)\, y_{322}^{2}$$
$$= (a_1, a_2, a_3), \text{ say.}$$

With this fuzzy number, the safety constraint for Models 2b(i), 2b(ii), 2b(iii) are obtained following (a), (b) and (c, d) respectively.

For model – 2c

Safety constraint with interval safety factors:

$$S = \sum_{q=1}^{2} \sum_{i=1}^{3} \sum_{j=1}^{2} \sum_{k=1}^{2} \left[s_{ijkL}^{q}, s_{ijkR}^{q} \right] y_{ijk}^{q} \geq [B_L, B_R]$$

$$\Rightarrow \sum_{q=1}^{2} \sum_{i=1}^{3} \sum_{j=1}^{2} \sum_{k=1}^{2} \left(\frac{s_{ijkL}^{q} + s_{ijkR}^{q}}{2} \right) y_{ijk}^{q} \geq \frac{B_L + B_R}{2}. \text{ [L. Sahoo (8)]}.$$

For Model – 2d

Chance constraint programming:

$$\sum_{q=1}^{2}\sum_{k=1}^{2}\sum_{j=1}^{2}\sum_{i=1}^{3} \hat{s}_{ijk}^{q}\, y_{ijk}^{q} \geq B$$

$$\Rightarrow \quad \text{prob} \sum_{q=1}^{2}\sum_{k=1}^{3}\sum_{j=1}^{3}\sum_{i=1}^{3} \hat{s}_{ijk}^{q}\, y_{ijk}^{q} \geq B \geq \varepsilon$$

$$\Rightarrow \quad \sum_{q=1}^{2}\sum_{k=1}^{2}\sum_{i=1}^{3} E\left(\hat{s}_{ijk}^{q}\right) y_{ijk}^{q} - k_{\alpha}\sum_{q=1}^{2}\sum_{k=1}^{2}\sum_{i=1}^{3} \text{var}\left(\hat{s}_{ijk}^{q}\right) y_{ijk}^{q} \geq B.$$

6. RESULT AND DISCUSSION

The above constrained optimization problems are executed using LINGO 13.0 and the results of Models – 1, 2a, 2b(i), 2b(ii), 2b(iii), 2c and 2d are presented in Table 3

Optimal Result for Different Models

The above constrained optimization problems are executed using LINGO 13.0 and the results of Models – 1, 2a, 2b(i), 2b(ii), 2b(iii), 2c and 2d are presented below:

Model 1:

$$x_{121}^{1} = 6,\; x_{311}^{1} = 16,\; x_{222}^{1} = 8,\; x_{112}^{2} = 21,$$
$$x_{311}^{2} = 2,\; x_{321}^{2} = 2,\; x_{222}^{2} = 12,$$

and rest of all are zero.

$$\text{Min} < m(z),\, w(z) > \; = \; < 2096.82,\, 82 >,\; S = 3.6.$$

Model 2a:

$$x_{121}^{1} = 6,\; x_{221}^{1} = 1,\; x_{222}^{1} = 7,\; x_{311}^{1} = 17,$$
$$x_{112}^{2} = 21,\; x_{222}^{2} = 14,\; x_{311}^{2} = 2,$$

and rest of all are zero.

$$\text{Min} < m(z),\, w(z) > \; = \; < 2119.5,\, 79.5 >,\; S = 4.3.$$

Model 2b (i):

$$x_{121}^{1} = 9.4,\; x_{211}^{1} = 0.45,\; x_{212}^{1} = 0.2,\; x_{221}^{1} = 7.35,$$
$$x_{222}^{1} = 0.01,\; x_{311}^{1} = 16.32,\; x_{312}^{1} = 0.05,\; x_{321}^{1} = 0.01,$$
$$x_{112}^{2} = 28.9,\; x_{122}^{2} = .51,\; x_{222}^{2} = 12,\; x_{312}^{2} = 0.06,$$
$$x_{312}^{2} = 1.6,\; x_{322}^{2} = 0.32$$

and rest of all are zero.

$$\text{Min} < m(z), w(z) > = < 2677.6, 103.6 >, \tilde{S} = (7.8, 8.5, 9.7).$$

Model 2b(ii):

$$x^1_{121} = 6.2, \; x^1_{211} = .17, \; x^1_{221} = 7.98, \; x^1_{311} = 16.7,$$

$$x^2_{111} = .15, \; x^2_{112} = 29.5, \; x^2_{122} = .21, \; x^2_{211} = .16,$$

$$x^2_{221} = .7, \; x^2_{222} = 11.6, \; x^2_{311} = .48, \; x^2_{312} = .11,$$

$$x^2_{321} = 1.13, \; x^2_{322} = .4$$

and rest of all are zero.

$$\text{Min} < m(z), w(z) > = < 2478.42, 100 >, \tilde{S} = (7.8, 8.1, 8.8).$$

Model 2b (iii):

$$x^1_{121} = 8.47, \; x^1_{212} = .31, \; x^1_{221} = 7.6, \; x^1_{222} = .67,$$

$$x^1_{311} = 16.3, \; x^1_{321} = .085, \; x^2_{111} = .27, \; x^2_{112} = 23.6,$$

$$x^2_{121} = .02, \; x^2_{211} = .02, \; x^2_{221} = 1.35, \; x^2_{222} = 10.2,$$

$$x^2_{311} = .02, \; x^2_{321} = 1.13, \; x^2_{322} = 1.48$$

and rest of all are zero.

$$\text{Min} < m(z), w(z) > = < 2184.03, 53.46 >, \tilde{S} = (7.45, 8.2, 8.95).$$

Model 2c:

$$x^1_{121} = 6, \; x^1_{222} = 8, \; x^1_{311} = 17, \; x^2_{122} = 21,$$

$$x^2_{222} = 13, \; x^2_{311} = 2, \; x^2_{321} = 1$$

and rest of all are zero.

$$\text{Min} < m(z), w(z) > = < 2102, 81 >, S = [3.1, 4].$$

Model 2d:

$$x^1_{111} = 1.9, \; x^1_{112} = 4.1, \; x^1_{211} = 2, \; x^1_{221} = 6,$$

$$x^1_{321} = 16, \; x^2_{112} = 21, \; x^2_{211} = .002, \; x^2_{222} = 9.7,$$

$$x^2_{311} = 4.3$$

and rest of all are zero.

$$\text{Min} < m(z), w(z) > = < 2882.86, 96.4 >.$$

7. DISCUSSION

As the natures of the input data are different, the optimum results cannot be compared. For the same input data, the models 2b(i), 2b(ii), 2b(iii) have been executed. As 'Possibility' and 'Necessity' represent pessimism and optimism respectively and 'credibility' is lies in between these two, the optimum results of these models are as per expectations. Optimum results of models 2b(i) and 2b(ii) are highest and lowest whereas model 2b(iii) gives figure in between these two. Out of all the models, the model-1 with crisp data gives the minimum cost. Note that in this model, total safety measure constraint is not taken into account and along the predicted routes, the total safety measure is 3.6. In other models DM wants to have more total safety measure for the system i.e., 4.3. For this reason the transportation routes are rearranged so that total safety measure is greater or equal to 4.3. As a result, total mean cost is observed to increase. These additional costs are incurred against the increased total safety measure.

8. CONCLUSION

The main objects of this paper are to present a solution procedure for Multi-item Solid Transportation Problem with safety measure. Multi-item Solid Transportation Problem with Safety Measure gives an idea about a new factor, 'Safety Measure', to transport commodities by means of different conveyances. Five Models have been proposed as Multi-item Solid Transportation Problem: (i) Without Safety Factor, (ii) With Safety Factor, (iii)With Safety Factor (Under Possibility/Necessity/Credibility), (iv) With safety Factor as interval, (v) With Stochastic Safety Factors

In the present study mathematical problems were solved using LINGO 13.0 Software. The present study aimed at presenting two types of constraints: deterministic and uncertain considering both fuzzy and stochastic sense. A analytical software, LINGO 13.0 was used while solving mathematical problem. In context with real field problem, the technique could be used as very effective and promising and in view of a practical importance.

References

1. Charnes, A., and Cooper, W.W., 1953: *Chance-constrained and normal deviates*, Journal of American Statistics Association 57, 134-118.

2. Charnes, A. and Cooper, W.W., 1963: *Deterministic equivalents for optimizing and satisfying under chance constraints*, Operations Research 11, 18-39.

3. Das, B., Maity, K., and Maiti, M., 2007: *A Warehouse Supply-Chain model under Possibility/Necessity/Credibility Measure*, Mathematical and Computer Modeling 46 398-409.

4. Dubois, D., and Prade, H., 1988: *Possibility Theory: An Approach to Computerized processing of Uncertain*, Plenum Press, New York.

5. Hitchcock, F.L., 1941: *The distribution of a product from several sources to numerous localities*, Journal of Mathematical Physic 20224–230.

6. Jimenez, F., and Verdegay, J. L., 1998: *Uncertain Solid Transportation Problem*, Fuzzy Sets and Systems - Elsevier, Volume 100, Issues 1–3, Pages 45-57.

7. Liu, B., and Liu, Y. K., 2002: *Expected value of fuzzy variable and fuzzy expected value model*, IEEE Transactions on fuzzy Systems, Vol. 10. No. 4.445-450.

8. Sahoo, L., Bhunia, A.K., and Kapur, P.K., 2012: *Genetic algorithm based multi-objective reliability optimization in interval environment*, Computers and Industrial Engineering, 62, 152-160.

9. Sengupta, A., Pal, T.K., and Chakraborty, D., 2001: *Interpretation of inequality constraints involving interval coefficients and a solution to interval linear programming*, Fuzzy Sets and Systems, 119, 129-138.

10. Sengupta, A., and Pal, T.K., 2000: *Oncomparing interval numbers*, European Journal of Operation Research 127, 28-43.

11. Sengupta, A., and Pal, T.K., 2004: *Interval-Valued Transportation problem with Multiple Penalty Factors*, Vidyasagar University Journal of physical Science, Page No. 71 to 81.

12. Zadeh, L.A., 1978: *Fuzzy set as a basis for a theory of possibility*, Fuzzy Sets and Systems. Vol. 1, 3-28.

Algebra of Negation Fragment of a Logic with Graded Notion of Consequence

Soma Dutta

Institute of Mathematical Sciences, Chennai
E-mail: somadutta9@gmail.com

ABSTRACT

In this paper we shall study the properties of the algebraic structure which can be considered as the model for the graded version of classical conditions viz., overlap, dilution, cut, law of explosiveness and reasoning by cases.

1. INTRODUCTION

"Until now the construction of superficial many-valued logics, that is, logics with an arbitrary number (bigger than two) of truth values but always incorporating a binary consequence relation, has prevailed in investigations of logical many-valuedness."

This observation is made by Pelta in 2004 [8]. Much before that, in 1987, Chakraborty [1] introduced the notion of graded consequence to address the idea that, 'if α_1, α_2, ..., α_n, and α receive many-valued truth values (i.e. values other than the least (0) and the top (1)) then, generally α_1, α_2, ..., α_n | α may also receive many-valued truth value.

The notion of graded consequence is a generalization of the notion of two-valued consequence relation [6], in many-valued context. A graded consequence relation, say |~ is a fuzzy relation between $P(F)$, the power set of formulae and F, the set of formulae, where for each set of formulae X and a single formula α, gr $(X|\!\sim\!\alpha)$ denotes the degree to which α is a consequence is X. Then generalizing the classical conditions for overlap, dilution and cut, |~ has been axiomatized by the following axioms.

    ($GC1$) If $\alpha \in X$ then gr $(X|\!\sim\!\alpha) = 1$.

    ($GC2$) If $X \subseteq Y$ then gr $(X|\!\sim\!\alpha) \leq$ gr $(Y|\!\sim\!\alpha)$.

    ($GC3$) $\inf_{\beta \in Y}$ gr $(X|\!\sim\!\beta) *$ gr $(X \cup Y|\!\sim\!\alpha) \leq$ gr $(X|\!\sim\!\alpha)$.

In graded context, the meta-level sentences, like $X|\sim\alpha$ get many-valued value, and it is denoted by gr $(X|\sim\alpha)$. The meta-linguistic *'and'* and *'for all'* present in the classical form of cut condition are computed by '*' and 'inf' respectively. When semantics and algebra come into play, we shall see below, the value set L for the object level formuale along with *, its residuum $\rightarrow$, and 'inf' forms the algebraic structure for the meta language of $|\sim$; in this context this is a complete residuated lattice [7].

Classically, the notion of semantic consequence, which is represented by a binary relation, say $|=$ between any set X of formulae and a single formula α, formalizes the idea that 'α is a semantic consequence of X iff for all state of affairs, if for all formulae of X receive the truth value true, then α also receives the truth value true'. This idea has been generalized by Shoesmith and Smiley [9], by relativizing the notion of semantic consequence with respect to any arbitrary collection of states of affairs. According to [9],

$'X |= \alpha$ iff given any collection of state of affairs $\{T_i\}_{i \in I}$ if for all formulae of X receive the truth value true, then α also receives the truth value true' (Σ)

In the context of graded consequence, $\{T_i\}_{i \in I}$ present in the sentence (Σ) is an arbitrary collection of fuzzy sets assigning values to object level formulae from L. The degree to which α is a semantic consequence of X, denoted by gr $(X|\approx \alpha)$, is given below.

$$gr\ (X|\approx \{Ti\}_{i \in I}\ \alpha) = \inf_i\{\inf_{\gamma \in X} T_i(\gamma) \rightarrow T_i(\alpha)\}$$

It is to be noted, that to compute the meta-linguistic 'for all' and 'if-then' present in (Σ) the operators inf and $\blacklozenge$ of the meta level algebra $(L, *, \rightarrow, 0, 1)$ are used. Then assuming $(L, *, \rightarrow, 0, 1)$ to be a complete residuated lattice, the following theorems establishing the connection between graded consequence relation and its semantic counterpart are proved in [2].

Representation Theorem

(i) For any collection of fuzzy sets $\{T_i\}_{i \in I}$ over formulae, the fuzzy relation $|\approx$ generated by $\{T_i\}_{i \in I}$ is a graded consequence relation.

(ii) Given a graded consequence relation $|\sim$ there exists a collection of fuzzy sets say $\{T_i\}_{i \in I}$ such that $|\approx$ generated by $\{T_i\}_{i \in I}$ coincides with $|\sim$.

Generally, negation fragment of a logic determines the connection between 'consequence' and 'inconsistency', and classically these are equivalent to each other. In many-valued logics, the notion of inconsistency is defined in terms of the notion of consequence too; but whether the notions of consequence and inconsistency are interwoven, as found in the classical case [10], has not been come into the focus of the study. In the context of graded consequence this effort has been made and as an outcome some results come into the fore establishing the equivalence between the notions graded consequence and graded inconsistency [3]. Extending the language with object level negation ($\neg$) following axioms have been appended to the previous set of axioms for graded consequence relation.

(*GC*4) There is some $k > 0$ such that, gr $(\{\alpha, \neg\alpha\}|\sim\beta) = k$.

(*GC*5) gr $(X \cup \{\alpha\}|\sim\beta)$ * gr $(X \cup \{\neg\alpha\}|\sim\beta) \leq$ gr $(X|\sim\beta)$.

In [5], we found that this form of reasoning by cases demands $x \vee \neg_o x = 1$ as a sufficient condition to have $GC5$, where $\neg_o$ is the operator for computing object level negation $\neg$. But this restriction does not seem commensurate with many-valued scenario where many algebraic structures are available violating this classical condition, known as the law of excluded middle. This leads us towards the following, more general version for reasoning by cases.

(GC^M5) There is some $c > 0$,

$$\text{gr } (X \cup \{\alpha\}|\sim\beta) * \text{gr } (X \cup \{\neg\alpha\}|\sim\beta) * c \leq \text{gr } (X|\sim\beta).$$

In the context of classical logic, Boolean algebra acts as the algebraic counterpart where all the logical laws happen to be true. In graded context, we already found a complete residuated lattice as the meta level algebraic structure for the notion of graded consequence. But in presence of the axioms $GC4$ and (GC^M5), what other additional features need to be added with the structure, becomes a matter of interest.

So, in this paper, we shall focus on mainly studying the properties of the algebraic structure which might be considered as a model for the axioms $GC1$ to (GC^M5).

2. ALGEBRA FOR GRADED CONSEQUENCE IN PRESENCE OF NEGATION IN THE LANGUAGE: A BRIEF BACKGROUND

$GC4$ is actually the graded version of the classical law of explosive condition which states 'for any formula α, $\{\alpha, \neg\alpha\}$ together yield any formula β'. Classically, α and $\neg\alpha$ cannot be true together and this by default allows any formula to follow. But in many-valued context it is not the case that $x \wedge \neg_o x$ is false always. In [4, 5], we have extensively studied the necessary and sufficient conditions for $GC4$, GC^M5 to hold. Let us consider $(L, \wedge, \vee, *, \rightarrow, 0, 1)$, a complete residuated lattice to be the meta-level algebra for a graded consequence relation $|\sim$ and $\neg_o$ to be the operator for object level negation $\neg$. Besides, in this context we need to impose one condition on the collection of fuzzy sets $\{T_i\}_{i \in I}$ too; each T_i of the collection should obey the truth functionality condition and maintain the constraints at classical values i.e. on the set $\{0, 1\}$.

Sufficient condition for GC4: Let (i) there be some k' ($\neq 1$), such that $x \wedge \neg_o x \leq k'$ for any b in L and (ii) if $k' \neq 0$ then there exist z ($\neq 0$) such that $k' * z = 0$. Then for any graded consequence relation with the above structure for the meta language, GC4 holds.

Sufficient condition for $GC^M 5$: Let (i) there be an element c ($\neq 0$) in L such that $x \leq c$ or $x > c$ for all x in L, (ii) $x \vee \neg_o x \geq c$ for all $x \in L$ and (iii) $c * z = z$ for all $z \leq c$. Then for any graded consequence relation $\vDash_{\{Ti\}i \in I}$, with the above structure for the meta language $GC^M 5$ holds.

These two lead towards a sufficient condition for $GC4$ and $GC^M 5$ to hold simultaneously.

Sufficient condition for GC4 and $GC^M 5$: Let (i) there be an element c ($\neq 0$) $\in L$ such that for all $x \in L$, $x \leq c$ or $x > c$, (ii) $c * z = z$ for all $z \leq c$, (iii) $x \vee \neg_o x \geq c$ for all $x \in L$, and (iv) there be an element k' ($k' \leq 0 < c$) in L such that $x \wedge \neg_o x \leq k'$ for all $x \in L$, and if $k' \neq 0$ then $k' * z = 0$ for some z ($\neq 0$) in L. Then for any $\vDash_{\{Ti\}i \in I}$ defined with respect to the structure $(L, \wedge, \vee, *, \rightarrow, 0, 1)$, GC4 and $GC^M 5$ hold simultaneously.

In the Introduction it has been mentioned that irrespective of the language, assuming $\{T_i\}_{i \in I}$ to be a collection of fuzzy sets assigning values to object level formulae, one needs a complete residuated lattice to come up with a graded consequence relation which is exactly same as $\vDash_{\{Ti\}i \in I}$, the graded consequence consequence relation generated by $\{T_i\}_{i \in I}$. In presence of negation ($\neg$) in the object language, an algebraic structure, say $(L, \wedge, \vee, \neg_o, 0, 1)$, called object level algebra, comes into play. The above mentioned theorem on sufficient condition for simultaneous presence of $GC4$, $GC^M 5$ leads us towards an interrelation between the meta level algebra $(L, \wedge, \vee, *, \rightarrow, 0, 1)$ and the object level algebra $(L, \wedge, \vee, \neg_o, 0, 1)$. In the next section, we shall present a detailed study on the properties of the extended algebraic structure. We shall call $(L, *, \rightarrow, \neg_o, 0, 1)$, the algebra for graded consequence with negation, in short GC($\neg$) algebra.

3. GC(¬) ALGEBRA: A SURVEY

While studying sufficient condition for $GC4$ and $GC^M 5$, it is found that apart from the least (0) and top (1) element of L two other elements, viz., k' and c need to be distinguished. Let us consider the structure $(L, \wedge, \vee, *, \rightarrow, \neg_o, k', c, 0, 1)$ where

    (i) $(L, \wedge, \vee, *, \rightarrow, 0, 1)$ is a complete residuated lattice,

    (ii) $0 \leq k' < c \leq 1$,

    (iii) $L = L_1 \cup U_1$, where $L_1 = \{x \in L/x \leq c\}$, $U_1 = \{x \in L/x \geq c\}$,

    (iv) $(L_1, *, c)$ is a commutative monoid,

(v) $x \vee \neg_\sigma x \geq c$ for all $x \in L$,

(vi) $x \wedge \neg_\sigma x \leq k'$ for all $x \in L$, and

(vii) if $k' \neq 0$ then $k' * z = 0$ for some z $(\neq 0)$ in L.

Proposition 1 $a \in L_1$ and $b \in U_1$ imply $a * b^n = a$ for any positive integer n.

Corollary 2 $a \in L_1$ and $b \in U_1$ imply $a * b = a$.

Proposition 3 $(U_1, *, 1)$ is a monoid.

Proposition 4 $(L_1, \wedge, \vee, 0, c)$ is a complete lattice.

Proof Let $\{a_i\}_{i \in I}$ be a collection of elements of L_1. Then $a_i \leq c$ for each $i \in I$.

As L is a complete lattice $\vee_i a_i$ exists.

Therefore $a_i \leq c$ for each $i \in I$ implies $\vee_i a_i \leq c$ and hence $\vee_i a_i \in L_1$.

Also $\wedge_i a_i$ exists in L and $\wedge_i a_i \leq \vee_i a_i \leq c$ imply $\wedge_i a_i \in L_1$.

Proposition 5 $(U_1, \wedge, \vee, c, 1)$ is a complete lattice.

Theorem 6 $(L_1, *, \rightarrow_{L1}, 0, c)$ forms a complete residuated lattice, where $x \rightarrow_{L1} y = \max\{z \in L_1: x * z \leq y\}$.

Proof As by (iv) we know $(L_1, *, c)$ is a monoid, we have to prove that $(*, \rightarrow_{L1})$ forms an adjoint pair i.e. $x * z \leq y$ iff $x \leq z \rightarrow_{L1} y$ holds for any $x, y, z \in L_1$.

That is we have to show that for any $x, y \in L_1$, $x \rightarrow_{L1} y$ i.e. $\max\{z \in L_1: x * z \leq y\}$ exists.

Let us consider two cases viz., (A) $x \nleq y$ and (B) $x \nleq y$.

(A) Let $x \nleq y$. Now as $x * 0 = 0 \leq y$, $0 \in \{z \in L_1: x * z \leq y\}$.

Therefore, $\{z \in L_1: x * z \leq y\}$ is non-empty and by proposition 4, $\sup\{z \in L_1: x * z \leq y\}$ exists in L_1. Let $z' \leq \{z \in L_1: x * z \leq y\}$.

As $x * z' \leq y$, by residuation property of the value set L, we have $z' \leq x \rightarrow y$. [By (i)]

That is, $x \rightarrow y$ is an upper bound of all z's belonging to $\{z \in L_1: x * z \leq y\}$.

We want to show $x \rightarrow y \leq c$. If not, then $c < x \rightarrow y$. [by (iii)]

Hence $x = x * c \leq x * (x \rightarrow y) \leq y$. This contradicts the assumption $x \nleq y$.

Therefore, $x \rightarrow y \leq c$ i.e. $x \rightarrow y \in L_1$. Hence $x \rightarrow y = \max\{z \in L_1: x * z \leq y\}$. That is, $x \rightarrow_{L1} y$ exists and equals to $x \rightarrow y = \max\{z \in L_1: x * z \leq y\}$.

(B) Let $x \leq y$. Therefore, $x * c = x \leq y$ implies $c \in \{z \in L_1: x * z \leq y\}$.

Also, for all $z \in L_1$ i.e. $z \leq c$, $x * z \leq x * c \leq x \leq y$. Then $\max\{z \in L_1: x * z \leq y\} = c$.

Hence $x \rightarrow_{L1} y = c$ for $x \leq y$.

Theorem 7 $(U_1, x, \to U_1, c, 1)$ forms a complete residuated lattice, where $x \to_{U1} y = \max\{z \in U_1 : x * z \le y\}$.

Proof   To prove this theorem we have to show that for any $x, y \in U_1, x \to U_1 y$ exists.

Now two cases arise. (A) $x \nleq y$ (B) $x \nleq y$.

(A) Let $x \nleq y$. As $x * c \le c \le y$ we have $c \in \{z \in U_1 : x * z \le y\}$. Therefore the set is non-empty and hence $\sup\{z \in U_1 : x * z \le y\}$ exists. [by Proposition 5]

Now for any $z \in \{z \in U_1 : x * z \le y\}$, $x * z \le y$ implies $z \le x \to y$ [by (i)]

We want to show $c \le x \to y$ i.e., $x \to y \in U_1$.

If not, then $x \to y < c$ [By (iii)]

As L is a complete residuated lattice, we have, i.e., $y \le x \to y < c$. This contradicts the assumption that $y \in U_1$. Therefore, $c \le x \to y$, or in other words $x \to y \in U_1$.

Hence $\max\{z \in L_1 : x * z \le y\}$ i.e. $x \to U_1 y$ exists and $x \to U_1 y = x \to y$.

(B) $x \le y$ i.e., $x = x * 1 \le y$. Hence $x \to U_1 y = \max\{z \in L_1 : x * z \le y\} = 1$.

Theorem 8   If for some $x \ne 0$, there is $z \ne 0$ such that $x * z = 0$ then $z < c$.

Proof   Now for any $z \in L$, two cases arise – $z \ge c$ and $z < c$.

If $z \ge c$ then $x * z \ge x * c$, i.e. $0 \ge x * c$ implies $x * c = 0$. Now three subcases arise.

(A) $x \in L_1 - \{c\}$ (B) $x \in U_1 - \{c\}$ (C) $x = c$.

(A) $x \in L_1 - \{c\}$ implies $x * c = x = 0$. This contradicts the assumption that $x \ne 0$.

(B) $x \in U_1 - \{c\}$ implies $x * c \in U_1$ [by Proposition 3, (U_1 is closed under *)] So, $x * c$ cannot be 0.

(C) $x = c$ implies $0 = x*c = c*c = c$. This contradicts the condition (ii) of the initial assumptions. Hence $z < c$.

Corollary 9   If for some $z (\ne 0) \in L$ has a divisor of zero then $z < c$.

Proposition 10   If $k' * z = 0$ for some $z (\ne 0) \le k'$, then for any $x, y (\ne 0) \le z$, $x * y = 0$.

Proposition 11   If for some $z \ne 0$, $k' * z = 0$, then $x * z = 0$ for all $x \le k'$.

Theorem 12   Let for some $a (< c) \in L_1$, (I) there is no z such that $a < z < c$ and

(II) for every $x < a$ there is an element y non-comparable to a such that $a * y = x$.

Then a is an idempotent element of L.

Proposition 13 If $x \in U_1$ then $\neg_\sigma x \leq k'$.

Theorem 14 For $x \leq k'$ either $\neg_\sigma x \in U_1$ or $\neg_\sigma x$ is non-comparable to x and $x \vee \neg_\sigma x = c$.

Proof As $x \leq k'$, $\neg_\sigma x$ is not less equal to k' [For $\neg_\sigma x \leq k'$, $x \vee \neg_\sigma x \leq k'$, which contradicts (v)]

Therefore either (I) $\neg_\sigma x > k'$ or (II) $\neg_\sigma x$ is non-comparable to k'.

(I) For $\neg_\sigma x > k'$ two cases arise $- c > \neg_\sigma x > k' \leq x$ and $\neg_\sigma x \geq c > k' \geq x$.

If $c > \neg_\sigma x > k' \geq x$, $x \vee \neg_\sigma x = \neg_\sigma x < c$. This contradicts condition (v).

So, $\neg_\sigma x \geq c > k' \geq x$, $x \wedge \neg_\sigma x = x \leq k'$ and $x \vee \neg_\sigma x = \neg_\sigma x \geq c$. Therefore, $\neg_\sigma x \in U_1$.

(II) Let $\neg_\sigma x$ be non-comparable to k'. As, $x \leq k'$, $x \wedge \neg_\sigma x \leq k'$. Also, as $k' < c$ [by (ii)], $x < c$.

Now either $x \leq \neg_\sigma x$ or x, $\neg_\sigma x$ are non-comparable. [Since $\neg_\sigma x < x \, (\leq k')$ cannot be the case]

Now if $x \leq \neg_\sigma x$, then $x \vee \neg_\sigma x = \neg_\sigma x \geq c > k'$. [By (v) $x \vee \neg_\sigma x \geq c$]

This contradicts the assumption $\neg_\sigma x$ is non-comparable to k'.

Hence x and $\neg_\sigma x$ are non-comparable.

As $x < c$ and x, $\neg_\sigma x$ are non-comparable, $\neg_\sigma x \notin U_1$.

Hence $\neg_\sigma x < c$ which implies $x \vee \neg_\sigma x \leq c$. Now, as by condition (v), $x \vee \neg_\sigma x \geq c$, we have $x \vee \neg_\sigma x = c$.

Theorem 15 If $k' < x < c$ then x and $\neg_\sigma x$ are non-comparable and $x \vee \neg_\sigma x = c$.

Proof First we want to prove that x and $\neg_\sigma x$ are non-comparable.

If not, then either $x \leq \neg_\sigma x$ or $\neg_\sigma x \leq x$.

For $x \leq \neg_\sigma x$, we have $x \wedge \neg_\sigma x = x > k'$. This contradicts the assumption (vi).

For $\neg_\sigma x \leq x$, we have $x \vee \neg_\sigma x = x < c$. This contradicts the assumption (v).

Hence x and $\neg_\sigma x$ are non-comparable.

Now as $k' < x < c$ and x, $\neg_\sigma x$ are non-comparable, neither $\neg_\sigma x \leq k'$ nor $c \leq \neg_\sigma x$.

Because if $\neg_\sigma x \leq k'$, then $\neg_\sigma x \leq x$ and if $c \leq \neg_\sigma x$ then $x < c \leq \neg_\sigma x$. Both the cases contradict the fact that x and $\neg_\sigma x$ are non-comparable.

Then either (I) $k' < \neg_\sigma x < c$ or (II) k' and $\neg_\sigma x$ are non-comparable and $\neg_\sigma x < c$.

(I) For $k' < x < c$ and $k' < \neg_\sigma x < c$, we have $x \vee \neg_\sigma x \leq c$. So, using (v), we have $x \vee \neg_\sigma x = c$.

(II) In this case also x, $\neg_\sigma x < c$. So, as in the case (I), it can be proved that $x \vee \neg_\sigma x = c$.

Proposition 16   If L is a linearly ordered complete residuated lattice then z ($\neq 0$) has a divisor of zero impies $\neg_o z \geq c$.

Proposition 17   If L is a complete linearly ordered residuated lattice then for k' ($\neq 0$), $\neg_o c \leq k'$.

Notation   Let $U_1 = \{x \in L/x \geq c\}$, $L_N = \{x \in L/x \leq k'\}$ and $M = \{x \in L/k' < x < c$ or $x < c$ but x is non-comparable to $k'\}$.

Theorem 18   If for some $x \in L$, $k' < x < c$, then M is not a linearly ordered structure and M is not a singleton set.

Proof   Let $x \in M$ and $k' < x < c$. Then by Theorem 15, $x, \neg_o x$ are non-comparable and $x \vee \neg_o x = c$. As $x < c$ and $x, \neg_o x$ are non-comparable, $\neg_o x \notin U_1$.

Now, if $\neg_o x \in L_N$, then $\neg_o x \leq k' < x < c$. This contradicts the fact that $x, \neg_o x$ are non-comparable. Therefore. Hence $\neg_o x \in M$.

So, $x, \neg_o x \in M$ and they are non-comparable to each other. Hence the theorem.

4. CONCLUSION

From the properties of GC($\neg$) algebra certain distinguishing features appear more prominently. These indicate some points of departure of negation fragment of a logic with graded notion of consequence from that of the classical logic. The points of departure are as follows.

  (i) Classical conditions for the law of contradiction and the law of excluded middle get relaxed in this graded context. Here, in graded context, an element which is not necessarily 0, acts as the threshold for the law of contradiction, and an element c which is not necessarily 1, acts as the threshold for the law of excluded middle.

 (ii) Classically, the value which corresponds to the law of excluded middle is the top as well as the identity element for the value set. In this graded context, the threshold c for the law of excluded middle works as an identity element for those elements who lie below it.

(iii) Classically, 0 (may be called the zero element) corresponds to the value for the law of contradiction. While in the context of graded consequence k', which is not necessarily the zero element (i.e. 0), gives a threshold value for the law of contradiction, and if k' is non-zero then it must have a divisor of zero.

 (iv) As the definitions suggest, U_1 can be considered as the zone from where any instance of the law of excluded middle can assume values, and L_1 is complementary to it. For $x \leq y$ if $x, y \in U_1$, $(x, y \in L_1)$, then $x \rightarrow_{U1} y = x \rightarrow y$ ($x \rightarrow_{L1} y = x \rightarrow y$). That is, for $x \leq y$ taken from U_1, the residuum for the whole value set remains the same as the residuum for U_1, and similar result holds for L_1 also.

(v) L_N can be considered as the zone from where any instance of the law contradiction can assume values. Classically, 0 (zero) is the only element of L_N. But in this context, any element of L_N must have a divisor of zero.

(vi) Classically, U_1 contains non-zero element. And here, from graded perspective no element of U_1 can have divisor of zero.

(vii) Classically, the value set is a linear structure where, there is no element lying between the zones for the law of excluded middle (i.e. U_1) and the law of contradiction (i.e. L_N). From the perspective of GC($\neg$) algebra we can see that either the middle zone (denoted as M) is empty or it must have non-comparable elements. More specifically, for $x \in M$, $\neg_o x$ is non-comparable to x and $\neg_o x \in M$.

(viii) In a linear structure with respect to the operator $\neg_o$, images of the elements of U_1 belong to L_N and conversely images of the elements of L_N belong to U_1. The same holds in the classical case also.

The above mentioned points may have some deeper logico-philosophical insight which needs to be investigated further.

References

1. Chakraborty, M.K., 1988: *Use of fuzzy set theory in introducing graded consequence in multiple valued logic*, in M.M. Gupta and T. Yamakawa (eds.), Fuzzy Logic in Knowledge-Based Systems, Decision and Control, Elsevier Science Publishers, B.V. (North Holland) 247-257.

2. Chakraborty, M.K., 1995: *Graded Consequence: further studies*, Journal of Applied Non-Classical Logics, 5(2), 127-137.

3. Chakraborty, M.K., and Basu, S., 1997: *Graded Consequence and some Metalogical Notions Generalized*, Fundamenta Informaticae 32, 299-311.

4. Chakraborty, M.K., and Dutta, S., 2010: *Graded Consequence Revisited, Fuzzy Sets and Systems*, 161.

5. Dutta, S., 2011: *Graded Consequence as the foundation of Fuzzy Logic*, Ph.D thesis, University of Calcutta.

6. Gentzen, G., 1969: *Investigations into Logical Deductions*, in the collected papers of G. Gentzen, M.E. Szabo, ed., 68-131, North Holland Publications, Amsterdam.

7. Hajek, P., 1998: *Metamathematics of Fuzzy Logic*, Kluwer Academic Publishers.

8. Pelta, C., 2004: Wide sets, *Deep many-valuedness and Sorites arguments*, Mathware and Soft computing, 11, 5-11.

9. Shoesmith, D.J., and Smiley, T.J., 1978: *Multiple Conclusion Logic*, Cambridge University Press.

10. Surma, S.J., 1981: *The growth of logic out of the foundational research in mathematics, in Modern logic - A survey*, E. Agazzi (ed), D. Reidel Publishing co., Dordrecht, 15-33.

Rough Sets, Fuzzy Sets and Soft Computing
Editor: S. Bhattacharya Halder

Generalized Disjunctive Syllogism in Inverse Approximate Reasoning[*]

Banibrata Mondal[*] and Swapan Raha[#]

Department of Mathematics
Visva-Bharati, Santiniketan, West Bengal
E-mail: mbanibrata@gmail.com[*]; swapan.raha@visva-bharati.ac.in[#]

ABSTRACT

This paper presents a method of inverse approximate reasoning using generalized disjunctive syllogism inference mechanism. Approximate reasoning is a methodology which can deduce a specific information from general knowledge and specific observation. It is dependent on the form of general knowledge and the corresponding deductive mechanism. In ordinary approximate reasoning, we derive B' from $A \to B$ and A' by some mechanism. In inverse approximate reasoning, we conclude A' from $A \to B$ and B' using an altogether different mechanism. We first briefly discuss the existing approximate reasoning method based on generalised disjunctive syllogism with example. Then we fit this method in the case of inverse approximate reasoning. An example is cited to illustrate the method.

Keywords: Approximate reasoning, Generalised disjunctive syllogism, Inverse approximate reasoning

1. INTRODUCTION

Reasoning is a mental activity that allows us to derive new premises from given premises with some degree of confidence. It is a fact that human beings are better in reasoning than machines as they have the capability of taking effective decisions on the basis of imprecise linguistic information. A collection of imprecise information given by human experts often forms the basis of a fuzzy system which is represented by fuzzy sets and fuzzy relations. The task of a fuzzy system is to exploit knowledge acquired by experts over time and model the world with it. In the rule based system, from a given rule and an observed

---

[*] This research has been partially supported by the UGCSAP(DRS) Phase-II programme of the Department of Mathematics, Visva-Bharati. Authors appreciate UGC's financial support.

data on the antecedent, we conclude something by applying some method of inference which, we call Forward Approximate Reasoning. The problem with the forward method is that many rules may be applicable for a particular observation and the whole process is not directed towards a goal. In backward reasoning, methods work towards a final state and work at a working memory for a goal. When observed data is given for the consequent part of a given rule to get the best member from the set of input data for the forward reasoning process then the method of inference will be called Inverse Approximate Reasoning. T. Arnould, S. Tano, Y. Kato and T. Miyoshi [1] have considered the problem of inverse approximate reasoning as backward reasoning. They concentrated on finding all fuzzy input data which would produce a conclusion at some interval as the set of input data may be quite large for a given rule. J.Y. Dieulot and P. Borne's [2] simple algorithm for the inversion of fuzzy relational equation with sum-product composition, based on some assumptions is applied to the inversion of a discrete fuzzy linguistic model. But there concluding fuzzy set is restricted at some definite interval to get the solution to input data. E. Eslami and J.J. Buckley [3] used the generalized modus ponens (GMP) in forward approximate reasoning and have investigated the generalized modus tollens (GMT) for the method of inverse approximate reasoning. But though the GMT is frequently applied to the given conclusion, the best member is not in the set of input data. So this method is not a reliable one for inverse approximate reasoning. Considering this limitation they have proposed to investigate the method of inverse approximate reasoning applying the principle of maximum entropy[4]. Here they choose the best member from the set of input data for the given data by maximizing the measure of entropy of the member of input set. Choosing the best member in this way is maximally non-committal with regard to missing information. However, for large cardinality of discrete fuzzy sets, it is quite complicated to produce the input sets. For rules with more complicated antecedents, with more constraints and different measures of entropy the method is also not verified. Mellouli and Bouchon-Meunier [5, 6] have termed the inverse Approximate Reasoning as Fuzzy Abductive Reasoning. Their approach of abduction aims at finding condition on antecedent A of the rule 'If X is A then Y is B' (A and B are fuzzy sets over the universes of discourse U and V respectively, and X, Y are the linguistic variables over U, V respectively) so that observation B' over V is satisfied. They reversed the GMP for S-implication and R-implication considering semantic of the rule, as described by Duboisand Pradein [7]. The reversal of the GMP for R-implication resulted a 'maximal explanation' AG such that any explanation A' should be included in AG given by

$$\forall u \in U,\ \mu A_G(u) = \inf_{v \in V} I_T\left(I_T\left(\mu_A(u),\ \mu_B(v)\right),\ \mu_{B'}(v)\right)$$

where IT is a right continuous R-implication associated with a t-norm T. However, their studies were limited to a particular class of modifiers and to

Gödel implication. Revault d'allones, H. Akdag, and Bouchon-Meunier in [8, 9] generalised these results to other hedges and implications. One of these results is shown to be contradiction with GMP. Finally, particularizing the observation B' bounded by the implication operator, they studies GMP on possible shapes of information. They sometimes find it diffcult to choose an implication a simplications of different classes sometimes generate similar shapes. The choice of GMP operator should be taken into account in the semantic interpretation processes. This drawback of inverse approximate reasoning motivates us to think about an alternative way to represent it. In [10](1993), the authors have presented a purely semantic approach to approximate reasoning based on the concept of the law of disjunctive syllogism. They have formulated it based on generalised disjunctive syllogism. We try to adjust this model in the case of inverse approximate reasoning. To build the model we first transformed the conditional rule in the form of disjunction, if semantic of the rule demands this type of transformation. We then apply the law of disjunctive syllogism in the generalised form.

2. PRELIMINARIES

In this section we recall some definitions and basic concept related to terms that are frequently used to develop the literature.

2.1 Similarity Indices

The similarity between two objects suggests the degree to which properties of one may be inferred from those of the other. Providing a measure of similarity depends mostly on the perceptions of different observers. Emphasis should also be given to different members of the set, so that no one member can influence the ultimate result. Many measures of similarity have been proposed in the existing literature. A careful analysis of the different similarity measures reveals that it is impossible to single out one particular similarity measure that works well for all purpose.

Suppose U be an arbitrary finite set, and $F(U)$ be the collection of all fuzzy subsets of U. Suppose $A, B \in F(U)$, and a similarity index between the pair $\{A, B\}$is denoted as $S(A, B; U)$ or simply $S(A, B)$which can also be considered as a function $S: F(U)^2 \rightarrow [0, 1]$. In order to provide a definition for similarity index, a number of factors must be considered. We expect that a similarity measure (A, B) should satisfy the following axioms.

P1. $S(B, A) = S(A, B)$.

P2. $0 \leq S(A, B) \leq 1$.

P3. $S(A, B) = 1$ if and only if $A = B$.

P4. For two fuzzy sets A and B, simultaneously not null, if $S(A, B) = 0$ then $\min(\mu_A(u), \mu_B(u)) = 0$ for all $u \in U$, i.e., $A \cap B = \Phi$.

P5. If either $A \subseteq B \in C$ or $A \supseteq B \supseteq C$ then $S(A, C) \le \min\{S(A, B), S(B, C)\}$.

A similarity measure between two fuzzy sets satisfying these axioms can also be termed as a *f*-near-degree defined in [11]. For $0 \le \varepsilon \le 1$, if $S(A, B) \ge \varepsilon$, we say that the two fuzzy sets A and B are ε-similar. We now take a definition of measure of similarity which has been proposed in [12].

Definition: $S(A, B) = 1 - \left(\dfrac{\sum\limits_{u} |\mu_A(u) - \mu_B(u)|^q}{n} \right)^{1/q}$

where n is the cardinality of the universe of discourse and q is the family parameter. It is easy to say that the similarity measure given by *Definition* satisfies axioms **P1**, **P2**, **P3**, **P4** and **P5**.

2.2 Generalised Disjunctive Syllogism

First of all, we have to clear how deductive inference patterns, like Modus Ponens or Modus Tollens, or Abductive reasoning patterns behave in classical logic.

Modus ponens:

$$\frac{\begin{array}{c} a \to b \\ a \end{array}}{b}$$

and Modus Tollens:

$$\frac{\begin{array}{c} a \to b \\ \neg\, b \end{array}}{\neg\, a}$$

are deductive inference patterns (they preserve the truth), while a pattern like

$$\frac{\begin{array}{c} a \to b \\ b \end{array}}{a}$$

is a pattern of Abductive inference, it does not preserve the truth (i.e., '$a \to b$' and 'b' may be true while a may be false). In other words, given '$a \to b$', 'a' is not a logical consequence of 'b', only a pluasible explanation.

We describe generalizations of these classical inference rules based on so called Compositional Rule of Inference (CRI).

Placed in a fuzzy contex, where linguistic variables X and Y are defined over U and V respectively, and fuzzy sets A, A' defined on U and fuzzy sets B, B' defined on V, the Generalised Modus Ponens is expressed by the schema given in Table 1.

Table 1 GMP

p: If X is A then Y is B
q: X is A'
r: Y is B'

In this schema, B' is calculated by

$$\mu_{B'}(v) = \sup_{u \in U} T(\mu_{A'}(u), \mu_R(u, v)) \; \forall v \in V \tag{1}$$

where T is a GMP operator and, $\mu_R(u, v)$ is the membership function of fuzzy implication modelling the rule. Observe that it becomes the classical modus ponens when the sets are crisp and $A' = A$, $B' = B$. This equation, which can also be written in the matrix form as

$$B' = A' \circ R$$

is called Composition Rule of Inference under composition.

Equation (1) has a non empty solution if the necessary (not suffcient) condition

$$\mu_{B'}(v) \leq \sup_{u \in U} \mu_{A'}(u) \; \forall v \in V$$

[13] is satisfied.

Another inference rule in fuzzy logic, which is a Generalised Modus Tollens, is expressed by the schema given in Table 2.

Table 2 GMT

p : If X is A then Y is B
q : Y is B'
r : X is A'

In this schema, the CRI has the form

$$\mu_{A'}(u) = \sup_{u \in U} T(\mu_{B'}(v), \mu_R(u, v)) \; \forall u \in u \tag{2}$$

where T is a t-norm. When the sets are crisp and $A' = \neg A$, $B' = \neg B$, we obtain the classical modus tollens.

In two valued logic the law of disjunctive syllogism can be stated as 'Given a disjunction and a negation of any of the disjuncts, the other can be inferred'. Symbolically,

$$a \vee b$$

$$\dfrac{\neg\, a}{b}$$

In using this rule the user must be sure that the disjunct as appears in the second premise is exactly the negation of the one that appears in the first premise, i.e., they must be contradictory pairs. No doubt, such condition is a restriction so long as the real life problems are concerned. Here we often find certain pairs of information which are not completely contradictory but certainly close to the same. In this case, the above framework is not admissible. For this let us remove this restriction on exactness and generalize the concept to the case where the disjuncts are inexact or imprecise in nature and hence the 'degree of contradiction' is not absolutely specified. But as the first premise is a restriction of the disjuncts, and the user may have some possibly inexact knowledge about any of the disjuncts, whatever the level of contradiction may be, it is always possible to infer the shape of the other disjunct. Thus we get a schema given in Table 3.

Table 3   Generalised disjunctive syllogism

$$p:\ X \text{ is } A \text{ or } Y \text{ is } B$$
$$\dfrac{q:\ X \text{ is } A'}{r:\ Y \text{ is } B'}$$

where B' is close to B as A' is close to not A. Mathematical formulation is done in the next section.

3. APPROXIMATE REASONING USING DISJUNCTIVE SYLLOGISM

In [10] authors used disjunctive syllogism to model approximate reasoning. We first analyze it briefly. Let us consider a Schema of approximate reasoning given in Table 4.

Table 4   Approximate reasoning

$$p:\ X \text{ is } A \text{ or } Y \text{ is } B$$
$$\dfrac{q:\ X \text{ is } A^{*}}{r:\ Y \text{ is } B^{*}}$$

Here, X, Y are two linguistic variables that define objects over the universe U, V and A, B, A^{*}, B^{*} are inexact concepts which are approximated by fuzzy sets over U, V, U, V respectively. If A^{*} is complementary to A then by disjunctive syllogism B^{*} is close to B.

We use dissimilarity/similarity measure to interpret the concepts – $A*$ is complementary to A or $B*$ is close to B. If $S(A, A*) \leq 0.5$ then A and $A*$ is complementary. However, if $S(B, B*) > 0.5$ then $B*$ is close to B.

3.1 Mathematical Formulation

Two very important operations on fuzzy sets and fuzzy relations are projection and cylindrical extension. One is the inverse operation to the other, in some sense. The projection operation squeezes the dimension of a relation. Whereas, cylindrical extension extends it. These are almost always combined with each other. Cylindrical extension is the least specific of all relations compatible with the projection. It guarantees that no information not included in the projection is employed in determining the extended relation. Hence, cylindrical extension is totally unbiased. We, simply, define two operators as in [14, 15].

Let R be a fuzzy relation on $U = \times^n$, $(i_1, \ldots, i_k)$ be a subsequence of $(1, \ldots, n)$ and $(j_1, \ldots, j_l)$ be the complementary subsequence of $(1, \ldots, n)$. Let $V = \times^k_{m=1} U_{i_m}$.

Definition–Projection: The projection of R on V is defined by

$$proj\ R \text{ on } V = \int_{V} \sup_{x_{j1}, \ldots, x_{j1}} \mu_R(x_1, \ldots, x_n)/(x_{i_1}, \ldots, x_{i_k}).$$

If R be defined on $U \times V$:

$$proj\ R \text{ on } U = \int_{U} \sup_{v} \mu_R(u, v)/v \qquad (3)$$

Instead of the supremum, maximum operation is used when domains are discrete.

Now we define cylindrical extension. Let E be a fuzzy relation on $V = X^k_{m=1} U_{i_m}$ and $U = X^n_{i=1} U_i$.

Definition–Cylindrical Extension: The cylindrical extension of E into U is defined by

$$ce(E) = \int_{U} \mu_E \left(x_{i_1}, \ldots, x_{i_k}\right)/(x_1, \ldots, x_n)$$

Let B be a fuzzy set defined on V. The cylindrical extension of B on $U \times V$ is the set of all tuples $(u, v) \in U \times V$ with membership degree equal to $\mu_B(v)$, i.e.,

$$ce(B) = \int_{U \times V} \mu_B (v)/(u, v) \qquad (4)$$

Hence, *proj* $ce(E)$ on $V = E$, but, in general, $ce(proj\ R$ on $V) \neq R$.

We now consider the schema given in Table 4 and investigate the schema for Generalised disjunctive syllogism. We describe this method simply by an algorithm.

ALGORITHM–GDSAPPR

Step 1. To translate the rule into a fuzzy relation R, i.e., a fuzzy subset of the Cartesian product $U \times V$ such that

$$\mu_R(u, v) = \min \{1 - \mu_A(u), \mu_B(v)\}$$

Step 2. To take the cylindrical extension of fuzzy subset A^* in U on $U \times V$ by the definition in (4) andsay it R'.

Step 3. To construct $R^* = R \cap R'$, where $\cap$ is defined by any conjunction operator.

Step 4. To obtain $B^* = proj\ R^*$ on V, defined by definition in (3).

Mathematically, we get,

$$\mu_{B*}(v) = proj_{u \in U}\ R^*(u, v)$$

$$= \sup_{u \in U} T(\mu_{R'}(u, v), \mu_R(u, v)),\ T \text{ is a t-norm}$$

$$= \sup_{u \in U} T(\mu_{A*}(u), \mu_R(u, v)),\ \text{by (4)}$$

$$= \sup_{u \in U} (\min (1 - \mu_A(u), \mu_B(v)) \wedge (1 - \mu_A(u))$$

$$= \sup_{u \in U} (\min (1 - \mu_A(u), \mu_B(v))$$

$$= \mu_B(v) \text{ if and only if } \mu_B(v) \leq \sup (1 - \mu_A(u)).$$

This can be achieve if and only if fuzzy sets A and B are chosen in such a way that the membership values attains at least once both the bounds 0 and 1 at some points within their respective domains. In case these bounds are not attained even then any essential information will not be lost in the ultimate inference. Only the inferred results will be a very close approximation of the desired one. The phenomenon is also true in case of GMP.

4. INVERSE APPROXIMATE REASONING

Before going to the definition of inverse approximate reasoning, we first cite an example to make out what inverse approximate does mean. Let there be a fuzzy rule: 'If O_2 flow rate is LOW. Then heating power is LOW'. Now, suppose we observe 'heating power is rather LOW'. Then, by the method of inverse approximate reasoning with single rule, we may construct hypotheses which would explain the causes of the observation by fuzzy mathematical model. There may be another type of example cited in [6]. Let there be two fuzzy rules:

If the traffic is crowded Then the flow is low;

If the visibility is weak Then the flow is low;

Suppose, we observe 'the flow is very low'. According to these two linguistic rules and to the corresponding obsevation, we first construct hypotheses by abduction such as the 'the traffic is very crowded' or 'the visibility is very weak' or 'the traffic is very crowded and the visibility is very weak'. It is very difficult to make a decision which are possible explanations of this observation. N. Mellouli and B. Bouchon-Meunier have used GMP to construct abductive hypotheses and used the measure of similitude to construct the best possible explanation. We do not consider these types of rules through out our present paper. We simply consider single rule in the method of inverse approximate reasoning and discuss how we tackle multiple rules with different consequences in rule based system.

Definition–Inverse Approximate Reasoning: Let

$$\text{If } X \text{ is } A \text{ then } Y \text{ is } B \tag{5}$$

be a given rule, where A and B are fuzzy subsets of the universes of discourse U and V respectively; and X and Y are variables taking values in the sets U and V respectively. From a given fact 'X is A'', where A' is a fuzzy subset of U we can conclude that 'Y is B', where B' is a fuzzy subset of V, by applying some method of approximate reasoning. This is called forward approximate reasoning from the fuzzy sets of U into the fuzzy sets of V. Now for given 'Y is B^*', we consider $\Omega(B^*)$ be the set of all fuzzy subsets of A^* of U such that for given 'X is A^*' we can conclude 'Y is B^*' by some specific method of approximate reasoning by the given rule. We have to choose the best members of $\Omega(B^*)$(not empty) in some sense and have to define some inverse mapping from fuzzy subsets of V into fuzzy subsets of U, which we refer here as inverse approximate reasoning.

4.1 Method

Let a Schema of inverse approximate reasoning given in Table 5.

Table 5 Inverse approximation reasoning

p: If X is A then Y is B
q: Y is B^*
r: X is A^*

where two premises p, q and conclusion r are fabricated by fuzzy sets A, B, B^* and A^* respectively. A and A^* are defined over the universe of discourse $U = \{u_1, u_2, ..., u_m\}$; B and B^* are defined over the same universe of discourse $V = \{v_1, v_2, ..., v_n\}$; X and Y are two linguistic variables take values from U and V respectively. Also, B^* could be obtained from A^* by applying some forward reasoning method. Now our main aim is to construct the shape of $A*$ for a simple rule in (5).

In the case of inverse approximate reasoning schema given in Table 5, when observations are slightly different from the expected conclusion governed by the rule, fuzzy abduction can explain the observations. These explanations have been worked out by Bouchon-Meunier with other researchers sequentially in [5][2000], [6][2003], [8][2007] and in [9][2009], considering the semantic of the rule and GMP operators in the frame work of fuzzy abductive reasoning method. We try to investigate the inverse approximate reasoning method when observations are far different from the expected conclusion governed by the rule and we expect $A*$ which are far different from A in the given rule. Again, the observed $B*$ could be derived from not $A*$, by using some forward reasoning method with suitable GMP operator or t-norm. So in our inverse approximate reasoning schema, we can use generalized disjunctive syllogism schema to obtain $A*$ in the antecedent part of the given rule.

The logic operation of implication is essential for approximate reasoning as it is for reasoning with classical logic. In a relational context, if a fuzzy implication I is modelled the rule by a fuzzy relation from U to V then we quantify the relational degree between premise 'X is A' and conclusion 'Y is V'. We distinguish three classes of fuzzy implications.

In classical logic, $a \rightarrow b = \neg a \vee b$ is extended to fuzzy logic in the form of

$$I(a, b) = T*(n(a), b)$$

for all $a, b \in [0, 1]$, where T is a fuzzy union (t-conorm) and n is fuzzy complement. This class of fuzzy implication is referred to as S-implication in literature.

Another way of defining implication in classical logic is of the form $a \wedge (\neg a \vee b) = a \wedge b$. We remark that the truth value of $a \wedge (\neg a \vee b)$ is not greater than the truth value of b, and then we substitute logical operators by fuzzy operators as

$$I(a, b) = \sup (z \in [0, 1] |\ T (a, z) \leq b)$$

for all $a, b \in [0, 1]$, where T denotes continuous fuzzy intersection (t-norm). They are usually called R-implications, as they are closely connected to residuated semigroup.

Moreover, classical implication may also be rewritten, due to the law of absorption of negation in classical logic, as $a \wedge (\neg a \vee b) = \neg (a \neg (a \wedge b))$ and we substitute logical operators by fuzzy operators as

$$I(a, b) = n(T(a, n (T(a, b)))),$$

where n, T are defined as earlier. These fuzzy implications are called QL-implications, since they are originally employed in quantum logic.

While the definitions of implications in classical logic are equivalent, their extensions in fuzzy logic are not equivalent. In addition to the three classes of fuzzy implications, which are predominant in the literature, other fuzzy implications are possible.

Fuzzy rule based systems have been mainly used as a convenient tool to handle control laws from data. To apply disjunctive syllogism we have to deal with the rule, the semantic of which demands the form of S-implication of the rule. Therefore, if $a \rightarrow b$ is transformed to $\neg a \vee b$, the schemes look like simply in classical logic and in fuzzy logic respectively, as

<table>
<tr><td>prem1: $\neg a \vee b$</td><td>prem1: $\neg a \vee b$</td></tr>
<tr><td>prem2: $\neg b$</td><td>prem2: b'</td></tr>
<tr><td>——————</td><td>——————</td></tr>
<tr><td>conl: $\neg a$</td><td>conl: a'</td></tr>
</table>

where a' is exactly $\neg a$ whenever b' is identical with $\neg b$.

Hence the schem a given in Table 5 be comes an equivalent schema as in Table 6.

Table 6 An equivalent form: Inverse approximate reasoning

<table>
<tr><td>p: X is not A or Y is B</td></tr>
<tr><td>q: Y is B^*</td></tr>
<tr><td>—————————</td></tr>
<tr><td>r: X is A^*</td></tr>
</table>

We, here, expect A^* is sufficiently dissimilar to A whenever B^* is sufficiently dissimilar to B.

ALGORITHM–GDSINV

Step 1. To translate the rule into a fuzzy relation or implication operator R such that

$$\mu_R(u, v) = \min (1 - \mu_A(u), \mu_B(v)) \text{ or}$$

$$\mu_R(u, v) = T^*(1 - \mu_A(u), \mu_B(v)), \text{ where } T^* \text{ is a t-conorm}$$

or

$$\mu_R(u, v) = I(\mu_A(u), \mu_B(v)), \text{ where } I \text{ is an S-implication}$$

as the semantic of the rule demands.

Step 2. To take the cylindrical extension of fuzzy subset B^* in V on $U \times V$ by the definition in (4) ands ay it R^*.

Step 3. To construct $R^* = R \cap R^*$, where $\cap$ is defined by any conjunction operator.

Step 4. To obtain $A^* = proj\ R^*$ on U, defined by definition in (3).

Mathematically, we get,

$$\mu_{A*}(u) = proj_{v \in V} R^*(u, v)$$

$$= \sup_{v \in V} T(\mu_{R*}(u, v), \mu_R(u, v)), \ T \text{ is a t-norm}$$

$$= \sup_{v \in V} T(\mu_{B*}(v), \mu_R(u, v)), \text{ by definite in (4)}$$

which is same as the equation in (2) and establish the CRI in the form of GMT. We have to select an appropriate fuzzy implication or fuzzy relation derived from the requirement that the GMT coincide with classical modus tollens. Also standard negation of the resultant fuzzy set gives the given observation by GMP. Hence, mathematical formulation of the above algorithm is:

$$\mu_{A*}(u) = \sup_{v \in V} T(\mu_{B*}(v), \mu_R(u, v)) \tag{6}$$

We shall illustrate our methods in a fuzzy rule based system by citing an example.

5. EXAMPLE

To illustrate our result we consider a simple fuzzy model adopted from [2] which describes-how the heating power of a burner depends on the oxygen supply. Fuzzy sets:

$$A = \{\text{LOW, OK, HIGH}\}$$

$$B = \{\text{LOW, MEDIUM, HIGH}\}$$

The figures of membership functions of A and B are given in Fig. 1 and Fig. 2 respectively.

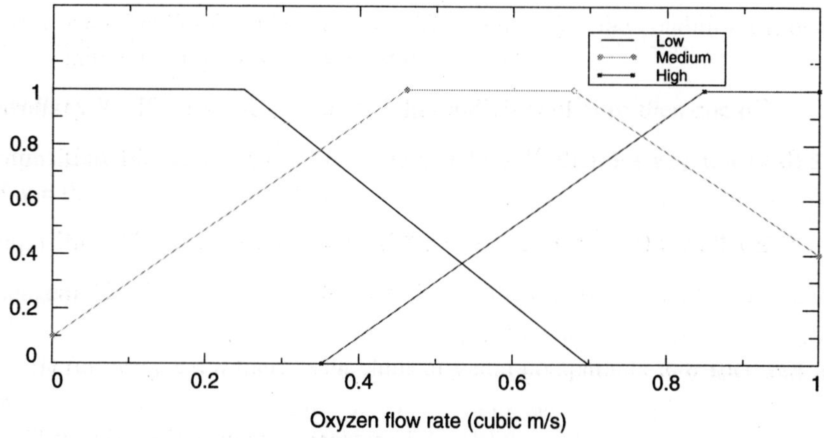

Fig. 1 Membership functions of A

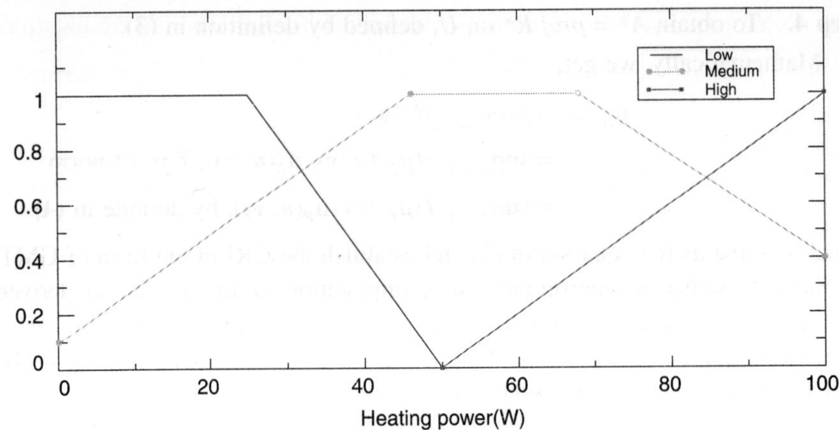

Fig. 2 Membership functions of B

The three rules are:

IF O_2 flow rate is LOW THEN heating power is LOW

IF O_2 flow rate is OK THEN heating power is MEDIUM

IF O_2 flow rate is HIGH THEN heating power is HIGH.

The respective universes of discourse of A and B are:

$$U = \{0, 0.5, 1\}$$

and
$$V = \{0, 25, 50, 75, 100\}.$$

From Fig. 1 and Fig. 2, we have

$$\mu_{LOW}(u) = \{1, 0.4, 0\} = a_1 \quad \mu_{LOW}(v) = \{1, 0.7, 0, 0, 0\} = b_1$$

$$\mu_{OK}(u) = \{0.1, 1, 0.4\} = a_2 \quad \mu_{MEDIUM}(v) = \{0.1, 0.6, 1, 0.7, 0.2\} = b_2$$

$$\mu_{HIGH}(u) = \{0, 0.3, 1\} = a_3 \quad \mu_{HIGH}(v) = \{0, 0, 0, 0.5, 1\} = b_3$$

If given data $b^* = \{0.808, 0.986, 0.73, 1, 0.51\}$ corresponding to a 'rather LOW' heating and taking the first rule as consideration then by the method of J.Y. Dielot and P. Borne [2] $a = \{0.5, 0.6, 0.2\}$ is obtained which corresponds to a 'rather LOW' flow rate.

However, our method is different from it. Our method says: if the given data is sufficiently dissimilar to the consequent part of a given rule then one may conclude that the resulting fuzzy set is sufficiently dissimilar to the antecedent part of the rule. Again, the negation of this resulting fuzzy set also produces the fuzzy set of observation by some forward reasoning method.

Rule Selection: Similarity measures between b^* and the other fuzzy sets involved in the consequences of the rules calculated separately are: $s_1 = S(b^*, b_1)$ = 0.3817, $s_2 = S(b^*, b_2) = 0.5736$, $s_3 = S(b^*, b_3) = 0.2723$ and $\min(s_1, s_2, s_3)$ = 0.2723 = s_3. So for the given b^*, the third rule p_3 is required to fire.

Rule Execution: (i) Here equivalent rule R is defined by t-norm 'min' as

$$R(v, u) = \text{not HIGH} \times \text{not HIGH}$$

$$= \begin{pmatrix} 1 & .7 & 0 \\ 1 & .7 & 0 \\ 1 & .7 & 0 \\ .5 & .5 & 0 \\ 0 & 0 & 0 \end{pmatrix}$$

Defining fuzzy complement by strong negation we get the membership grades of 'not HIGH' in V are

$$\text{not HIGH} = \{1, 1, 1, 0.5, 0\}$$

Let s be the similarity between two fuzzy sets described by 'not HIGH' and b^* in V.

$$s = 1 - \sqrt{\frac{|1 - .808|^2 + |1 + .986|^2 + |1 - .73|^2 + |1 - .5|^2 + |1 - .51|^2}{5}}$$

$$\approx 0.6478$$

Now the modified conditional relation R^* is obtained by (i.e., $\mu_{R*}(v, u) = 1 - (1 - \mu_R(v, u)).s$)

$$R^* = \begin{pmatrix} 1 & 0.805 & 0.352 \\ 1 & 0.805 & 0.352 \\ 1 & 0.805 & 0.352 \\ 0.676 & 0.676 & 0.352 \\ 0.352 & 0.352 & 0.352 \end{pmatrix}$$

Therefore, a^* will be obtained using

$$a^* = \sup \mu_{R*}(v, u) = \{1, 0.805, 0.352\}$$

which corresponds to a 'rather LOW' flow rate. We can calculate the measure of similarity s between two fuzzy sets describing 'not HIGH' and a^* in U.

$$s \approx 0.7877$$

This results shows if b^* is more similar to not b in V, i.e., more dissimilar to b then we get more similar result a to not a, i.e., more dissimilar result a^* to a provided that not a^* yields not b^* in forward method of approximate reasoning. Again, if the defuzzification scheme in [12] be applied to the fuzzy sets involved in the given data and in the derived output data, we get the defuzzified values of b^* and a are 46.39 and 0.35 respectively. Therefore, we may conclude: to obtain 46.39 W heating power 0.35 m^3 sec^{-1} O_2 flow rate is necessary.

(ii) Here, rule is defined by disjunction operator which is considered as t-conorm operator T^*. Let us consider $T^*(a, b) = a + b - a.b$. Therefore,

$$\mu_R(u, v) = T^*(1 - \mu_A(u), \mu_B(v))$$

$$= 1 - \mu_A(u) + \mu_A(u) \cdot \mu_B(v)$$

$$R = \begin{pmatrix} 1 & 1 & 1 & 1 & 1 \\ 0.70 & 0.70 & 0.70 & 0.85 & 1 \\ 0 & 0 & 0 & 0.80 & 1 \end{pmatrix}$$

Cylindrical extension R' of b^* is calculated by

$$R' = \begin{pmatrix} 0.81 & 0.99 & 0.73 & 1 & 0.51 \\ 0.81 & 0.99 & 0.73 & 1 & 0.51 \\ 0.81 & 0.99 & 0.73 & 1 & 0.51 \end{pmatrix}$$

Next, we construct $R^* = R * R^*$, where $*$ is a conjunction operator. Here, we consider Lukaseiwicz t-norm as conjunction and we get,

$$R^* = \begin{pmatrix} 0.81 & 0.99 & 0.73 & 1 & 0.51 \\ 0.51 & 0.69 & 0.43 & 0.85 & 0.51 \\ 0 & 0 & 0 & 0.50 & 0.51 \end{pmatrix}$$

Therefore, a^* will be obtained applying,

$$a^* = \sup_v \mu_{R*}(u, v) = \{1, 0.85, 0.51\},$$

which also corresponds to a 'rather LOW' flow rate. The measure of similarity between a^* and a_3 in U, i.e., $S(a, a^*) \approx 0.2829$ which, again, establishes by our method of inverse approximate reasoning using cylindrical extension and projection that if b^* is more dissimilar to b in V then a^* is more dissimilar to a in V. By defuzzification of a we get 0.40. Hence, by another method, we show that 0.40 m^3 sec^{-1} O_2 flow rate is necessary to turn out 46.39 W heating power, which proves reasonableness of our method.

6. CONCLUSION

Inverse approximate reasoning is an important topic of research for goal directed reasoning. It is used in searching method to obtain the goal. Many researchers, scientists and engineers solved the problem of inverse approximate reasoning in different ways. The inference mechanism of generalised disjunctive syllogism plays a vital role to model inverse approximate reasoning. We have shown when the rule in the scheme of inverse approximate reasoning demands the form of an S-implication we can apply generalized disjunctive syllogism to obtain a solution. Also, we have shown if the rule can be transformed into the form of a relation proposed by us the solution will be obtained.

References

1. Arnould, T., Tano, S., Kato, Y., and Miyoshi, Y., 1993: Backward chaining with fuzzy ifthen..rules. In: Proc. 2nd IEEE International Conf. on Fuzzy Systems, San Francisco, 548–553

2. Dieulot, J.Y., and Borne, P., 2005: Inverse fuzzy sum-product composition and its application to fuzzy linguistic modelling. Studies in Informatics and Control 14(2).

3. Eslami, E., and Buckley, J.J., 1997: Inverse approximate reasoning. Fuzzy Sets and Systems, 87, 155-158.

4. Eslami, E., and Buckley, J.J., 1997: Inverse approximate reasoning II: Maximize entropy. Fuzzy Sets and Systems 87, 291-295.

5. Mellouli, N., and Bouchon-Meunier, B., 2003: Abductive reasoning and measure of similitude in the presence of fuzzy rules. Fuzzy Sets and Systems, 15, 177-188.

6. Mellouli, N., and Bouchon-Meunier, B., 2000: Fuzzy approaches of abductive inference. In: Proc. of the 8th Int. Workshop Non-Monotonic Reasoning NMR'2000, Breikenridge, USA.

7. Ughetto, L., Dubois, D., and Prade, H., 1999: Implicative and conjunctive fuzzy rule–a tool for reasoning from knowledge and examples. In: AAAI'99/ IAAI'99 Proc. of the 16th national Conference on Artificial Intelligence and the 11 th innovative applications of Artificial Intelligence Conference, American Association for Artificial Intelligence, MenloPark, CA, USA, 214-219.

8. d'Allones, A.R., Akdag, H., and Bouchon-Meunier, B., 2007: Selecting implications in fuzzy ab-ductive problems. In: Proceedings of the IEEE International Symposium on Foundations of Computational Intelligence, 597-602.

9. d'Allones, A.R., Akdag, H., and Bouchon-Meunier, B., 2009: For a data-driven interpretation of rules, wrt gmp conclusions, in abductive problems. Journal of Uncertain Systems, 3(4), 280-297.

10. Raha, S., and Ray, K.S., 1993: Approximate reasoning based on generalised disjunctive syllogism. Fuzzy Sets and Systems, 61(2), 143-151.

11. Xudong, L., and Zhang, C., 1999: An axiom founadation for uncertain reasonings in rule based expert system: Nt-algebra. Knowledge and Information Systems: an international journal, 1(4), 415-433.

12. Mondal, B., Mazumdar, D., Raha, S., 2006: Similarity in approximate reasoning. International Journal of Computational Cognition 4(3), 46-56.

13. Baets, B., 1994: A primer on solving fuzzy relational equations on the unit interval. International Journal of Uncertainity, Fuzziness and Knowledge-based Systems 2(2), 205–225

14. Zadeh, L.A., 1975: The concept of linguistic varible and its application to approximate reasoning: Part I and part II. Information Science 8.

15. Drainkov, D., Hellendoorn, H., and Reinfrank, M., 1996: An Introduction to Fuzzy Control. Narosa Publishing House, New Delhi, India

A Study on Fuzzy Competition Graphs

Sovan Samanta[1*] and Anita Pal[2#]
[1]Department of Applied Mathematics with Oceanology and Computer Programming, Vidyasagar University, Midnapore
[2]Department of Mathematics, National Institute of Technology Durgapur
E-mail: ssamantavu@gmail.com[*]; anita.buie@gmail.com[#]

ABSTRACT

Fuzzy competition graph as the generalization of competition graph is introduced here. Two generalizations of fuzzy competition graph as fuzzy k-competition graphs and p-competition fuzzy graphs are also defined. These graphs are related to fuzzy digraphs. Fuzzy neighbourhood graphs, which are related to fuzzy graphs are defined here. Some relations between fuzzy competition graphs and fuzzy neighbourhood graphs have been established. Also several results to find strong edges of stated graphs are established.

Keywords: Fuzzy graphs, fuzzy competition graphs, fuzzy k-competition graphs, fuzzy neighbourhood graphs, p-competition fuzzy graphs.

1. INTRODUCTION

In 1968 Cohen introduced the notion of competition graphs in connection with a problem in ecology. Let $\overleftarrow{D} = \left(V, \overleftarrow{E}\right)$ be a digraph, which corresponds to a food web. A vertex $x \in V\left(\overleftarrow{D}\right)$ represents a species in the food web and an arc $\overleftarrow{(x, s)} \in \overleftarrow{E}\left(\overleftarrow{D}\right)$ means that x preys on the species s. If two species x and y have a common prey s, they will compete for the prey s. Based on this analogy, Cohen defined a graph which represents the relations of competition among the species in the food web. The competition graph $C\left(\overleftarrow{D}\right)$ of a digraph $\left(\overleftarrow{D}\right) = \left(V, \overleftarrow{E}\right)$ is an undirected graph $G = (V, E)$ which has the same vertex set V and has an edge between two distinct vertices $x, y \in V$ if there exists a vertex $s \in V$ and arcs $\overleftarrow{(x, s)}, \overleftarrow{(y, s)} \in \overleftarrow{E}\left(\overleftarrow{D}\right)$. The competition graph is also applicable in channel assignment, coding, modelling of complex economic and energy systems, etc.

A lot of works have been done on competition graphs and its variations. In all these works, it is assumed that the vertices and edges of the graphs are precisely defined. But, in reality we observe that sometimes the vertices and edges of a graph can not be defined precisely. For example, in ecology, species may be of different types like vegetarian, non-vegetarian, strong, weak, etc. Similarly in ecology, preys may be tasty, digestive, harmful, etc. The terms tasty, digestive, harmful, etc. have no precise meanings. They are fuzzy in nature and hence the species and preys may be assumed as fuzzy sets and inter-relationship between the species and preys can be designed by a fuzzy graph. This motivates the necessity of fuzzy competition graphs.

The concept of fuzzy graph was introduced by Rosenfeld in 1975. Fuzzy graph theory has a vast area of applications. It is used in evaluation of human cardiac function, fuzzy neural networks, etc. Fuzzy graphs can be used to solve traffic light problem, time table scheduling, etc. In fuzzy set theory, there are different types of fuzzy graphs which may be a graph with crisp vertex set and fuzzy edge set or fuzzy vertex set and crisp edge set or fuzzy vertex set and fuzzy edge set or crisp vertices and edges with fuzzy connectivity, etc. A lot of works have been done on fuzzy graphs [3, 7, 8, 9].

The competition graphs and fuzzy graphs are well known topics. In this article, fuzzy competition graphs are defined as motivated from fuzzy food web. Also, the generalization of it, the fuzzy k-competition graphs and the p-competition fuzzy graphs are introduced. Fuzzy neighbourhood graphs and their properties are investigated in this paper.

2. PRELIMINARIES

A directed graph (digraph) $\overleftarrow{G}$ is a graph which consists of non-empty finite set $V\left(\overleftarrow{G}\right)$ of elements called vertices and a finite set $\overleftarrow{E}\left(\overleftarrow{G}\right)$ of ordered pairs of distinct vertices called arcs. We will often write $\overleftarrow{G} = \left(V, \overleftarrow{E}\right)$. For an arc (u, v), u is the tail and v is the head. The order (size) of $\overleftarrow{G}$ is the number of vertices (arcs) in $\overleftarrow{G}$. The out-neighbourhood [14] of a vertex v is the set $N^+(v) = \left\{u \in V - v\colon \overleftarrow{(v, u)} \in \overleftarrow{E}\right\}$. Similarly, the in-neighbourhood [14] $N^-(v)$ of a vertex v is the set $\left\{w \in V - v\colon \overleftarrow{(w, v)} \in \overleftarrow{E}\right\}$. The open neighbourhood of a vertex is the union of out-neighbourhood and in-neighbourhood of the vertex. A walk in $\overleftarrow{G}$ is an alternating sequence $W = x_1 \overleftarrow{e_1} x_2 \overleftarrow{e_2} \ldots x_{k-1} \overleftarrow{e_k} x_k$ of vertices x_i and arcs $\overleftarrow{e_i}$ of $\overleftarrow{G}$ such that tail of $\overleftarrow{e_i}$ is $\overleftarrow{x_i}$ and head is x_{i+1} for every $i = 1, 2, \ldots, k - 1$. A walk is closed if $x_1 = x_k$. A trail is a walk in which all arcs are distinct. A path is a walk in which all vertices are distinct. A path $x_1, x_2, \ldots, x_k$ with $k \geq 3$ is a cycle if $x_1 = x_k$.

Definition 1: The competition graph $C\left(\overleftarrow{G}\right)$ of a digraph $\overleftarrow{G} = \left(V, \overleftarrow{E}\right)$ is an undirected graph $G = (V, E)$ which has the same vertex set V and has an edge between distinct two vertices $x, y \in V$ if there exist a vertex $a \in V$ and arcs $\overleftarrow{(x, a)}, \overleftarrow{(y, a)} \in \overleftarrow{E}$ in $\overleftarrow{G}$. We say that a graph G is a competition graph if there exists a digraph $\overleftarrow{G}$ such that $C\left(\overleftarrow{G}\right) = G$.

Many variations of competition graph have been available in literature. One of the important graphs, known as p-competition graphs, is defined below.

Definition 2: If p is a positive integer, the p-competition graph $C_p\left(\overleftarrow{G}\right)$ corresponding to the digraph $\overleftarrow{G}$ is defined to have a vertex set V with an edge between x and y in V if and only if, for some distinct vertices $a_1, a_2, ..., a_p$ in V, $\overleftarrow{(x, a_1)}, \overleftarrow{(y, a_1)}, \overleftarrow{(x, a_2)}, \overleftarrow{(y, a_2)}, ..., \overleftarrow{(x, a_p)}, \overleftarrow{(y, a_p)}$ are arcs in $\overleftarrow{G}$.

If $\overleftarrow{G}$ is thought of as a food web whose vertices are the species in some ecosystem, (x, y) is an edge of $C_p\left(\overleftarrow{G}\right)$ if and only if x and y have at least p common preys. So $C_1\left(\overleftarrow{G}\right)$ is the competition graph.

A fuzzy set A on a set X is characterized by a mapping $m: X \to [0, 1]$, which is called the membership function. A fuzzy set is denoted by $A = (X, m)$.

Definition 3: A fuzzy graph $\xi = (V, \sigma, \mu)$ is a non-empty set V together with a pair of functions $\sigma: V \to [0, 1]$ and $\mu: V \times V \to [0, 1]$ such that for all $x, y \in V$, $\mu(x, y) \leq \sigma(x) \wedge \sigma(y)$ and μ is a symmetric fuzzy relation on σ. Here $\sigma(x)$ and $\mu(x, y)$ represent the membership values of the vertex x and of the edge (x, y) in ξ.

Since μ is well defined, a fuzzy graph has no multiple edges. A loop at a vertex x in a fuzzy graph is represented by $\mu(x, x) = 0$. The fuzzy set (V, σ) is called fuzzy vertex set of ξ and the elements of the fuzzy set are called fuzzy vertices. $(V \times V, \mu)$ is called the fuzzy edge set of ξ and the elements of the fuzzy set are called fuzzy edge. An edge is non-trivial if $\mu(x, y) \neq 0$. The fuzzy graph $\xi' = (V', \tau, \nu)$ is called a fuzzy subgraph [21] of ξ if $\tau(x) \leq \sigma(x)$ for all $x \in V'$ and $\nu(x, y) \leq \mu(x, y)$ for all $x, y \in V'$ where $V' \subset V$.

The underlying crisp graph of $\xi = (\sigma, \mu)$ is denoted by $G^* = (V, E)$, where $V(G^*) = \{u \in V: \sigma(u) > 0\}$ and $E(G^*) = \{(u, v) \in V \times V: \mu(u, v) > 0\}$. The order of a fuzzy graph ξ is $|\xi| = \sum_{x \in V} \sigma(x)$. An edge (u, v) of a fuzzy graph is called an effective edge [40] if $\mu(u, v) = \min \{\sigma(u), \sigma(v)\}$. An effective incident degree of a fuzzy graph is defined as number of effective incident edges on a vertex. A pendant vertex [28] in a fuzzy graph is defined as a vertex of an effective incident degree as one.

For the fuzzy graph $\xi = (V, \sigma, \mu)$, an edge (x, y), $x, y \in V$ is called strong if $\frac{1}{2} \min \{\sigma(x), \sigma(y)\} \leq \mu(x, y)$ and it is called weak otherwise.

Like crisp digraph, fuzzy digraph has the following definition in literature.

Definition 4: Directed fuzzy graph (fuzzy digraph) $\overleftarrow{\xi} = \left(V, \sigma, \overleftarrow{\mu}\right)$ is a non-empty set V together with a pair of functions $\sigma: V \to [0, 1]$ and $\overleftarrow{\mu}: V \times V \to [0, 1]$ such that for all $x, y \in V$, $\overleftarrow{\mu}(x, y) \leq \sigma(x) \wedge \sigma(y)$.

Since $\overleftarrow{\mu}$ is well defined, a fuzzy digraph has at most two directed edges (which must have opposite directions) between any two vertices. Here $\overleftarrow{\mu}(u, v)$ is denoted by the membership value of the edge $\overleftarrow{(u, v)}$. The loop at a vertex x is represented by $\overleftarrow{\mu}(x, x) \neq 0$. Here $\overleftarrow{\mu}$ need not be symmetric as $\overleftarrow{\mu}(x, y)$ and $\overleftarrow{\mu}(y, x)$ may have different values. The underlying crisp graph of directed fuzzy graph is the graph similarly obtained except the directed arcs are replaced by undirected edges.

3. FUZZY COMPETITION GRAPHS

Now, we come to our main objective of the paper, the fuzzy competition graph. Like crisp graph, fuzzy out-neighbourhood and fuzzy in-neighbourhood of a vertex in directed fuzzy graph are defined below.

Definition 5: Fuzzy out-neighbourhood of a vertex v of a directed fuzzy graph $\overleftarrow{\xi} = \left(V, \sigma, \overleftarrow{\mu}\right)$ is the fuzzy set $N^+(v) = (X_v^+, m_v^+)$ where $X_v^+ = \left\{ u \mid \overleftarrow{\mu}(v, u) > 0 \right\}$ and $m_v^+: X_v^+ \to [0, 1]$ defined by $m_v^+(u) = \overleftarrow{\mu}(v, u)$. Similarly, fuzzy in-neighbourhood of a vertex v of a directed fuzzy graph $\overleftarrow{\xi} = \left(V, \sigma, \overleftarrow{\mu}\right)$ is the fuzzy set $N^-(v) = (X_v^-, m_v^-)$ where $X_v^- = \left\{ u \mid \overleftarrow{\mu}(u, v) > 0 \right\}$ and $m_v^-: X_v^- \to [0, 1]$ defined by $m_v^-(u) = \overleftarrow{\mu}(u, v)$.

Now we define fuzzy competition graph.

Definition 6: The fuzzy competition graph $C\left(\overleftarrow{\xi}\right)$ of a fuzzy digraph $\overleftarrow{\xi} = \left(V, \sigma, \overleftarrow{\mu}\right)$ is an undirected fuzzy graph $\xi = (V, \sigma, \mu)$ which has the same fuzzy vertex set as in $\overleftarrow{\xi}$ and has a fuzzy edge between two vertices $x, y \in V$ in $C\left(\overleftarrow{\xi}\right)$ if and only if $N^+(x) \cap N^+(y)$ is non-empty fuzzy set in $\overleftarrow{\xi}$ and the edge membership value between x and y in $C\left(\overleftarrow{\xi}\right)$ is $\mu(x, y) = (\sigma(x) \wedge \sigma(y)) \, h(N^+(x) \cap N^+(y))$.

Example 1:   Let $\overleftarrow{\xi}$ be a directed fuzzy graph. Let the vertices with membership values of $\overleftarrow{\xi}$ be $(a, 0.3), (b, 0.6), (c, 0.4), (d, 0.5), (e, 0.4)$ with membership values of arcs be $\overleftarrow{\mu}(b, a) = 0.2, \overleftarrow{\mu}(c, b) = 0.35, \overleftarrow{\mu}(d, c) = 0.2, \overleftarrow{\mu}(e, d) = 0.2, \overleftarrow{\mu}(e, a) = 0.25, \overleftarrow{\mu}(a, d) = 0.3$. It is shown in Fig. 1(a). The corresponding fuzzy competition graph is shown in Fig. 1(b).

Edges of fuzzy competition graphs represent that two vertices (species) compete for at least one prey. So the strength of the edges is important to the character of competitions.

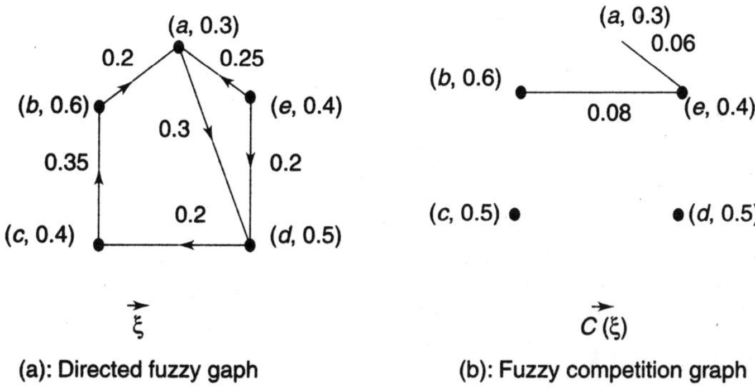

(a): Directed fuzzy gaph   (b): Fuzzy competition graph

Fig. 1   Example of fuzzy competition graphs

Now an extension of fuzzy competition graph, called fuzzy k-competition graph is defined in the following.

Definition 7:   Let k be a non-negative number. The fuzzy k-competition graph $C_k\left(\overleftarrow{\xi}\right)$ of a fuzzy digraph $\overleftarrow{\xi} = \left(V, \sigma, \overleftarrow{\mu}\right)$ is an undirected fuzzy graph $\xi = (V, \sigma, \mu)$ which has the same fuzzy vertex set as $\overleftarrow{\xi}$ and has a fuzzy edge between two vertices $x, y \in V$ in $C_k\left(\overleftarrow{\xi}\right)$ if and only if $|N^+(x) \cap N^+(y)| > k$. The edge membership value between x and y in $C_k\left(\overleftarrow{\xi}\right)$ is $\mu(x, y) = \dfrac{(k' - k)}{k'} [\sigma(x) \wedge \sigma(y)]\, h(N^+(x) \cap N^+(y))$

where $k' = |N^+(x) \cap N^+(y)|$.

So fuzzy k-competition graph is simply fuzzy competition graph when $k = 0$. An example of fuzzy 0.2-competition graph is given below.

Example 2:   Let $\overleftarrow{\xi}$ be a directed fuzzy graph. Let vertices with membership values of $\overleftarrow{\xi}$ be $(x, 0.4), (y, 0.6), (a, 0.6), (b, 0.7), (c, 0.8), (d, 0.65)$ and the membership values of arcs be $\overleftarrow{\mu}(x, a) = 0.3$, $\overleftarrow{\mu}(x, b) = 0.35$, $\overleftarrow{\mu}(x, c) = 0.36$, $\overleftarrow{\mu}(x, d) = 0.4$, $\overleftarrow{\mu}(y, a) = 0.6$, $\overleftarrow{\mu}(y, b) = 0.5$, $\overleftarrow{\mu}(y, c) = 0.45$, $\overleftarrow{\mu}(y, d) = 0.35$. It is shown in Fig. 2(a). The corresponding fuzzy 0.2-competition graph is shown in the Fig. 2(b).

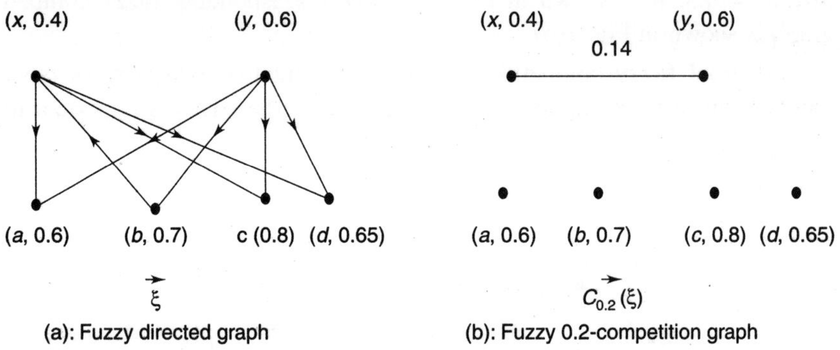

(a): Fuzzy directed graph   (b): Fuzzy 0.2-competition graph

Fig. 2   Example of 0.2-fuzzy competition graph

Theorem 1:   Let $\overleftarrow{\xi} = \left(V, \sigma, \overleftarrow{\mu}\right)$ be a fuzzy digraph. If $N^+(x) \cap N^+(y)$ contains single element of $\overleftarrow{\xi}$, then the edge (x, y) of $C\left(\overleftarrow{\xi}\right)$ is strong if and only if $|N^+(x) \cap N^+(y)| > 0.5$.

Proof.   Here $\overleftarrow{\xi} = \left(V, \sigma, \overleftarrow{\mu}\right)$ is a fuzzy digraph. Let $N^+(x) \cap N^+(y) = \{(a, m)\}$, where m is the membership value of the element a. Here, $|N^+(x) \cap N^+(y)| = m = h(N^+(x) \cap N^+(y))$. So, $\mu(x, y) = m \times \sigma(x) \wedge \sigma(y)$. Hence the edge (x, y) in $C\left(\overleftarrow{\xi}\right)$ is strong if and only if $m > 0.5$.

If all the edges of a fuzzy digraph are strong, then all the edges of the corresponding fuzzy competition graph may not be strong. This result is illustrated below. Let us consider two vertices x, y with $\sigma(x) = 0.3$, $\sigma(y) = 0.4$ in a fuzzy digraph such that the vertices have a common prey z with $\sigma(z) = 0.2$.

Let $\overleftarrow{\mu}(x, z) = 0.2$, $\overleftarrow{\mu}(y, z) = 0.15$. Clearly, the edges $\overleftarrow{(x, z)}$ and $\overleftarrow{(y, z)}$ are strong. But membership value of the edge (x, y) in corresponding competition graph is $0.3 \times 0.15 = 0.045$. Hence the edge is not strong as $\dfrac{0.045}{0.3} = 0.15 < 0.5$. But if all the edges are strong of a fuzzy digraph, then a result can be found from the following theorem.

Theorem 2:   If all the edges of a fuzzy digraph $\overleftarrow{\xi} = \left(V, \sigma, \overleftarrow{\mu}\right)$ be strong, then $\dfrac{\mu(x, y)}{(\sigma(x) \wedge \sigma(y))^2} > 0.5$ for all edge (x, y) in $C\left(\overleftarrow{\xi}\right)$.

Proof. Let $\overleftarrow{\xi} = \left(V, \sigma, \overleftarrow{\mu}\right)$ be a fuzzy digraph and every edge of $\overleftarrow{\xi}$ be strong

i.e., $\dfrac{\overleftarrow{\mu}(x, y)}{\sigma(x) \wedge \sigma(y)} > 0.5$ for all edge (x, y) in $\overleftarrow{\xi}$. Let the corresponding fuzzy

competition graph be $C\left(\overleftarrow{\xi}\right) = (V, \sigma, \mu)$.

Case 1: Let $N^+(x) \cap N^+(y)$ be a null set for all $x, y \in V$. Then there exist no

edge in $C\left(\overleftarrow{\xi}\right)$ between x and y.

Case 2: $N^+(x) \wedge N^+(y)$ is not a null set. Let $N^+(x) \wedge N^+(y) = \{(a_1, m_1), (a_2, m_2),$
..., $(a_z, m_z)\}$, where m_i, $i = 1, 2, ..., z$ are the membership values of a_i, $i = 1, 2, ..., z$,
respectively. So

$$m_i = \overleftarrow{\mu}(x, a_i), \overleftarrow{\mu}(y, a_i)\}, \ i = 1, 2, ..., z.$$

Let $\quad h(N^+(x) \cap N^+(y)) = \max \{m_i, i = 1, 2, ..., z\} = m_{\max}.$

$$\mu(x, y) = (\sigma(x) \wedge \sigma(y)) \, h(N^+(x) \cap N^+(y))$$

$$= m_{\max} \times \sigma(x) \wedge \sigma(y).$$

Hence $\quad \dfrac{\mu(x, y)}{(\sigma(x) \wedge \sigma(y))^2} = \dfrac{m_{\max}}{\sigma(x) \wedge \sigma(y)} > 0.5.$

We have seen that if height of intersection between two out neighbourhoods of two vertices of a fuzzy digraph is greater than 0.5, the edge between the two vertices in corresponding fuzzy competition graph is strong. This result is not true in corresponding fuzzy k-competition graph. A related result is proved below.

Theorem 3: Let $\overleftarrow{\xi} = \left(V, \sigma, \overleftarrow{\mu}\right)$ be a fuzzy digraph. If $h(N^+(x) \cap N^+(y)) = 1$ and $|N^+(x) \cap N^+(y)| > 2k$, then the edge (x, y) is strong in $C_k\left(\overleftarrow{\xi}\right)$.

Proof. Let $\overleftarrow{\xi} = \left(V, \sigma, \overleftarrow{\mu}\right)$ be a fuzzy digraph and $C_k\left(\overleftarrow{\xi}\right) = (V, \sigma, \mu)$ be the corresponding fuzzy k-competition graph. Also let, $h(N^+(x) \cap N^+(y)) = 1$ and $|N^+(x) \cap N^+(y)| > 2k$.

Now, $\mu(x, y) = \dfrac{k' - k}{k'} \sigma(x) \wedge \sigma(y) \, h(N^+(x) \cap N^+(y))$, where $k' = |N^+(x) \cap N^+(y)|$.

So, $\mu(x, y) = \dfrac{k' - k}{k'} \sigma(x) \wedge \sigma(y)$. Hence $\dfrac{\mu(x, y)}{\sigma(x) \wedge \sigma(y)} = \dfrac{k' - k}{k'} > 0.5$ as $k' > 2k$.

Hence the edge (x, y) is strong.

For any positive integer p, we define a p-competition fuzzy graph as follows.

Definition 8: Let p be a positive integer. The p-competition fuzzy graph $C^p\left(\overleftarrow{\xi}\right)$ of a fuzzy digraph $\overleftarrow{\xi} = \left(V, \sigma, \overleftarrow{\mu}\right)$ is an undirected fuzzy graph $\xi = (V, \sigma, \mu)$ which has same fuzzy vertex set as $\overleftarrow{\xi}$ and has a fuzzy edge between two vertices x and $y \in V$ in $C^p\left(\overleftarrow{\xi}\right)$ if and only if $|\text{supp}\,(N^+(x) \cap N^+(y))| \geq p$. The edge membership value between x and y in $C^p\left(\overleftarrow{\xi}\right)$ is $\mu(x, y) = \dfrac{(n-p)+1}{n}\,[\sigma(x) \wedge \sigma(y)]$ $h(N^+(x) \cap N^+(y))$ where $n = |\text{supp}\,(N^+(x) \cap N^+(y))|$.

The following example illustrates the 2-competition fuzzy graph.

Example 3: Let a fuzzy graph (see Figure 3(a)) with vertices $\{x, y, z, a, b, c\}$ where $\sigma(x) = 0.7$, $\sigma(y) = 0.8$, $\sigma(z) = 0.9$, $\sigma(a) = 0.75$, $\sigma(b) = 0.85$, $\sigma(c) = 0.95$ and membership values of arcs be $\overleftarrow{\mu}\,(x, a) = 0.7$, $\overleftarrow{\mu}\,(x, b) = 0.7$, $\overleftarrow{\mu}\,(y, a) = 0.6$, $\overleftarrow{\mu}\,(y, b) = 0.8$, $\overleftarrow{\mu}\,(y, c) = 0.8$, $\overleftarrow{\mu}\,(z, c) = 0.9$. The corresponding 2-competition fuzzy graph is shown in the Fig. 3(b).

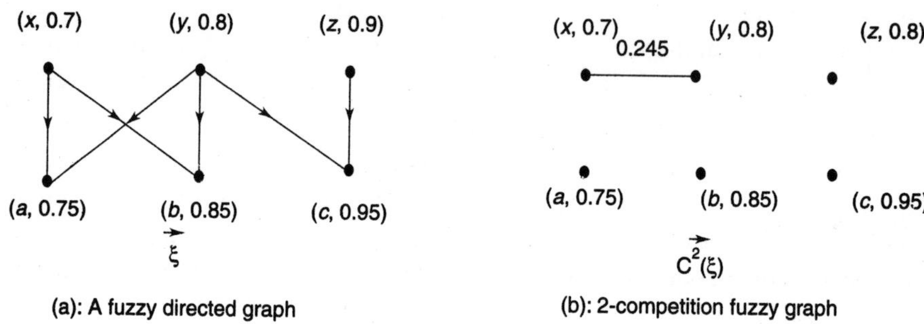

(a): A fuzzy directed graph

(b): 2-competition fuzzy graph

Fig. 3 Example of 2-competition fuzzy graph

Fuzzy k-competition graph and p-competition fuzzy graph are two generalizations of fuzzy competition graph, where in first fuzzy graph a real positive number k is related to the cardinality of a fuzzy set. In the second fuzzy graph i.e. p-competition fuzzy graph, p is related to the cardinality of a crisp set.

Theorem 4: Let $\overleftarrow{\xi} = \left(V, \sigma, \overleftarrow{\mu}\right)$ be a fuzzy digraph. If $h(N^+(x) \cap N^+(y)) = 1$ in $C^{\left[\frac{n}{2}\right]}\left(\overleftarrow{\xi}\right)$, then the edge (x, y) is strong where $n = |\text{supp}\,(N^+(x) \cap N^+(y))|$. (Note that for any real number x, $[x]$ = greatest integer not exceeding x).

Proof. Here $\overleftarrow{\xi} = (V, \sigma, \overleftarrow{\mu})$ is a fuzzy digraph. Let the corresponding $\left[\frac{n}{2}\right]$-fuzzy competition graph be $\xi = (V, \sigma, \mu)$ where $n = |\text{supp}\,(N^+(x) \cap N^+(y))|$. Also assume that $h(N^+(x) \cap N^+(y)) = 1$, $x, y \in V$. Now, $\mu(x, y) = \dfrac{n - \left[\frac{n}{2}\right] + 1}{n}\ \sigma(x) \wedge \sigma(y)$.

This gives the result, $\dfrac{\mu(x, y)}{\sigma(x) \wedge \sigma(y)} = \dfrac{n - \left[\frac{n}{2}\right] + 1}{n} > 0.5$. Hence the edge (x, y) is strong.

4. CONCLUSIONS

All real world competitions can be designed by fuzzy competition graphs, introduced here. This study extends two subclass of fuzzy competition graphs. One of the graphs is fuzzy k-competition graph which is one kind of intersection graph of fuzzy out neighbourhood of vertices of a fuzzy digraph. Other is p-competition fuzzy graph which is, also, an intersection graph of supports of fuzzy out neighbourhood of vertices of a fuzzy digraph. The positive real number k in fuzzy k-competition graphs and the positive integer p in p-competition fuzzy graphs measure the strength of competitions of the corresponding fuzzy competition graphs. Some results have been proved to find the strong edges of the fuzzy competition graphs. These results will help to find strong competitors in all kinds of business markets. m-step fuzzy competition graphs along with many topics can be introduced as an extension of this study.

References

1. Cho, H.H., Kim, S.R. and Nam, Y., 2000: The m-step competition graph of a diagraph, Discrete Applied Mathematics, 105(1-3), 115-127.

2. Cohen, J.E., 1968: Interval graphs and food webs: a finding and a problem, Document 17696-PR, RAND Corporation, Santa Monica, CA.

3. Eslahchi, C., and Onaghe, B.N., Vertex Strength of Fuzzy Graphs, International Journal of Mathematics and Mathematical Sciences, Volume 2006, Article ID 43614, Pages 1-9, DOI 10.1155/IJMMS/2006/43614.

4. Isaak, G., Kim, S.R., McKee, T.A., McMorris F.R., and Roberts, F.S., 1992: 2-competition graphs, SIAM J. Disc. Math., 5(4), 524-538.

5. Jenson, J.B., and Gutin, G.Z., 2009: Digraphs: Theory, Algorithms and Applications, Springer-verlag.

6. Kim, S.R., McKee, T.A., McMorris F.R., and Roberts, F.S., 1995: p-competition graphs, Linear Algebra and its Application, 217, 167-178.

7. Mordeson, J.N., and Nair, P.S., 2000: Fuzzy graphs and hypergraphs, Physica Verlag.

8. Samanta, S., and Pal, M., 2011: Fuzzy tolerance graphs, International Journal of Latest Trends in Mathematics, 1(2), 57-67.

9. Samanta, S., and Pal, M., 2011: Fuzzy threshold graphs, CIIT International Journal of Fuzzy Systems, 3(12), 360-364.

Rough Sets, Fuzzy Sets and Soft Computing
Editor: S. Bhattacharya Halder
Copyright © 2015, Narosa Publishing House, New Delhi

Fuzzy *k*-Competition Graph and *p*-Competition Fuzzy Graph

Sovan Samanta[1], Anita Pal[2] and Madhumangal Pal[1]
[1]Department of Applied Mathematics with Oceanology and Computer Programming, Vidyasagar University, Midnapore
[2]Department of Mathematics, National Institute Technology Durgapur

ABSTRACT

In this paper, we define fuzzy competition graphs, fuzzy *k*-competition graphs, fuzzy neighbourhood graphs, fuzzy *k*-neighbourhood graphs and *p*-competition fuzzy graphs. Also some basic theorems related to the stated graphs have been presented.

Keywords: Fuzzy graph, fuzzy competition graph, fuzzy neighbourhood graph.

1. INTRODUCTION

Cohen [2] introduced the notion of competition graph in connection with a problem in ecology in 1968. The competition graphs have various applications to channel assignments, coding, and modeling of complex economic and energy systems.

A directed graph (digraph) $\vec{G}$ consists of non empty finite set $V\left(\vec{G}\right)$ of elements called vertices and a finite set $E\left(\vec{G}\right)$ of ordered pairs of distinct vertices called arcs. We will often write $\vec{G} = \left(V, \vec{E}\right)$. The order (size) of $\vec{G}$ is the number of vertices (arcs) in $\vec{G}$. For an arc $\overrightarrow{(u, v)}$, u is tail and v is head and denote it by $u \to v$. The out-neighbourhood $N^+(v)$ of a vertex v is the set $\left\{ u \in V - v : \overrightarrow{(v, u)} \in \vec{E} \right\}$. Similarly, the in-neighbourhood $N^-(v)$ of a vertex v is the set $\left\{ w \in V - v : \overrightarrow{(w, v)} \in \vec{E} \right\}$. The open neighbourhood of v, $N(v) = N^+(v) \cup N^-(v)$. A walk in $\vec{G}$ is an alternating sequence $W = x_1 \, \vec{e_1} \, x_2 \, \vec{e_2} \, ... \, x_{k-1} \, \vec{e_k} \, x_k$ of vertices x_i and arcs $\vec{e_i}$ of $\vec{G}$ such that tail of $\vec{e_i}$ is x_i and head is x_{i+1} for every $i = 1, 2, ...,$

$k - 1$. A walk is closed if $x_1 = x_k$. A trail is a walk in which all arcs are distinct. A path is a walk in which all vertices are distinct. A path $x_1, x_2, ..., x_k$ with $k \geq 3$ is a cycle if $x_1 = x_k$.

Open neighbourhood graph $N(G)$ of G is a graph whose vertex set is same as G and has an edge between two vertices x and y in $N(G)$ if and only if $N(x) \cup N(y) \neq \phi$ in G. Closed neighbourhood $N[v]$ of v is the set $N(v) \cup \{v\}$. Closed neighbourhood graph $N[G]$ of a graph G is similarly defined, except has an edge in $N[G]$ if and only if $N[x] \cap N[y] \neq \phi$ in G. (p)-neighbourhood graph (read as open p-neighbourhood graph), $N_p(G)$ of a graph G is a graph whose vertex set is same as G and has an edge between two vertices x and y if and only if $|N(x) \cap N(y)| \geq p$ in G. Similarly $[p]$-neighbourhood graph (closed p-neighbourhood graph) $N_p[G]$ is defined except has an edge if and only if $|N[x] \cap N[y]| \geq p$ in G.

The competition graph $C\left(\vec{G}\right)$ of a diagraph $\vec{G} = \left(V, \vec{E}\right)$ is an undirected graph $G = (V, E)$ which has the same vertex set V and has an edge between distinct two vertices $x, y \in V$ if there exists a vertex $a \in V$ and arcs $\overrightarrow{(x, a)}, \overrightarrow{(y, a)} \in \vec{E}$ in $\vec{G}$. We say that a graph G is a competition graph if there exists a diagraph $\vec{G}$ such that $C\left(\vec{G}\right) = G$.

Many variations of ordinary competition graph have been introduced. One of such a graph is p-competition graph [5]. If p is a positive integer, the p-competition graph $C_p\left(\vec{G}\right)$ corresponding to diagraph $\vec{G}$ is defined to have a vertex set V with an edge between x and y in V if and only if, for some distinct vertices $a_1, a_2, ..., a_p$ in V, the pairs $\overrightarrow{(x, a_1)}, \overrightarrow{(y, a_1)}, \overrightarrow{(x, a_2)}, \overrightarrow{(y, a_2)}, ..., \overrightarrow{(x, a_p)}, \overrightarrow{(y, a_p)}$ are arcs. If $\vec{G}$ is thought of as a food web whose vertices are the species in some ecosystem, (x, y) is an edge of $C_p\left(\vec{G}\right)$ if and only if x and y have at least p common prey. So $C_1\left(\vec{G}\right)$ is ordinary competition graph.

The notion of edge clique covering plays an important role in the study of ordinary competition graphs. An edge clique covering (ECC) of G is a collection of cliques such that every edge of G is in at least one of these cliques. The minimum size of a ECC of G is called the ECC number of the graph G, and is denoted by $\theta_e(G)$. In the case of p-competition graph, a related notion plays an analogous role. Suppose G is a graph and $F = \{S_1, S_2, ..., S_r\}$ is a family of subsets of the vertex set of G, repetitions allowed. We say that F is a p-edge clique covering or p-ECC if every set $i_1, i_2, ..., i_p$ of p distinct subscripts, $T = S_{i_1} \cap S_{i_2} ... \cap S_{i_p}$ either is empty or induces a clique of G, and the collection of sets of the form T covers all edges of G. Let $\theta_e^p(G)$ be the smallest r for which there is a p-ECC.

1.1 An Example of Competition Graph

Let $\vec{G} = \left(V, \vec{E}\right)$ be a diagraph, which corresponds to a food web. A vertex $x \in V$ in $\vec{G}$ stands for species in the food web. An arc $\overrightarrow{(x, a)} \in \vec{E}$ in $\vec{G}$ means that the species x preys on the species a. If two species x and y have a common prey a, they will compete for the prey a. Cohen defined a graph which represents the relations of competition among the species in the food web.

1.2 Fuzzy sets

A fuzzy set A on a set X is characterized by a mapping $m: X \rightarrow [0, 1]$, called the membership function. We shall denote a fuzzy set as $A = (X, m)$. The support of A is supp $A = \{x \in X \mid m(x) \neq 0\}$. The core of A is the crisp set of all members whose membership values are 1. A is non trivial if supp A is nonempty. The height of A is $h(A) = \max \{m(x) \mid x \in X\}$. A is normal if $h(A) = 1$. The membership function of the intersection of two fuzzy sets A and B with membership functions m_A and m_B respectively is defined as the minimum of the two individual membership functions. $m_{A \cap B} = \min(m_A, m_B)$. We write $A = (X, m) \leq B = (X, m')$ (fuzzy subset) if $m(x) \leq m'(x)$ for all $x \in X$. The family of all fuzzy subset is denoted by $F(x)$.

1.3 Fuzzy Graphs

A fuzzy graph $\xi = (V, \sigma, \mu)$ is a non empty set V together with a pair of functions $\sigma: V \rightarrow [0, 1]$ and $\mu: V \times V \rightarrow [0, 1]$ such that for all $x, y \in V$, $\mu(x, y) \leq \sigma(x) \cap \sigma(y)$. μ is a symmetric fuzzy relation on σ. Since μ is well defined, a fuzzy graph has no multiple edges. An edge is nontrivial if $\mu(x, y) \neq 0$. A loop at x is represented by $\mu(x, x) = 0$. The underlying crisp graph of $\xi = (\sigma, \mu)$ is denoted by $G^* = (V, E)$, where $V = \{u \in V: \sigma(u) > 0\}$ and $E = \{(u, v) \in V \times V: \mu(u, v) > 0\}$. Also the underlying crisp graph of directed fuzzy graph is the graph similarly obtained except the directed arcs are replaced by undirected edges. The order of a fuzzy graph ξ is $|\xi| = \sum_{x \in V} \sigma(x)$. The strength of connectedness between two vertices u and v is $\mu^\infty(u, v) = \sup \{\mu^k(u, v) \mid k = 1, 2, ...\}$ where $\mu^k(u, v) = \sup\{\mu(u, u_1) \wedge \mu(u_1, u_2) \wedge ... \wedge \mu(u_{k-1}, v) \mid u_1, u_2, ..., u_{k-1} \in V\}$.

Directed fuzzy graph (fuzzy digraph) $\vec{\xi} = \left(V, \sigma, \vec{\mu}\right)$ is a nonempty set V together with a pair of functions $\sigma: V \rightarrow [0, 1]$ and $\vec{\mu}: V \times V \rightarrow [0, 1]$ such that for all $x, y \in V$, $\vec{\mu}(x, y) \leq \sigma(x) \cap \sigma(y)$. Since $\vec{\mu}$ is well defined, a fuzzy digraph has at most two directed edges (which must have opposite directions) between any two vertices. A fuzzy multigraph is a multivalued symmetric mapping $\mu: V \times V \rightarrow [0, 1]$. A fuzzy multigraphs can be considered to be the "disjoint union" or "disjoint sum" of a collection of simple fuzzy graphs, as is done with crisp multigraphs. The same holds for multidigraphs. The fuzzy graph $\xi' = (V', \tau, v)$ is called a fuzzy subgraph of ξ if $\tau(x) \leq \sigma(x)$ for all $x \in V'$ and $v(x, y) \leq \mu(x, y)$ for all $x, y \in V'$ where $V' \subset V$.

A path ρ in fuzzy graph is a sequence of distinct nodes $x_0, x_1, ..., x_n$ such that $\mu(x_{i-1}, x_i) > 0$, $1 \leq i \leq n$. Here $n \, ^{3*} \, 0$ is called the length of the path. The consecutive pairs (x_{i-1}, x_i) are called the arcs of the path. The strength of ρ is defined as $\overset{n}{\underset{i=1}{\Lambda}} \mu(x_{i-1}, x_i)$. In other words, strength of a path is the weight of the weakest arc of the path. If the path has length 0, then its strength is $\sigma(x_0)$. We call ρ a cycle if $x_0 = x_n$ and $n \geq 3$. We recall that a graph without cycle is called acyclic or forest, a connected forest is tree. We call a fuzzy graph a forest if the graph consisting of its nonzero arcs is a forest. We call fuzzy graph a fuzzy forest if it has a fuzzy spanning subgraph $\xi' = (\tau, v)$ which is a forest, where for all arcs (x, y) not in the subgraph ξ', we have $\mu(x, y) < v^{\infty}(x, y)$. Thus if $(x, y) \in \xi$ but $(x, y) \notin \xi'$, there is a path in ξ' between x and y whose strength is greater than $\mu(x, y)$. (σ, μ) is a cycle if and only if $(\text{supp}(\sigma), \text{supp}(\mu))$ is a cycle. (σ, μ) is a fuzzy cycle if and only if $(\text{supp}(\sigma), \text{supp}(\mu))$ is a cycle and there does not exist unique $(x, y) \in \text{supp}(\mu)$ such that $\mu(x, y) = \wedge\{\mu(u, v) | (u, v) \in \text{supp}(\mu)\}$. (σ, μ) is a clique if $(\text{supp}(\sigma), \text{supp}(\mu))$ is a clique. (σ, μ) is fuzzy clique [12] if it is a clique and every cycle in it is a fuzzy cycle.

1.4 Review of Previous Works

Mathew and Sunitha [6] described the types of arcs in a fuzzy graph. Nagoorgani and Malarvizhi [9] established the isomorphism properties of strong fuzzy graphs. Nagoorgani and Radha [8] defined regular fuzzy graphs. Nagoorgani and Vadivel [10] showed relations between the parameters of independent domination and irredundance in fuzzy graphs. Nagoorgani and Vijayalaakshmi [11] defined insentive arc in domination of fuzzy graph. Nair and Cheng [12] defined cliques and fuzzy cliques in fuzzy graphs. Nair [13] established the definition of Perfect and precisely perfect fuzzy graphs. Natarajan [14] et. al. showed strong (weak) domination in fuzzy graphs.

2. OUR WORKS

We now define fuzzy competition graphs and related notions and prove some basic results.

2.1 Fuzzy Competition Graph

Out-neighbourhood $N^+(v)$ of a vertex v of a directed fuzzy graph $\vec{\xi} = \left(V, \sigma, \vec{\mu}\right)$ is the set $\left\{u \in V - v : \overrightarrow{(v, u)} \text{ is a strong arc in } \xi\right\}$. Similarly, in-neighbourhood

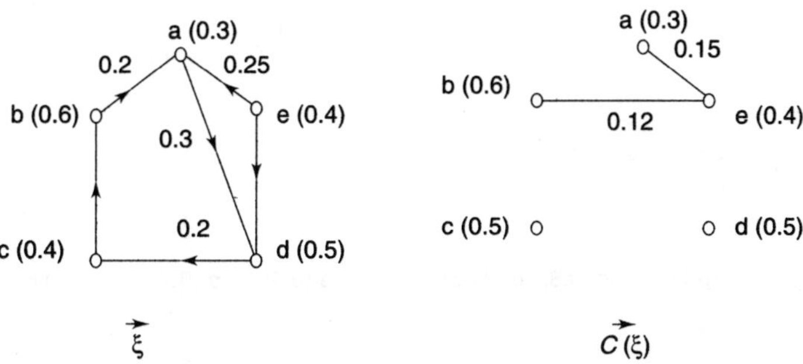

Fig. 1 Example of fuzzy competition graph

$N^-(v)$ of a vertex v of directed fuzzy graph $\vec{\xi} = \left(V,\ \sigma,\ \vec{\mu}\right)$ is the set $\left\{ w \in V - v \colon \overrightarrow{(w,\ v)} \text{ is a strong arc in } \vec{\xi} \right\}$.

Definition 1: The fuzzy competition graph $C\left(\vec{\xi}\right)$ of a fuzzy diagraph $\vec{\xi} = \left(V,\ \sigma,\ \vec{\mu}\right)$ is an undirected fuzzy graph $\xi = (V,\ \sigma,\ \mu)$ which has same vertex set as in $\vec{\xi}$ and has an edge between two vertices $x,\ y \in V$ in $C\left(\vec{\xi}\right)$ if and only if $N^+(x) \cap N^+(y)$ is non empty fuzzy set in $\vec{\xi}$ and the edge membership value between x and y in $C\left(\vec{\xi}\right)$ is $\mu(x,\ y) = (\sigma(x) \wedge \sigma(y))\ h(N^+(x) \cap N^+(y))$.

Example 1: Let $\vec{\xi}$ be a directed fuzzy graph. Let the vertices with membership values of $\vec{\xi}$ be $a(0.3)$, $b(0.6)$, $c(0.4)$, $d(0.5)$, $e(0.4)$ with strong arcs $\overrightarrow{(b,\ a)}$, $\overrightarrow{(c,\ b)}$, $\overrightarrow{(d,\ c)}$, $\overrightarrow{(e,\ d)}$, $\overrightarrow{(e,\ a)}$, $\overrightarrow{(d,\ a)}$. It is shown on left side in Fig. 1. The corresponding competition graph is shown on the right of the same figure.

Definition 2: Let k be a non negative number. The fuzzy k-competition graph $C_k\left(\vec{\xi}\right)$ of a fuzzy diagraph $\vec{\xi} = \left(V,\ \sigma,\ \vec{\mu}\right)$ is an undirected fuzzy graph $\xi = (V,\ \sigma,\ \mu)$ which has the same vertex set as $\vec{\xi}$ and has an edge between two vertices x and $y \in V$ in $C_k\left(\vec{\xi}\right)$ if and only if $|N^+(x) \cap N^+(y)| > k$. The edge membership value between x and y in $C_k\left(\vec{\xi}\right)$ is $\mu(x,\ y) = (k' - k)/k'\ (\sigma(x) \wedge \sigma(y))\ h(N^+(x) \cap N^+(y))$ where $k' = |N^+(x) \cap N^+(y)|$.

So fuzzy k-competition graph is ordinary fuzzy competition graph for $k = 0$.

Example 2: Let $\vec{\xi}$ be a directed fuzzy graph. Let vertices with membership values of $\vec{\xi}$ be $x(0.4)$, $y(0.6)$, $a(0.6)$, $b(0.7)$, $c(0.8)$, $d(0.65)$ and the strong arcs

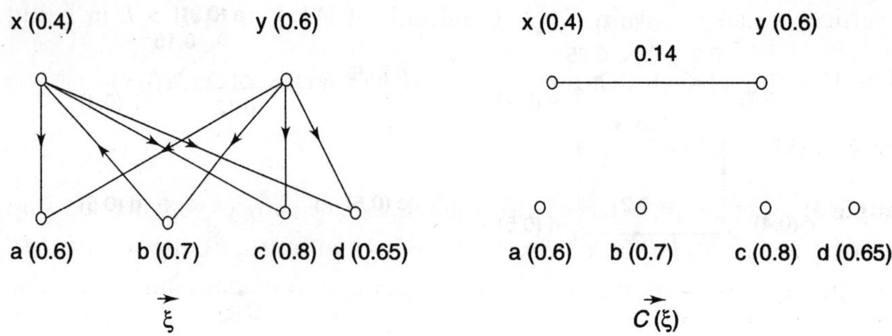

Fig. 2 Example of fuzzy 1.25-competition graph

be $\overrightarrow{(x, a)}$, $\overrightarrow{(x, b)}$, $\overrightarrow{(x, c)}$, $\overrightarrow{(x, d)}$, $\overrightarrow{(y, a)}$, $\overrightarrow{(y, b)}$, $\overrightarrow{(y, c)}$, $\overrightarrow{(y, d)}$. It is shown on left side in Fig. 2. The corresponding fuzzy 1.25-competition graph is shown on the right of the figure.

2.2 Fuzzy Neighbourhood Graphs

In this section fuzzy open neighbourhood graphs are defined and then fuzzy closed neighbourhood graphs and fuzzy k-neighbourhood graphs of open and closed types are defined.

Definition 3: Let $\xi = (V, \sigma, \mu)$ be a fuzzy graph. Open neighbourhood graph of ξ is a fuzzy graph $N(\xi) = (V, \sigma, \mu')$ whose vertex set is same as ξ and has an edge between two vertices x and $y \in V$ in $N(\xi)$ if and only if $N(x) \cap N(y)$ is non empty fuzzy set in ξ and $\mu': V \times V \to [0, 1]$ such that $\mu'(x, y) = (\sigma(x) \wedge \sigma(y))\, h(N(x) \cap N(y))$.

Definition 4: Let $\xi = (V, \sigma, \mu)$ be a fuzzy graph. Closed neighbourhood graph of ξ is a fuzzy graph $N[\xi] = (V, \sigma, \mu')$ whose vertex set is same as ξ and has an edge between two vertices x and $y \in V$ in $N[\xi]$ if and only if $N[x] \cap N[y]$ is non empty fuzzy set in ξ and $\mu': V \times V \to [0, 1]$ such that $\mu'(x, y) = (\sigma(x) \wedge \sigma(y))\, h(N[x] \cap N[y])$.

Definition 5: Let $\xi = (V, \sigma, \mu)$ be a fuzzy graph. Fuzzy (k)-neighbourhood fuzzy graph (read as open fuzzy k-neighbourhood graph) of ξ is a fuzzy graph $N_k(\xi) = (V, \sigma, \mu')$ whose vertex set is same as ξ and has an edge between two vertices x and $y \in V$ in $N_k(\xi)$ if and only if $|N(x) \cap N(y)| > k$ in ξ and

$$\mu': V \times V \to [0, 1] \text{ such that } \mu'(x, y) = \frac{(k' - k)}{k'} (\sigma(x) \wedge \sigma(y))\, h(N(x) \cap N(y))$$

where $k' = |N(x) \cap N(y)|$.

Definition 6: Let $\xi = (V, \sigma, \mu)$ be a fuzzy graph. fuzzy $[k]$-neighbourhood graph (read as fuzzy closed k-neighbourhood graph) of ξ is a fuzzy graph $N_k[\xi] = (V, \sigma, \mu')$ whose vertex set is same as ξ and has an edge between two vertices x and $y \in V$ in $N_k[\xi]$ if and only if $|N[x] \cap N[y]| > k$ in ξ and

$$\mu': V \times V \to [0, 1] \text{ such that } \mu'(x, y) = \frac{(k' - k)}{k'} (\sigma(x) \wedge \sigma(y)) \, h(N(x) \cap N(y))$$

where $k' = |N[x] \cap N[y]|$.

Example 3: Let $\xi = (V, \sigma, \mu)$ be a fuzzy graph. Let the vertices with membership values of ξ be $a(0.3)$, $b(0.6)$, $c(0.4)$, $d(0.5)$, $e(0.4)$ with $\mu(a, b)$, $\mu(b, c)$, $\mu(c, d)$, $\mu(d, e)$, $\mu(e, a)$, $\mu(a, d) \in (0, 1]$ and other edges have 0 membership value. The fuzzy open neighbourhood graph $N(\xi)$, fuzzy closed neighbourhood graph $N[\xi]$, fuzzy (0.3)-neighbourhood graph $N_{0.3}(\xi)$ and fuzzy [0.3] neighbourhood graph $N_{0.3}[\xi]$ of the fuzzy graph ξ are shown in Fig. 3.

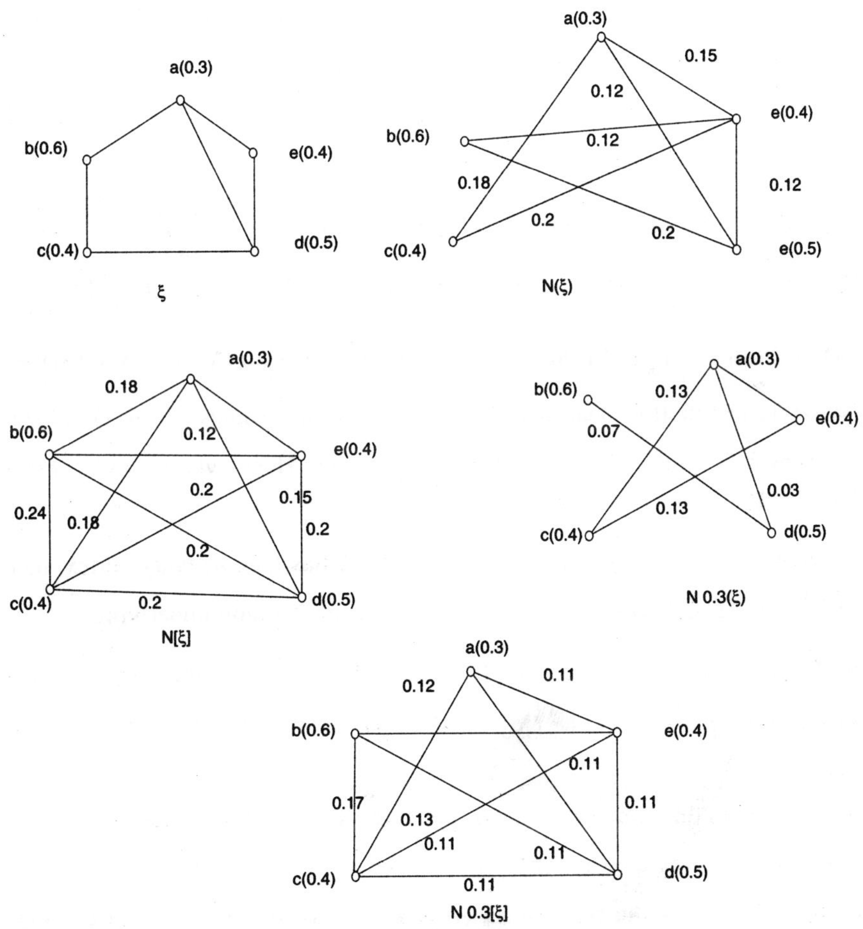

Fig. 3 Example of fuzzy neighbourhood graphs

Definition 7: Let $\vec{\xi} = (V, \sigma, \vec{\mu})$ be a fuzzy diagraph. The underlying fuzzy graph of $\vec{\xi}$ is denoted by $U(\xi)$ and is defined as $U(\xi) = (V, \sigma, \mu)$ where $\mu(u, v) = \min\{\vec{\mu}(u, v), \vec{\mu}(v, u)\}$ for all $u, v \in V$.

Theorem 1: If the symmetric fuzzy diagraph $\vec{\xi}$ is loop less, $C_k\left(\vec{\xi}\right) = N_k\left(U(\xi)\right)$ where $U(\xi)$ is the fuzzy graph underlying $\vec{\xi}$.

Proof. Let a fuzzy directed graph be $\vec{\xi} = (V, \sigma, \vec{\mu})$ and the corresponding underlying fuzzy graph be $U(\xi) = (V, \sigma, \mu)$. Also let $C_k\left(\vec{\xi}\right) = (V, \sigma, v)$ and $N_k(U(\xi)) = (V, \sigma, v')$. The vertex set of $\vec{\xi}$ is equal to $C_k\left(\vec{\xi}\right)$. Also an underlying fuzzy graph has the same vertex set as the directed fuzzy graph. Hence $N_k(U(\xi))$ has the same vertex set as $\vec{\xi}$. Now we need to show that $v(x, y) = v'(x, y)$ for all $x, y \in V$. If $v(x, y) = 0$ in $C_k\left(\vec{\xi}\right)$ then $|N^+(x) \cap N^+(y)| \le k$. As $\vec{\xi}$ is symmetric fuzzy set, $|N(x) \cap N(y)| \le k$ in $U(\xi)$. So $v'(x, y) = 0$.

If $|N^+(x) \cap N^+(y)| > k$ then $v(x, y) > 0$ in $C_k\left(\vec{\xi}\right)$. So $v(x, y) = \dfrac{(k'-k)}{k'}(\sigma(x) \wedge \sigma(y))$ $h(N^+(x) \cap N^+(y))$ where $k' = |N^+[x] \cap N^+[y]|$. As $\vec{\xi}$ is symmetric fuzzy set, $|N(x) \cap N(y)| > k$ in $U(\xi)$. So v' $\dfrac{(k'-k)}{k''}(\sigma(x) \wedge \sigma(y))$ $h(N(x) \cap N(y))$ where $k'' = |N[x] \cap N[y]|$. It is clear that $h(N^+(x) \cap N^+(y))$ in $\vec{\xi}$ equals to $h(N(x) \cap N(y))$ in $U(\xi)$ as $\vec{\xi}$ is symmetric. $k' = k''$ for similar reason. Hence $v(x, y) = v'(x, y)$ for all $x, y \in V$.

Theorem 2: If the symmetric fuzzy diagraph $\vec{\xi}$ has loop at every vertex, then $C_k\left(\vec{\xi}\right) = N_k[U(\xi)]$ where $U(\xi)$ is the loop less fuzzy graph underlying $\vec{\xi}$.

Proof: Let a fuzzy directed graph be $\vec{\xi} = (V, \sigma, \vec{\mu})$ and the corresponding underlying loop less graph be $U(\xi) = (V, \sigma, \mu)$. Also let $C_k\left(\vec{\xi}\right) = (V, \sigma, v)$ and $N_k[U(\xi)] = (V, \sigma, v')$. The vertex set of $\vec{\xi}$ is equal to $C_k\left(\vec{\xi}\right)$. Also an underlying fuzzy graph has the same vertex set as the directed fuzzy graph. Hence $N_k[U(\xi)]$ has the same vertex set as $\vec{\xi}$. Now we need to show that $v(x, y) = v'(x, y)$ for all $x, y \in V$. As the fuzzy directed graph $\vec{\xi}$ has loop at every vertex, out-neighbourhood of each vertex contains the vertex itself. Hence if $v(x, y) = 0$ in

$C_k\left(\vec{\xi}\right)$ then $|N^+(x) \cap N^+(y)| \le k$. As $\vec{\xi}$ is symmetric fuzzy set, $|N(x) \cap N(y)| \le k$ in $U(\xi)$. So $v'(x, y) = 0$.

If $|N^+(x) \cap N^+(y)| > k$ then $v(x, y) > 0$ in $C_k\left(\vec{\xi}\right)$. So $v(x, y) = \dfrac{(k' - k)}{k'}(\sigma(x) \wedge \sigma(y))\, h(N^+(x) \cap N^+(y))$ where $k' = |N^+[x] \cap N^+[y]|$. As $\vec{\xi}$ is symmetric fuzzy set, $|N(x) \cap N(y)| > k$ in $U(\xi)$. So $v' = \dfrac{(k' - k)}{k''}(\sigma(x) \cap \sigma(y))\, h(N(x) \cap N(y))$ where $k'' = |N[x] \cap N[y]|$. It is clear that $h(N^+(x) \cap N^+(y))$ in $\vec{\xi}$ equals to $h(N(x) \cap N(y))$ in $U(\xi)$ and $k' = k''$ as $\vec{\xi}$ is symmetric. Hence $v(x, y) = v'(x, y)$ for all $x, y \in V$.

We now define *p*-competition fuzzy graph and prove some related theorems.

Definition 8: Let p be a positive integer. The *p*-competition fuzzy graph $C^p\left(\vec{\xi}\right)$ of a fuzzy diagraph $\vec{\xi} = \left(V, \sigma, \vec{\mu}\right)$ is an undirected fuzzy graph $\xi = (V, \sigma, \mu)$ which has same vertex set as $\vec{\xi}$ and has an edge between two vertices x and $y \in V$ in $C^p\left(\vec{\xi}\right)$ if and only if $|\text{supp}(N^+(x) \cap N^+(y))| \ge p$. The edge membership value between x and y in $C^p\left(\vec{\xi}\right)$ is $\mu(x, y) = \dfrac{(n - p) + 1}{n}(\sigma(x) \wedge \sigma(y))\, h(N^+(x) \cap N^+(y))$ where $n = |\text{supp}(N^+(x) \cap N^+(y))|$.

Example 4: We now give an example of 2-competition fuzzy graph. Here the graph contains six vertices $x(0.7)$, $y(0.8)$, $z(0.9)$, $a(0.75)$, $b(0.85)$, $c(0.95)$ and strong arcs $\overrightarrow{(x, a)}$, $\overrightarrow{(x, b)}$, $\overrightarrow{(y, a)}$, $\overrightarrow{(y, b)}$, $\overrightarrow{(y, c)}$, $\overrightarrow{(z, c)}$. The corresponding 2-competition fuzzy graph is shown on the right of the Fig. 4.

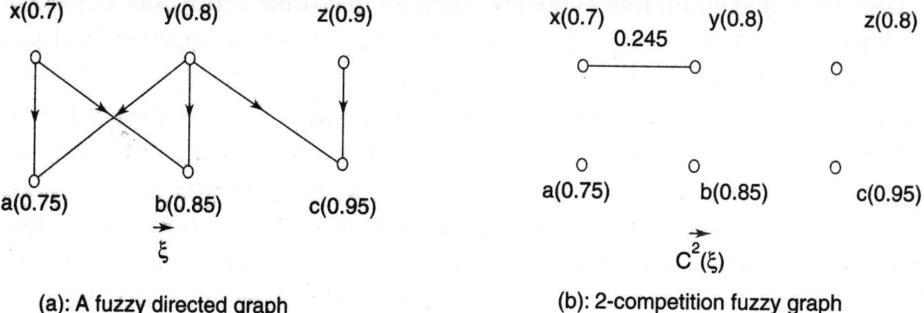

(a): A fuzzy directed graph

(b): 2-competition fuzzy graph

Fig. 4 Example of 2-competition fuzzy graph

2.3 ECC in Fuzzy Graph

Edge clique covering in fuzzy graph is collection of cliques such that every edge of ξ is in at least one of these cliques. The minimum size of the collection is called ECC number in fuzzy graphs and denoted by θ_e (ξ). In p-competition fuzzy graph a related notion p-ECC in fuzzy graphs plays an important role. Suppose ξ is a fuzzy graph and $F = \{S_1, S_2, ..., S_r\}$ is a family of subsets of the vertex set of ξ, repetitions allowed. We say that F is a p-edge clique covering or p-ECC in fuzzy graph if every set $i_1, i_2, ..., i_p$ of p distinct subscripts, $T = S_{i_1} \cap S_{i_2} \cap ... \cap S_{i_p}$ either is empty fuzzy set or induces a clique of ξ, and the collection of sets of the form T covers all edges of ξ. Let θ_e^p (ξ) be the smallest r for which there is a p-ECC.

Theorem 3:   A fuzzy graph ξ with n vertices is a p-competition fuzzy graph of an arbitrary fuzzy diagraph if and only if θ_e^p (ξ) $\leq n$, which is true if and only if ξ has a p-ECC consisting n sets.

Proof: Let p-competition graph $C^p\left(\vec{\xi}\right) = (V, \sigma, \mu)$ where $\vec{\xi} = \left(V, \sigma, \vec{\mu}\right)$. Let $V = \{v_1, v_2, ..., v_n\}$. For each i, let $S_i = v_j$: $\vec{\mu}(v_j, v_i) > 0\}$. It is clear that $S_1, S_2, ..., S_n$ make a family of fuzzy p-ECC.

Conversely, let ξ be a fuzzy graph and a p-ECC of this graph be $F = \{S_1, S_2, ..., S_r\}$ with $r \leq n$. Now we define $\vec{\xi} = \left(V, \sigma, \vec{\mu}\right)$ such that $\vec{\mu}(v_i, v_j) > 0$ if and only if $v_i \in S_j$. It is easy to verify that $\xi = C^p\left(\vec{\xi}\right)$.

Theorem 4:   Fuzzy cycle of length 4 is not a fuzzy 2-competition graph.

Proof: Let the fuzzy 4 cycle be (V, σ, μ). Let the membership values of the four vertices, in order as in cycle, be $\sigma(x_1) > 0$, $\sigma(x_2) > 0$, $\sigma(x_3) > 0$, $\sigma(x_4) > 0$ such that minimum of the values does not occur twice. So to prove fuzzy 4 cycle a fuzzy 2-competition graph we have to shows that there exists a fuzzy 2-ECC, F, consisting maximum four elements. Suppose that some vertex, say x_1, belongs to all four sets. As $\mu(x_2, x_3) > 0$ (see Fig. 5), so x_2, x_3 appear together in at least two set of F. Hence x_1, x_3 must appear at least two of sets of F which shows that $\mu(x_1, x_3) > 0$, contradicts $\mu(x_1, x_3) = 0$. Thus, each x_i is in at most three sets in F. But $\mu(x_1, x_2) > 0$ and $\mu(x_1, x_4) > 0$, so x_1 must be in at least two sets with x_2 and at least two sets with x_4 with x_2 and x_4 at most one set together. Thus we arrive at a condition that x_1 is in at least three sets, and so in exactly three sets of F. So each x_i is in exactly three sets of F then x_1 and x_3 are in two sets together, which is a contradiction. Thus fuzzy cycle of length 4 is not a fuzzy 2-competition graph.

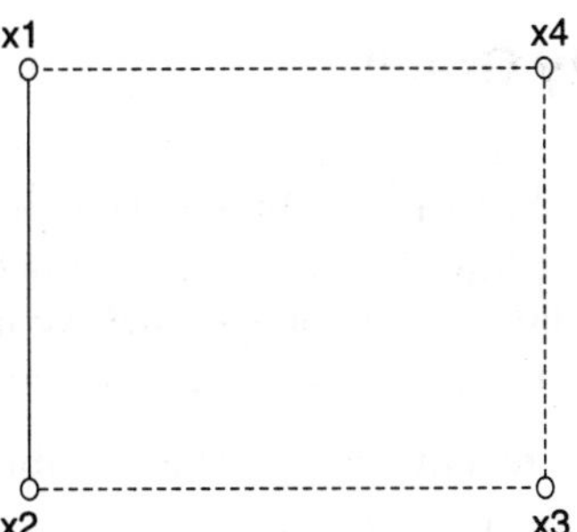

Fig. 5 Example of fuzzy four cycle

Theorem 5: Suppose ξ is a 2-competition graph of an arbitrary fuzzy diagraph. Then ξ' is a 2-competition graph of an arbitrary diagraph if ξ' is obtained from ξ by,

(a) adding a pendent edge at a vertex,

(b) adding a fuzzy path P_k with $k = 3$ between adjacent vertices x and y in ξ or adding a P_k with $k > 4$ between arbitrary distinct vertices in ξ,

(c) adding a path P_4 from x to y with a pendent edge, where x and y are distinct vertices in ξ.

Proof: Let $\xi = (V, \sigma, \mu)$ be a fuzzy graph such that V has n elements with $\mu(x_i) > 0$, $i = 1, 2, ..., n$. As it is a 2-competition graph it has 2-ECC $S_1, S_2, ..., S_n$ for ξ.

(a) Let ξ' be obtained from ξ by adding a pendent edge $\{x, u\}$ with $\mu(x, u) > 0$. We may assume that x_1 is in S_1. Now we can construct a 2-ECC for ξ' using $n + 1$ sets as $S'_1 = S_1 \cup \{u\}$, $S'_i = S_i$ for $i = 2, 3, ..., n$ and $S'_{n+1} = \{x, u\}$.

(b) Let ξ' be obtained from ξ by adding a fuzzy path P_k with $k = 3$ between adjacent vertices x and y in ξ or adding a P_k with $k > 4$ between arbitrary distinct vertices in ξ. Without loss of generality, we can assume that $x \in S_1$ and $y \in S_2$. If $\mu(x, y) > 0$ in ξ, then x, y are together in two sets S_i and S_j, and without loss of generality these are S_1 and S_2. If $\mu(x, y) = 0$, we may assume that $x \in S_1$. If y is in any $S_{j, j} \neq 1$, we may assume $y \in S_2$. We now built 2-ECC for ξ' which uses $n + k - 2$ sets by taking $S'_1 = S_1 \cup \{x_1\}$, $S'_2 = S_2 \cup \{x_{k-2}\}$, $S'_i = S_i$ for $i = 3, ..., n$, $S'_{n+1} = \{x, x_1, x_2\}$, $S'_{n+2} = \{x_1, x_2, x_3\}$, $S'_{n+3} = \{x_2, x_3, x_4\}$, ..., $S'_{n+k-2} = \{x_{k-3}, x_{k-2}, y\}$ (We should note that if this construction was used for $k = 4$ then $\{x_1, x_2\}$ could be in two sets S'_2, S'_{n+1}, which is not acceptable) suppose $k = 3$, and that $\{x, y\}$ is an edge of ξ. In this case there are two sets, S_1 and S_2 without loss of generality, which both contain x and y. We built a fuzzy

2-ECC for ξ' by taking $S_1' = S_1 \cup \{x_1\}$, $S_i' = S_i$ for $i = 2, ..., n$ and $S_{n+1}' = \{x, x_1, y\}$.

(c) Let ξ' be obtained from ξ by adding a path P_4 from x to y and a pendant edge, without loss of generality $\{x_2, v\}$. If x is not in any S_i, then we may place x in S_1 and not change the fact that we have a fuzzy 2-ECC. Thus by changing the subscripts if necessary, we may assume that x is in S_1.

Assume first that y is not in S_1. Then define a fuzzy 2-ECC in ξ' by taking $S'_1 = S_1 \cup \{x_1\}$, $S'_i = S_i$ for $i = 2, ..., n$ and $S_{n+1}' = \{x_1, x_2, y\}$, $S_{n+2}' = \{x, x_1, x_2, v\}$, $S_{n+3}' = \{y, x_2, v\}$ (note that this is a fuzzy 2-ECC, we need not in S_1, for otherwise y, x_1 are in two sets together).

Suppose next that x and y are both in S_1. Now we define a fuzzy 2-ECC for ξ' by taking $S_1' = S_1 \cup \{x_1, x_2\}$, $S_i' = S_i$ for $i = 2, ..., n$, $S_{n+1}' = \{x_1, x_2, y\}$, $S_{n+2}' = \{x, x_1, x_2, v\}$, $S_{n+3}' = \{y, x_2, v\}$.

Theorem 6: All unicyclic fuzzy graphs except fuzzy cycle of length 4 are fuzzy 2-competition graph.

Proof: We know complete fuzzy graph of four vertices K_4 is 2-competition fuzzy graph. Each n-cycle C_n with $n \neq 4$ a 2-competition fuzzy graph. Suppose that ξ is unicyclic and that the unique cyclic in ξ has length n. If $n \neq 4$, then ξ can be constructed from C_n by pendant edges, and so is a 2-competition fuzzy graph. If $n = 4$ and $\xi \neq C_4$, then ξ can be constructed from C_4 by successively adding pendant edges, and so is a 2-competition fuzzy graph.

References

1. Brigham, R.C., McMorris, F.R., and Vitray, R.P., 1995: Tolerance competition graphs, Linear Algebra and its Application, 217, 41-52.

2. Cohen, J.E., 1968: Interval graphs and food webs: a finding and a problem, Document 17696-PR, RAND Corporation (Santa Monica, CA).

3. Isaak, G., Kim, S.R., McKee, T.A., McMorris, F.R., and Roberts, F.S., 1992: 2-competition graphs, SIAM J. Disc. Math., 5(4), 524-538.

4. Kim, S.R., Graphs with one hole and competition number one, J. Korean Math. Soc, 42(6), 1251-1264.

5. Kim, S.R., McKee, T.A., McMorris, F.R., and Roberts, F.S., 1995: p-competition graphs, Linear Algebra and its Application, 217, 167-178.

6. Mathew, S., and Sunitha, M.S., 2009: Types of arcs in a fuzzy graph, Information Sciences, 179, 1760-1768.

7. Bhutani, K.R., Moderson J., and Rosenfeld, A., 2004: On degrees of end nodes and cut nodes in fuzzy graphs, 1(1), 57-64.

8. A. Nagoorgani and K. Radha, On regular fuzzy graphs, Journal of Physical Sciences, 12, 33-40, 2008.

9. Nagoorgani, A., and Malarvizhi, J., 2009: Isomorphism properties of strong fuzzy graphs, International Journal of Algorithms, Computing and Mathematics, 2(1), 39-47.

10. Nagoorgani, A., and Vadivel, P., Relations between the parameters of independent domination and irredundance in fuzzy graph, International Journal of Algorithms, Computing and Mathematics, 2(1), 15-19.

11. Nagoorgani, A., and Vijayalaakshmi, P., 2011: Insentive arc in domination of fuzzy graph, Int. J. Contemp. Math. Sciences, 6(26), 1303-1309.

12. Nair, P.S., and Cheng, S.C., 2001: Cliques and fuzzy cliques in fuzzy graphs, IFSA World Congress and 20th NAFIPS International Conference, 4, 2277-2280.

13. Nair, P.S., 2008: Perfect and precisely perfect fuzzy graphs, Fuzzy Information Processing Society, 19-22.

14. Natarajan, C., and Ayyasawamy, S.K., 2010: On strong (weak) domination in fuzzy graphs, World Academy of Science, Engineering and Technology, 67, 247-249.

Rough Sets, Fuzzy Sets and Soft Computing
Editor: S. Bhattacharya Halder

Statistically Pre-Cauchy Fuzzy Real-Valued Sequences Defined by Orlicz Function

Amar Jyoti Dutta
Institute of Advanced Study in Science and Technology
Pachim Boragaon, Garchuk, Guwahati, Assam
E-mail: amar_iasst@yahoo.co.in

ABSTRACT

In this article we have defined statistically pre-Cauchy sequence of fuzzy real numbers defined by Orlicz function. We have proved that a sequence $X = (X_k)$ of fuzzy real numbers is statistically pre-Cauchy if and only if

$$\lim_{n \to \infty} \frac{1}{n^2} \sum_{k,\, m \leq n} M\left(\frac{\overline{d}(X_k, X_m)}{\rho}\right) = 0.$$

We have also established some other results.

Keywords: Statistically convergent; statistically pre-Cauchy; Orlicz function; fuzzy real numbers.

AMS Classification No.: 40A05; 46A45; 46E30.

1. INTRODUCTION

The credit goes to Fast (1951) and Schoenberg (1959) who independently extended the concept of convergence of a sequence to statistical convergence with the help of natural density function on the set of positive integers. For a set $K \subset N$ the density is defined by

$$\delta(K) = \lim_{n \to \infty} \frac{1}{n} |\{k \in K: k \leq n\}|,$$

where $|\{k \in K: k \leq n\}|$ denote the numbers of element in the set $\{k \in K: k \leq n\}$.

The notion of statistically pre-Cauchy sequences for real sequences was introduced by Connor et al. (1994).

A real-valued sequence $x = (x_k)$ is said to be statistically convergent to a number L if for each $\varepsilon > 0$, $= 0$, where the vertical bars indicate the number of elements in the enclosed set.

The sequence $x = (x_k)$ is said to be statistically pre-Cauchy if

$$\lim_{n \to \infty} \frac{1}{n^2} |\{(k, p) \leq n: |x_k - x_p| \geq \varepsilon\}| = 0, \text{ for each } \varepsilon > 0.$$

An Orlicz function is defined as $M: [0, \infty) \to [0, \infty)$, which is continuous, non-decreasing and convex with $M(0) = 0$, $M(x) > 0$ for $x > 0$ and $M(x) \to \infty$ as $x \to \infty$. If the convexity of M is replaced by the subadditivity $M(x + y) \leq M(x) + M(y)$ then it is called the modulus function. An Orlicz function may be bounded or unbounded.

Studies on some important Orlicz sequence spaces from different point of view are found in (Esi and Et, 2000), (Esi, 2009), (Khan and Lohani, 2007), (Tripathy and Sarma, 2008).

A fuzzy real number X is a fuzzy set on R, more precisely a mapping $X: R \to I (= [0, 1])$, associating each real number t with its grade of membership $X(t)$, which satisfy the following conditions

(i) X is normal if there exists $t_0 \in R$ such that $X(t_0) = 1$.

(ii) X upper-semi-continuous if for each $\varepsilon > 0$, $X^{-1}([0, a + \varepsilon))$, is open in the usual topology of R, for all $a \in I$.

(iii) X is convex, if $X(t) \geq X(s) \wedge X(r) = min\ (X(s), X(r))$, where $s < t < r$.

The class of all upper-semi-continuous, normal, convex fuzzy real numbers is denoted by $R(I)$.

R can be embedded in $R(I)$, since each $r \in R$ can be regarded as a fuzzy number $\bar{r}$ defined by

$$\bar{r}(t) = \begin{cases} 1, \text{ for } t = r, \\ 0, \text{ otherwise} \end{cases}.$$

The additive identity and multiplicative identity of $R(I)$ are denoted by $\bar{0}$ and $\bar{1}$ respectively.

The α-level set of a fuzzy real number X is defined by

$$[X]_\alpha = \begin{cases} \{t \in R: X(t) \geq \alpha\}, \text{ for } 0 < \alpha \leq 1 \\ \overline{\{t \in R: X(t) > \alpha\}}, \text{ for } \alpha = 0 \end{cases}.$$

Let D be the set of all closed bounded intervals $X = [X^L, X^R]$ then we write $X \leq Y$ if and only if $X^L \leq Y^L$ and $X^R \leq Y^R$. We can write $d(X, Y) = max\{|X^L - Y^L|, |X^R - Y^R|\}$, where $X = [X^L, X^R]$ and $Y = [Y^L, Y^R]$.

It is can be easily verify that (D, d) is a complete metric space.

Consider the mapping $\bar{d}: R(I) \times R(I) \to R$ defined by

$$\bar{d}(X, Y) = \sup_{0 \le \alpha \le 1} d(X^\alpha, Y^\alpha), \text{ for } X, Y \in R(I).$$

Clearly $\bar{d}$ define a metric on $R(I)$.

A fuzzy real-valued sequence (X_k) is a function from the set of natural numbers into $R(I)$.

The set E^F of sequences taken from $R(I)$ is closed under addition and scalar multiplication defined as follows:

Let $(X_k), (Y_k) \in E^F, r \in R$ i.e. $X_k, Y_k \in R(I)$, and for all $k \in N$,

$$(X_k) + (Y_k) = (X_k + Y_k) \in E^F$$

and

$$r(X_k) = (rX_k) \in E^F,$$

where

$$rX_k(t) = \begin{cases} X_k (r^{-1} t), & \text{if } r \ne 0 \\ \bar{0}, & \text{if } r = 0 \end{cases}.$$

Definition 1. A sequence $X = (X_k)$ of fuzzy number is said to be convergent to a fuzzy number X_0 if for $\varepsilon > 0$ there exist a positive integer n_0 such that $\bar{d}(X_k, X_0) < \varepsilon$, for every $k > n_0$.

Definition 2. A fuzzy real-valued sequence (X_k) is said to be bounded if $\sup_k d(X_k, \bar{0}) < \infty$, equivalently, if there exist $\mu \in R(I)^*$, such that $|X_k| \le \mu$ for all $k \in N$, where $R(I)^*$ denotes the set of all positive fuzzy real numbers.

In the recent past different classes of sequences of fuzzy real numbers have been introduced and studied by Altinok et al. (2004), Savas (2000), Mursaleen and Basarir (2003), Esi (2008), Tripathy and Dutta (2007), Dutta (2011) and others.

In this article we have defined statistically pre-Cauchy sequence of fuzzy real numbers and defined by Orlicz function.

A fuzzy real valued sequence $X = (X_k)$ is said to be statistically pre-Cauchy if for each $\varepsilon > 0$,

$$\lim_{n \to \infty} \frac{1}{n^2} |i \le n, j \le n: \bar{d}(X_i, X_j) \ge \varepsilon| = 0.$$

2. MAIN RESULTS

Theorem 1. Let M be an Orlicz function, then the sequence (X_k) of fuzzy real numbers is statistically pre-Cauchy if and only if

$$\lim_{n \to \infty} \frac{1}{n^2} \sum_{k,\,m \leq n} M\left(\frac{\overline{d}(X_k,\,X_m)}{\rho}\right) = 0, \text{ for some } \rho > 0.$$

Proof. Let us suppose that $\lim_{n \to \infty} \dfrac{1}{n^2} \displaystyle\sum_{k,\,m \leq n} M\left(\dfrac{\overline{d}(X_k,\,X_m)}{\rho}\right) = 0$, for some $\rho > 0$.

We have

$$\lim_{n \to \infty} \frac{1}{n^2} \sum_{k,\,m \leq n} M\left(\frac{\overline{d}(X_k,\,X_m)}{\rho}\right) = \lim_{n \to \infty} \frac{1}{n^2} \sum_{\substack{k,\,m \leq n \\ \overline{d}(X_k,\,X_m) < \varepsilon}} M\left(\frac{\overline{d}(X_k,\,X_m)}{\rho}\right)$$

$$+ \lim_{n \to \infty} \frac{1}{n^2} \sum_{\substack{k,\,m \leq n \\ \overline{d}(X_k,\,X_m) \geq \varepsilon}} M\left(\frac{\overline{d}(X_k,\,X_m)}{\rho}\right)$$

$$\geq \lim_{n \to \infty} \frac{1}{n^2} \sum_{\substack{k,\,m \leq n \\ \overline{d}(X_k,\,X_m) \geq \varepsilon}} M\left(\frac{\overline{d}(X_k,\,X_m)}{\rho}\right)$$

$$\geq M(\varepsilon) \frac{1}{n^2} \,|(k,\,m):\, \overline{d}(X_k,\,X_m) > \varepsilon;$$

$$k,\,m \leq n|.$$

Since the L.H.S is zero, we get

$$\frac{1}{n^2}\, |\{(k,\,m):\, \overline{d}(X_k,\,X_m) \leq \varepsilon, \text{ for } k,\,m \leq n\}| = 0.$$

Thus (X_k) is statistically pre-Cauchy.

Conversely suppose that (X_k) is statistically pre-Cauchy. For a given $\varepsilon > 0$, choose δ such that $M(\delta) < \varepsilon/2$ and since M is continuous so there exists a positive integer K such that $M(X) < K/2$, for all X. We have

$$\lim_{n \to \infty} \frac{1}{n^2} \sum_{k,\,m \leq n} M\left(\frac{\overline{d}(X_k,\,X_m)}{\rho}\right) = \lim_{n \to \infty} \frac{1}{n^2} \sum_{\substack{k,\,m \leq n \\ \overline{d}(X_k,\,X_m) < \delta}} M\left(\frac{\overline{d}(X_k,\,X_m)}{\rho}\right)$$

$$+ \lim_{n \to \infty} \frac{1}{n^2} \sum_{\substack{k,\,m \leq n \\ \overline{d}(X_k,\,X_m) \geq \delta}} M\left(\frac{\overline{d}(X_k,\,X_m)}{\rho}\right)$$

$$\leq M(\delta) \lim_{n \to \infty} \frac{1}{n^2} \sum_{\substack{k, m \leq n \\ \bar{d}(X_k, X_m) \geq \delta}} M\left(\frac{\bar{d}(X_k, X_m)}{\rho}\right)$$

$$\leq \frac{\varepsilon}{2} + \frac{K}{2} \frac{1}{n^2} |(k, m): \bar{d}(X_k, X_m) > \delta;$$

for $k, m \leq n|$.

Since X is statistically pre-Cauchy, the R.H.S can be made less than ε. Thus we have

$$\lim_{n \to \infty} \frac{1}{n^2} \sum_{k, m \leq n} M\left(\frac{\bar{d}(X_k, X_m)}{\rho}\right) = 0.$$

This completes the proof.

Theorem 2. Let M be an Orlicz function, then the sequence (X_k) of fuzzy real numbers is statistically convergent to L if and only if

$$\lim_{n \to \infty} \frac{1}{n} \sum_{k=1}^{n} M\left(\frac{\bar{d}(X_k, L)}{\rho}\right) = 0.$$

Proof. Let us suppose that

$$\lim_{n \to \infty} \frac{1}{n} \sum_{k=1}^{n} M\left(\frac{\bar{d}(X_k, L)}{\rho}\right) = 0.$$

We have

$$\lim_{n \to \infty} \frac{1}{n} \sum_{k, m \leq n} M\left(\frac{\bar{d}(X_k, X_m)}{\rho}\right) = \lim_{n \to \infty} \frac{1}{n} \sum_{\substack{k, m \leq n \\ \bar{d}(X_k, X_m) < \varepsilon}} M\left(\frac{\bar{d}(X_k, X_m)}{\rho}\right)$$

$$+ \lim_{n \to \infty} \frac{1}{n} \sum_{\substack{k, m \leq n \\ \bar{d}(X_k, X_m) \geq \varepsilon}} M\left(\frac{\bar{d}(X_k, X_m)}{\rho}\right)$$

$$\geq \lim_{n \to \infty} \frac{1}{n} \sum_{\substack{k, m \leq n \\ \bar{d}(X_k, X_m) \geq \varepsilon}} M\left(\frac{\bar{d}(X_k, X_m)}{\rho}\right)$$

$$\geq M(\varepsilon) \frac{1}{n} |\{k \leq n: \bar{d}(X_k, X_m) \geq \varepsilon\}|.$$

Since $\displaystyle\lim_{n \to \infty} \frac{1}{n} \sum_{k=1}^{n} M\left(\frac{\overline{d}(X_k, L)}{\rho}\right) = 0$,

we get $\dfrac{1}{n} |\{k \leq n: \overline{d}(X_k, L) \geq \varepsilon\}| = 0$.

Thus (X_k) is statistically convergent to L. Converse part can be proved in a similar manner to the previous theorem.

Theorem 3. A statistically convergent fuzzy real valued sequence is statistically pre-Cauchy.

Proof. Let the fuzzy real-valued sequence (X_k) is statistically convergent. Let $A \subset N$ be such that $\delta(A) = 1$. Select $B \subset A$ such that $A\backslash B$ is finite and $B \times B \subset \{(j, k): \overline{d}(X_k, X_m) < \varepsilon\}$ for some $\varepsilon > 0$.

Now for $n \in N$, we have

$$[\delta(B)]^2 = \frac{1}{n^2} \sum_{k, m \leq n} \chi_{B \times B} (k, m) \leq \frac{1}{n^2} |\{(k, m): \overline{d}(X_k, X_m) < \varepsilon; k, m \leq n\}|.$$

Since, $\displaystyle\lim_{n \to \infty} \frac{1}{n} [\delta(B)] = 1$

$$\Rightarrow \lim_{n \to \infty} \frac{1}{n^2} |\{(k, m): \overline{d}(X_k, X_m) < \varepsilon; k, m \leq n\}| = 1.$$

This completes the proof.

References

1. Altinok, H., Altin, Y., and Et, M., 2004: *Lacunary almost statistical convergence of fuzzy numbers*, Thai Jour. Math, **2**(2), 265-274.

2. Connor, J.S., Fridy, J., and Kline, J., 1994: *Statistically Pre- Cauchy Sequences*, Analysis, **14**, 311-317.

3. Dutta, A.J., 2011: *Lacunary p-absolutely summable sequence of fuzzy real number*, Fasciculi Jour. Math., **46**, 58-64.

4. Esi, A., and Et, M., 2000: *Some new sequence spaces defined by a sequnce of Orliccz functions*, Indian Jour. Pure & Appl. Math., **31**(8), 967-972.

5. Et, M., 2001: *On some new Orlicz sequence spaces*, Jour. Analysis, **9**, 21-28.

6. Esi, A., 2008: *On some new classes of sequences of fuzzy numbers*, Int. Jour. Math. Anal, **2**(17), 837-844.

7. Esi, A., 2009: *Generalized Difference Sequence Spaces Defined by Orlicz Functions*, General Math., **17**(2), 53-66.

8. Fast, H., 1951: *Sur la Convergence Statistique*, Colloq. Math., **2**, 241-244.

9. Fridy, J.A., 1985: *On Statistical Convergence*, Anal., **5**, 301-313.

10. Khan, V.A., and Lohani, Q.M., 2007: *Statistically Pre-Cauchy Sequences and Orlicz Functions*, Southeast Asian Bull.Math., **31**, 1107-1112.

11. Mursaleen, M., and Basarir, M., 2003: *On some new sequence spaces of fuzzy numbers*. Indian Jour. Pure Appl Math, **34**(9), 1351-1357.

12. Savas, E., 2000: *A note on sequence of Fuzzy numbers*, Inf. Sc., **124**, 297-300.

13. Schoenberg, I.J., 1959: *The integrability of certain functions and related summability methods*, Amer. Math. Monthly, **66**, 361-375.

14. Tripathy, B.C., and Sarma, B., 2008: *Sequence spaces of fuzzy real numbers defined by Orlicz functions*, Math. Slov., **58**(5), 621-628.

15. Tripathy, B.C., and Dutta, A.J., 2007: *Statistically convergent and Cesàro summable double sequences of fuzzy real numbers*; Soochow Jour. Math., **33**(4), 835-848.

Rough Sets, Fuzzy Sets and Soft Computing
Editor: S. Bhattacharya Halder

Magnetic Field Effect on Double Diffusive Free Convection in a Partially Heated Enclosure

Salma Parvin[*], **Rehena Nasrin, M.A. Alim and N.F. Hossain**
*Department of Mathematics, Bangladesh University of
Engineering & Technology, Dhaka, Bangladesh
E-mail: salpar@math.buet.ac.bd*[*]

ABSTRACT

The double diffusive natural convection inside a partially heated square enclosure in presence of transverse magnetic field is investigated numerically. Energy and concentration equations take into account of Dufour and Soret effects respectively. The governing differential equations are transformed into a set of non-linear coupled ordinary differential equations and solved using similarity analysis with finite element technique using appropriate boundary conditions. The variations in the streamlines, isotherms and isoconcentrations for various Hartmann number are depicted graphically and analyzed in detail. The present work is compared with previously published work and found good agreement. Numerical results for average Nusselt and Sherwood numbers, average temperature and concentration and velocities in the chamber are presented as functions of the governing parameter mentioned above.

Keywords: MHD, double-diffusive natural convection, finite element method.

1. INTRODUCTION

The natural convection in enclosures continues to be a very active area of research during the past few decades. While a good number of works have made significant contributions for the development of the theory, an equally good number of works have been devoted to many engineering applications that include electronic or computer equipment, thermal energy storage systems and etc. We shall refer to a few important works that may serve as background for the present work.

Double diffusive convection of water has been studied by Bahloul et al. [1], Nithyadevi and Yang [2], Sezai and Mohamad [3], Sivasankaran and Kandaswamy [4, 5]. Yet, most work done considers flow inside closed enclosures, the applications included, such as pollution dispersion in lakes, chemical deposition, and melting and solidification process. Diffusion of matter caused by temperature gradients (Soret effect) and diffusion of heat caused by concentration gradients (Dufour effect) become very significant when the temperature and concentration gradients are very large. Generally these effects are considered as second order phenomenon. These effects may become important in some applications such as the solidification of binary alloys, groundwater pollutant migration, chemical reactors, and geosciences. The importance of these effects has also seen in Mansour et al. [6], Joly et al. [7], Platten [8] and Partha et al. [9].

Double diffusive and Soret induced convection in a shallow horizontal enclosure is analytically and numerically studied by Bahloul et al. [1] and also studied numerically by Mansour et al. [6]. They found that the Nusselt number has decreases in general with the Soret parameter while the Sherwood number increases or decreases with this parameter depending on the temperature gradient induced by each solution. Joly et al. [7] made the Soret effect on natural convection in a vertical enclosure. They analyzed the particular situation where the buoyancy forces induced by the thermal and solutal effects are opposing each other and of equal intensity.

In the above studies convection heat transfer is due to the imposed temperature gradient between the opposing walls of the enclosure taking the entire vertical wall to be thermally active. But in many naturally occurring situations and engineering applications it is only a part of the wall which is thermally active. For example in solar energy collectors due to shading, it is only the unshaded part of the wall that is thermally active. In order to have the results to possess applications, it is essential to study heat transfer in an enclosure with partially heated active walls. Only a few studies are reported in the literature concerning heat transfer in enclosures with partially heated side walls, by Oztop [10], Frederick and Quiroz [11] and Erbay et al. [12].

Natural convection in an enclosure with partially active walls is studied by Nithyadevi et al. [13, 14] and Kandaswamy et al. [15] without Soret and Dufour effects. Present study deals with the natural convection in a square enclosure filled with water and partially heated vertical walls for three different combinations of heating location in the presence of solute concentration with Soret and Dufour effects. The hot region is located at the top, middle and bottom of the left vertical wall of the enclosure. Effects of SoretDufour, chemical reaction and thermal radiation on MHD non-Darcy unsteady mixed convective heat and mass transfer over a stretching sheet was investigated by Pal and Mondal [16]. The author used shooting algorithm with Runge–Kutta–Fehlberg integration scheme to solve the governing equations. Parvin and Hossain [17] analyzed numerically the Joule-heating effect on magneto-hydrodynamic mixed convection, the flow and heat transfer characteristics in a lid-driven cavity with a sinusoidal wavy bottom surface. They found significant dependency of heat transfer on magnetic field and number of waves. Chamkha and Naser [18] performed analysis on Hydromagnetic Double-Diffusive Convection in a Rectangular Enclosure with Opposing Temperature and Concentration Gradients. Magnetohydrodynamic mixed convection heat transfer across vertical wavy (sinusoidal and triangular) isothermal channels are investigated numerically by Parvin et al.[19]. Their result reveals that the rate of heat transfer is more effective in a sinusoidal wavy channel than in a triangular one.

Present work analyzes numerically the effect of MHD on double-diffusive natural convection in a partially heated enclosure. The governing differential equations are solved by using Finite Element Method of Galerkin's weighted residual scheme. The effect of the Hartmann number (Ha) on the flow pattern, heat and mass transfer has been depicted. Comprehensive average Nusselt and Sherwood numbers, average temperature, concentration and velocity in the cavity are presented as functions of Ha.

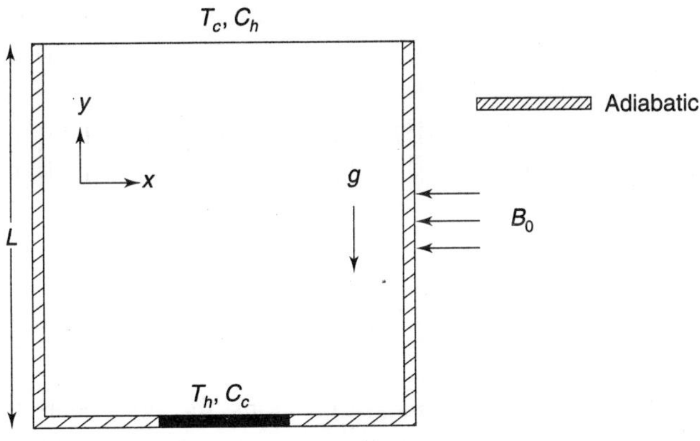

Fig. 1   Schematic diagram of the enclosure

2. PROBLEM FORMULATION

Figure 1 shows a schematic diagram of a partially heated inclined square enclosure of side L. The top wall has constant temperature T_c, while the bottom wall is partially heated with temperature T_h, where $T_h > T_c$ condition is sustained. The concentration at the top wall is maintained higher than bottom wall ($C_h > C_c$). The remaining bottom wall and the two side walls are considered adiabatic and impermeable. Uniform transverse magnetic field of strength B_0 is applied parallel to x-axis.

Two-dimensional, nonlinear, steady, MHD laminar flow of a viscous, incompressible and electrically conducting fluid inside a cavity in the presence of a transverse magnetic field is considered for investigation. The fluid properties are assumed constant except for the density variation which is treated according to Boussinesq approximation while viscous dissipation effects are considered negligible. The governing equations describing the present problem are as follows:

Continuity Equation:

$$\frac{\partial u}{\partial x} + \frac{\partial v}{\partial y} = 0 \tag{1}$$

Momentum Equations:

$$u \frac{\partial u}{\partial x} + v \frac{\partial u}{\partial y} = -\frac{1}{\rho} \frac{\partial p}{\partial x} + \nu \left(\frac{\partial^2 u}{\partial x^2} + \frac{\partial^2 u}{\partial y^2} \right) \tag{2}$$

$$u \frac{\partial u}{\partial x} + v \frac{\partial u}{\partial y} = -\frac{1}{\rho} \frac{\partial p}{\partial y} + \nu \left(\frac{\partial^2 u}{\partial x^2} + \frac{\partial^2 u}{\partial y^2} \right)$$

$$+ g \left[\beta_T \left(T - T_c \right) + \beta_c \left(c - C_c \right) \right] - \frac{\sigma B_0^2 v}{\rho} \tag{3}$$

Energy Equation:

$$u \frac{\partial T}{\partial x} + v \frac{\partial T}{\partial y} = \alpha \left(\frac{\partial^2 T}{\partial x^2} + \frac{\partial^2 T}{\partial y^2} \right) + \frac{Dk}{c_p \, c_s} \left(\frac{\partial^2 c}{\partial x^2} + \frac{\partial^2 c}{\partial y^2} \right) \tag{4}$$

Diffusion Equation:

$$u \frac{\partial C}{\partial x} + v \frac{\partial C}{\partial y} = D \left(\frac{\partial^2 c}{\partial x^2} + \frac{\partial^2 c}{\partial y^2} \right) + \frac{Dk}{T_m} \left(\frac{\partial^2 T}{\partial x^2} + \frac{\partial^2 T}{\partial y^2} \right) \tag{5}$$

where x and y are the distances measured along the horizontal and vertical directions respectively; u and v are the velocity components in the x and y directions respectively; T denote the fluid temperature, p is the pressure and ρ is the fluid density, g is the gravitational constant, β is the volumetric coefficient of thermal expansion, c_p is the fluid specific heat, k is the thermal conductivity of fluid.

The boundary conditions for the present problem are specified as follows:

at all solid boundaries $u = v = 0$

at the portion ($0.3L$-$0.7L$) of the bottom surface, $T = T_h$, $c = C_c$

at the upper wall, $T = T_c$, $c = C_h$

at the remaining boundaries $\dfrac{\partial T}{\partial n} = 0$, $\dfrac{\partial c}{\partial n} = 0$

The above equations are non-dimensionalized upon incorporating the following dimensionless variables:

$$X = \frac{x}{L}, \; Y = \frac{y}{L}, \; U = \frac{uL}{v}, \; V = \frac{vL}{v}, \; P = \frac{pL^2}{\rho v^2}, \; \theta = \frac{T - T_c}{T_h - T_c}, \; C = \frac{c - C_c}{C_h - C_c}$$

where X and Y are the coordinates varying along horizontal and vertical directions respectively, U and V are the velocity components in the X and Y directions respectively, θ is the dimensionless temperature, C is the dimensionless concentration and P is the dimensionless pressure.

After substituting the dimensionless variables into the equations (1-5), we get the following dimensionless equations

$$\frac{\partial U}{\partial X} + \frac{\partial V}{\partial Y} = 0 \tag{6}$$

$$U \frac{\partial U}{\partial X} + V \frac{\partial V}{\partial Y} = -\frac{\partial P}{\partial X} + \left(\frac{\partial^2 U}{\partial X^2} + \frac{\partial^2 U}{\partial Y^2} \right) \tag{7}$$

$$U \frac{\partial V}{\partial X} + V \frac{\partial V}{\partial Y} = -\frac{\partial P}{\partial Y} + \left(\frac{\partial^2 V}{\partial X^2} + \frac{\partial^2 V}{\partial Y^2} \right) + \frac{Ra_T}{Pr} (\theta - NC) - Ha^2 V \tag{8}$$

$$U \frac{\partial \theta}{\partial X} + V \frac{\partial \theta}{\partial Y} = \frac{1}{Pr} \frac{\partial^2 \theta}{\partial X^2} + \frac{\partial^2 \theta}{\partial Y^2} + D_f \left(\frac{\partial^2 C}{\partial X^2} + \frac{\partial^2 C}{\partial Y^2} \right) \tag{9}$$

$$U \frac{\partial C}{\partial X} + V \frac{\partial C}{\partial Y} = \frac{1}{Sc} \frac{\partial^2 C}{\partial X^2} + \frac{\partial^2 C}{\partial Y^2} + S_r \left(\frac{\partial^2 \theta}{\partial X^2} + \frac{\partial^2 \theta}{\partial Y^2} \right) \tag{10}$$

where $Pr = \dfrac{v}{\alpha}$ is the Prandtl number, $Sc = \dfrac{v}{D}$ is the Schmidt number, $Ra_T = \dfrac{g \, \beta_T \, L^3 \, (T_h - T_c)}{v^2}$ is the thermal Rayleigh number,

$Ra_c = \dfrac{g \, \beta_c \, L^3 \, (C_h - C_c)}{v^2}$ is the solutal Rayleigh number, $D_f = \dfrac{Dk \, (C_h - C_c)}{C_s \, C_p \, v \, (T_h - T_c)}$

is the Dufour and $S_r = \dfrac{Dk \, (T_h - T_c)}{T_m \, v \, (C_h - C_c)}$ is the Soret coefficients and $N = \dfrac{Ra_c}{Ra_T}$ is

the Buoyancy ratio number and Ha is the Hartmann number which is defined

as $Ha^2 = \dfrac{\sigma \, B_0^2 \, L^2}{\mu}$.

The dimensionless boundary conditions take the following forms:

at all solid boundaries $U = V = 0$

at the portion (0.3L-0.7L) of the bottom surface, $\theta = 1, C = 0$

at the top wall, $\theta = 0$, $C = 1$

at the remaining boundaries $\dfrac{\partial \theta}{\partial N} = 0$, $\dfrac{\partial C}{\partial N} = 0$

The average Nusselt and Sherwood numbers at the heated and concentrated surfaces of the enclosure may be expressed, respectively as

$$Nu = -\frac{1}{S}\int_0^S \frac{\partial \theta}{\partial N}\, dS$$

and

$$Sh = -\int_0^1 \frac{\partial C}{\partial N}\, dS.$$

where the dimensionless temperature and concentration gradients are $\dfrac{\partial \theta}{\partial N} = \sqrt{\left(\dfrac{\partial \theta}{\partial X}\right)^2 + \left(\dfrac{\partial \theta}{\partial Y}\right)^2}$ and $\dfrac{\partial C}{\partial N} = \sqrt{\left(\dfrac{\partial C}{\partial X}\right)^2 + \left(\dfrac{\partial C}{\partial Y}\right)^2}$ respectively and S is the dimensionless length of the heated or concentrated surface. The average temperature, concentration, and velocity in the cavity are defined as $\theta_{av} = \int \theta \, \dfrac{d\overline{V}}{\overline{V}}$, $C_{av} = \int C \, \dfrac{d\overline{V}}{\overline{V}}$ and $\mathbf{V}_{av} = \int \mathbf{V} \, \dfrac{d\overline{V}}{\overline{V}}$ respectively, where $\overline{V}$ is the cavity volume.

3. NUMERICAL IMPLEMENTATION

The Galerkin finite element method [20, 21] is used to solve the non-dimensional governing equations along with boundary conditions for the considered problem. The equation of continuity has been used as a constraint due to mass conservation and this restriction may be used to find the pressure distribution. The penalty finite element method [22] is used to solve the Eqs. (2) - (4), where the pressure P is eliminated by a penalty constraint. The continuity equation is automatically fulfilled for large values of this penalty constraint. Then the velocity components (U, V), temperature (θ) and concentration (C) are expanded using a basis set. The Galerkin finite element technique yields the subsequent nonlinear residual equations. Three points Gaussian quadrature is used to evaluate the integrals in these equations. The non-linear residual equations are solved using Newton–Raphson method to determine the coefficients of the expansions. The convergence of solutions is assumed when the relative error for each variable between consecutive iterations is recorded below the convergence criterion ε such that $|\psi^{n+1} - \psi^n| \leq 10^{-4}$, where n is the number of iteration and ψ is a function of U, V, θ and C.

4. MESH GENERATION

In the finite element method, the mesh generation is the technique to subdivide a domain into a set of sub-domains, called finite elements, control volume, etc. The discrete locations are defined by the numerical grid, at which the variables are to be calculated. It is basically a discrete representation of the geometric domain on which the problem is to be solved. The computational domains with irregular geometries by a collection of finite elements make the method a valuable practical tool for the solution of boundary value problems arising in various fields of engineering. Figure 2 displays the finite element mesh of the present physical domain.

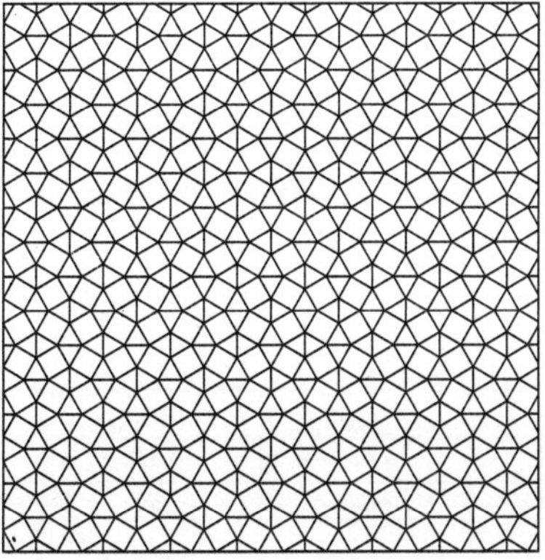

Fig. 2 Grid generation of the cavity

4.1 Grid Independent Test

An extensive mesh testing procedure is conducted to guarantee a grid-independent solution for $Ra = 10^4$, $Pr = 6.2$, $D_f = S_r = 0.5$, $Sc = 5$, $Ha = 20$ and $J = 1$ in a partially heated square enclosure. In the present work, we examine five different non-uniform grid systems with the following number of nodes and elements within the resolution field: 4136, 2569; 10369, 4730; 18626, 6516; 37341, 8457 and 47735, 10426. The numerical scheme is carried out for highly precise key in the average Nusselt (Nu) and Sherwood (Sh) numbers for the aforesaid nodes and elements to develop an understanding of the grid fineness as shown Table 1 and in Fig. 3. The scale of the average Nusselt and Sherwood numbers for 37341 nodes and 8457 elements shows a little difference with the results obtained for the other nodes and elements. Hence, considering the non-uniform grid system of 37341 nodes and 8457 elements is preferred for the computation.

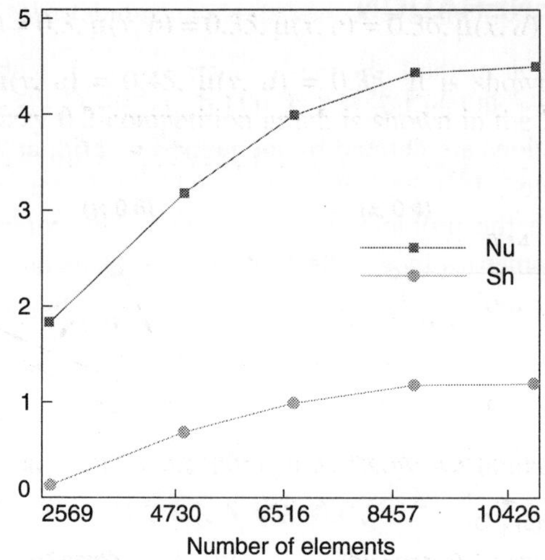

Fig. 3 Grid test for the geometry

Table 1 Grid Sensitivity Check at $Ra = 10^4$, $Pr = 6.2$, $D_f = S_r = 0.5$, $Sc = 5$ and $Ha = 20$

Nodes (elements)	4136 (2569)	10369 (4730)	18626 (6516)	37341 (8457)	47735 (10426)
Nu	1.82945	3.18176	3.98701	4.420435	4.490436
Sh	0.12945	0.68176	0.98701	1.173416	1.193416
Time (s)	225.261	289.587	382.531	412.283	519.572

5. RESULTS AND DISCUSSION

In this section, the effects of double diffusion and MHD with Joule heating on natural convection in a partially heated tilted square enclosure are explored. $Ra = Ra_T = Ra_c$ is assumed for the present numerical calculation. Numerical results of streamlines, isotherms and iso-concentrations for various Hartman number (Ha) are displayed. The considered values of the parameters are $Ra = 10^4$, $Pr = 6.2$, $S_r = D_f = 0.5$, $Sc = 5$ and $Ha = (0, 20, 40$ and $60)$. In addition, the values of the average Nusselt and Sherwood numbers as well as the average temperature, concentration and velocity in the domain have been calculated.

The influences of Hartmann number Ha on the streamlines, isotherms and iso-concentrations with $Ra = 10^4$, $Pr = 6.2$, $S_r = D_f = 0.5$ and $Sc = 5$ are shown in Fig. 4(a)–(c). At $Ha = 0$, the fluid flow is characterized by a primary rotating unicellular vortex of the size of the cavity generated by the movement of the fluid. At higher values of Ha, the flow pattern remains almost similar except the core region changes its shape from circular to oval. It is evident from the figures that the strength of the circulation reduces with higher values of *Ha*. This is due to the fact that the magnetic field creates the Lorentz force which acts perpendicular to the flow and tends to retard the fluid motion in the bulk of the cavity. It is observed that the thermal and concentrated boundary layer near the heated and concentrated deceased with elevating the magnetic field strength. In addition,

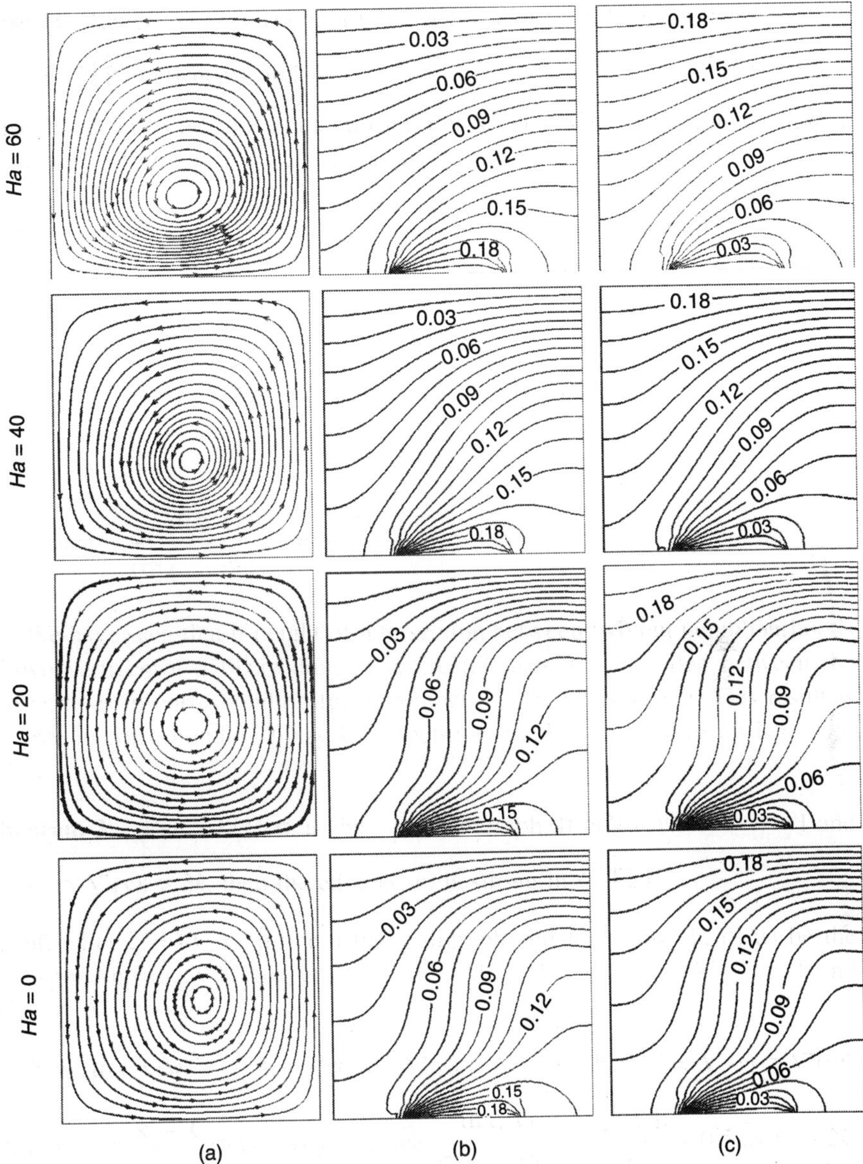

Fig. 4 Effect of *Ha* on (a) streamlines, (b) Isotherms and (c) Iso-concentration at $Ra = 10^4$, $Pr = 6.2$, $D_f = S_r = 0.5$ and $Sc = 5$.

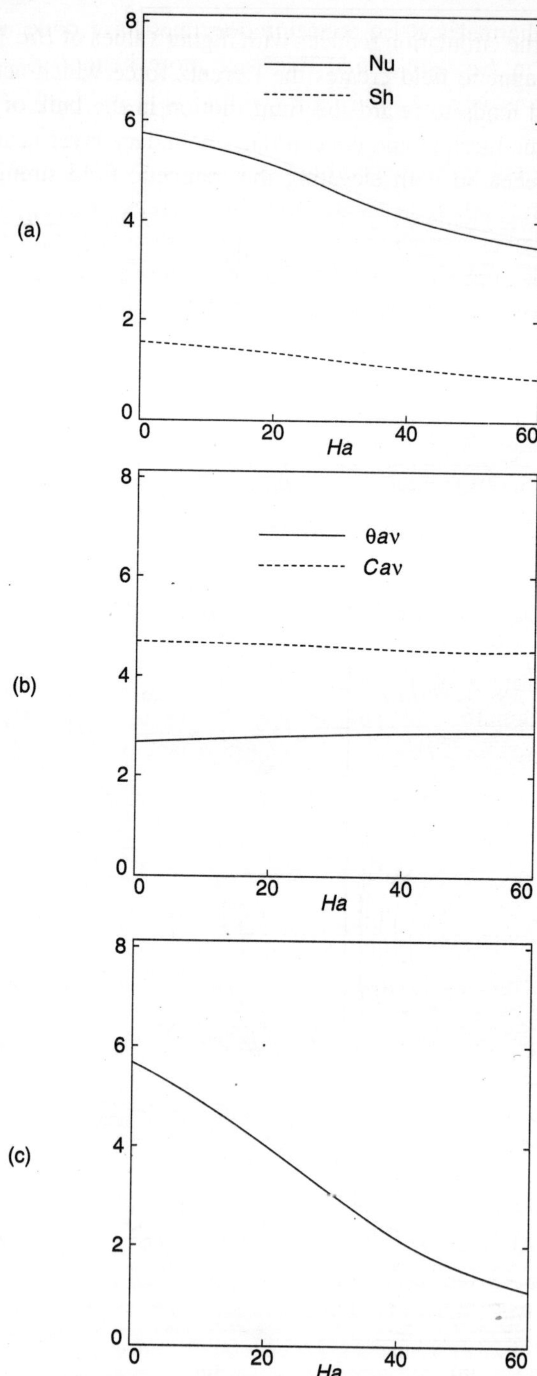

Fig. 5 Effect of *Ha* on (a) Nu and Sh, (b) θ_{av} and C_{av} and (c) V_{av} at $Ra = 10^4$, $Pr = 6.2$, $D_f = S_r = 0.5$ and $Sc = 5$.

pattern of isothermal and iso concentration lines have noticeable changes with the variation in *Ha*. Isothermal lines are more bended for lower Hartmann number where as they are almost parallel for higher values of Hartmann number. Iso concentration lines have the similar changes as isotherms due to variation of *Ha*.

From Fig 5 (a), it is found that the average Nusselt number (*Nu*) and Sherwood number (Sh) decreases mildly with increasing values of *Ha*. *Nu* and *Sh* are the uppermost at *Ha* = 0. However, Fig 5(b) shows that the mean temperature (θ_{av}) and concentration (C_{av}) of the fluid in the cavity has slight increment as *Ha* increases. Average velocity distributions inside the cavity for *Ha* variation is displayed in Fig 5(c). Velocity of the fluid particle deceased with greater magnetic field strength. At higher magnetic field, fluid flow is retarded and convection is suppressed. Conduction becomes the dominant mode of heat transfer and thereby higher magnetic field reduces both heat and mass transfer.

6. CONCLUSION

Effects of MHD on double diffusive natural convection in a partially heated cavity are studied numerically. Influences of Hartmann number is highlighted to study its impact on flow structure, heat and mass transfer characteristics while Rayleigh number, Prandtl number, Soret-Dufour coefficients and Schmidt number are kept fixed. It is noticed that the flow behavior and the heat transfer characteristics inside the cavity are strongly depending upon the strength of the magnetic field. In the absence of the magnetic force, the buoyancy force is dominant resulting better convective heat transfer performance. Increasing Hartmann number retards the fluid movement causing the lower temperature gradient as well as concentration gradient throughout the cavity. Therefore, the heat is transferred mainly by conduction. On the other hand, mean fluid temperature and concentration are observed least for the lowest value of magnetic parameter *Ha*.

References

1. Bahloul, A., Boutana, N., and Vasseur, P., 2003: *Double-diffusive and Soret-induced convection in a shallow horizontal porous layer*, J. Fluid Mech., Vol. 491, pp. 325-352.

2. Nithyadevi, N., and Yang, R.J., 2009: *Double diffusive natural convection in a partially heated enclosure with Soret and Dufour effects*. Int. J. Heat and Fluid Flow, Vol. 30, pp. 902-910.

3. Sezai, I., and Mohamad, A.A., 2000: *Double diffusive convection in a cubic enclosure with opposing temperature and concentration gradients*, Phys. Fluids, Vol. 12, pp. 2210–2223.

4. Sivasankaran, S., and Kandaswamy, P., 2006: *Double diffusive convection of water in a rectangular partitioned enclosure with temperature dependent species diffusivity.* Int. J. Fluid Mech. Res., Vol. 33, pp. 345-361.

5. Sivasankaran, S., and Kandaswamy, P., 2007: *Double diffusive convection of water in a rectangular partitioned enclosure with concentration dependent species diffusivity.* J. Korean Soc. Industrial Appl. Math., Vol. 11, pp. 71-83.

6. Mansour, A., Amahmid, A., Hasnaoui, M., and Bourich, M., 2006: *Multiplicity of solutions induced by thermosolutal convection in a square porous cavity heated from below and submitted to horizontal concentration gradient in the presence of Soret effect.* Num. Heat Trans., Vol. 49, pp. 69-94.

7. Joly, F., Vasseur, P., and Labrosse, G., 2000: *Soret-driven thermosolutal convection in a vertical enclosure.* Int. Commun. Heat Mass Trans., Vol. 27, pp. 755-764.

8. Platten, J.K., 2006: *The Soret effect: a review of recent experimental results.* J. Appl. Mech., Vol. 73, pp. 5-15.

9. Partha, M.K., 2006: Murthy, P.V.S.N., Raja Sekhar, G.P., *Soret and Dufour effects in a non-darcy porous medium.* J. Heat Trans., Vol. 128, pp. 605-610.

10. Oztop, H.F., 2007: *Natural convection in partially cooled and inclined porous rectangular enclosures.* Int. J. Thermal Sci., Vol. 46, pp. 149-156.

11. Frederick, R.L., and Quiroz, F., 2001: *On the transition from conduction to convection regime in a cubical enclosure with a partially heated wall.* Int. J. Heat Mass Trans., Vol. 44, pp. 1699-1709.

12. Erbay, B., Altac, Z., and Sulus, B., 2004: *Entropy generation in a square enclosure with partial heating from a vertical lateral wall.* Heat Mass Trans., Vol. 40, pp. 909-918.

13. Nithyadevi, N., Kandaswamy, P., and Sivasankaran, S., 2006: *Natural convection in a square cavity with partially active vertical walls: time periodic boundary condition.* Math. Prob. Eng., pp. 1-16.

14. Nithyadevi, N., Kandaswamy, P., and Lee, J., 2007: *Natural convection in a rectangular cavity with partially active side walls.* Int. J. Heat Mass Trans., Vol. 50, pp. 4688-4697.

15. Kandaswamy, P., Sivasankaran, S., and Nithyadevi, N., 2007: *Buoyancy-driven convection of water near its density maximum with partially active vertical walls.* Int. J. Heat Mass Trans., Vol. 50, pp. 942-948.

16. Pal, D., and Mondal, H., 2011: *Effects of SoretDufour, chemical reaction and thermal radiation on MHD non-Darcy unsteady mixed convective heat and mass transfer over a stretching sheet,* Communications in Nonlinear Science and Numerical Simulation, Vol. 16, no. 4, pp.1942-1958.

17. Salma Parvin, and Hossain, N.F., 2011: *Investigation on the Conjugate Effect of Joule heating and Magnetic Field on Combined Convection in a Lid-driven Cavity with Undulated Bottom Surface,* Journal of Advanced Science and Engineering Research, Vol. 1, pp. 210-223.

18. Chamkha, A.J., and Al-Naser, H., 2002: *Hydromagnetic Double-Diffusive Convection in a Rectangular Enclosure with Opposing Temperature and Concentration Gradients.* International Journal of Heat and Mass Transfer, Volume, Vol. 44, pp. 2465-2483.

19. Salma Parvin, Alim, M.A., and Hossain, N.F., 2011: *MHD mixed convection heat transfer through vertical wavy isothermal channels,* International Journal of Energy & Technology, Vol. 3 (34), pp. 1-9.

20. Taylor, C., and Hood, P., 1973: *A numerical solution of the Navier-Stokes equations using finite element technique,* Computer and Fluids, Vol. 1, pp. 73-89.

21. Dechaumphai, P., 1999: *Finite Element Method in Engineering,* 2nd ed., Chulalongkorn University Press, Bangkok.

22. Basak, T., Roy, S., and Pop, I., 2009: *Heat flow analysis for natural convection within trapezoidal enclosures based on heatline concept,* Int. J. Heat Mass Transfer, Vol. 52, pp. 2471–2483.

Nomenclature

c Dimensional concentration (kg m^{-3})

C Non-dimensional concentration

C_p Specific heat at constant pressure (kJ kg^{-1} K^{-1})

C_s Concentration susceptibility

D Solutal diffusivity (m^2 s^{-1})

D_f Dufour parameter

g Gravitational acceleration (m s^{-2})

h Local heat transfer coefficient (W m^{-2} K^{-1})

Ha Hartmann number

k Thermal conductivity (W m^{-1} K^{-1})

K_T Thermal diffusion ratio

L Lengh of the enclosure (m)

Nu Nusselt number

Pr Prandtl number

Sc Schmidt numbe

Sh Sherwood number

S_r Soret parameter

Ra Rayleigh number

T Dimensional temperature (°K)

u, v Dimensional x and y components of velocity (m s^{-1})

U, V Dimensionless velocities

X, Y Dimensionless coordinates
x, y Dimensional coordinates (m)

Greek Symbols

α Fluid thermal diffusivity (m^2 s^{-1})
β Thermal expansion coefficient (K^{-1})
θ Dimensionless temperature, $\theta = (T - T_c)/(T_h - T_c)$
μ Dynamic viscosity (N s m^{-2})
ν Kinematic viscosity (m^2 s^{-1})
ρ Density (kg m^{-3})

Subscripts

av average
c cold
h hot
m mean

Convective Flow in a Solar Collector: Effect of Aspect Ratio

Rehena Nasrin[*], Salma Parvin and M.A. Alim
Department of Mathematics, Bangladesh University of Engineering
& Technology, Dhaka, Bangladesh
E-mail: rehena@math.buet.ac.bd[*]

ABSTRACT

Solar energy is one of the best sources of renewable energy with minimal environmental impact. To investigate the natural convection inside a solar collector having the flat-plate cover and the sine-wave absorber a numerical study has been conducted. The governing differential equations with boundary conditions are solved by Finite Element Method using Galerkin's weighted residual scheme. The effect of physical parameter namely collector aspect ratio (Ar) on the natural convection heat transfer is simulated. Comprehensive average Nusselt number, average temperature and mean velocity field for working fluid within the collector are presented as functions of the parameter mentioned above. Comparison with the previously published work is made and found to be an excellent agreement. The numerical results show that the highest heat transfer rate is observed for the largest Ar. In addition, the design for enhancing the performance of the collector is determined by examining the above mentioned results.

Keywords: Natural convection, solar collector, finite element method, sinusoidal wavy absorber.

1. INTRODUCTION

The natural convection in enclosures continues to be a very active area of research during the past few decades. Because of the desirable environmental and safety aspects it is widely believed that solar energy should be utilized instead of other alternative energy forms, even when the costs involved are slightly higher. The flat-plate solar collector is commonly used today for the collection of low temperature solar thermal energy. Solar collectors are key elements in many applications, such as building heating systems, solar drying devices, etc. Solar energy has the greatest potential of all the sources of renewable energy especially when other sources in the country have depleted.

To increase the efficiency of the solar water heater there are so many methods introduced by Xiaowu and Hua [1] and Hussain [2]. But the novel approach is to introduce the nanofluids in solar water heater instead of conventional heat transfer fluids (like water). The poor heat transfer properties of these conventional fluids compared to most solids are the primary obstacle to the high compactness and effectiveness of the system. The essential initiative is to seek the solid particles having thermal conductivity of several hundred times higher than those of conventional fluids.

For shortwave solar radiation surface the absorptance of the collector depends on the nature and colour of the coating and on the incident angle. Usually black colour is used. Various colour coatings had been proposed by Orel et al. [3] and Tripanagnostopoulos et al. [4] mainly for aesthetic reasons. A low-cost mechanically manufactured selective solar absorber surface method had been proposed by Konttinen et al. [5]. A numerical experiment is performed for inclined solar collectors by Varol and Oztop [6]. Bég et al. [7] performed non-similar mixed convection heat and species transfer along an inclined solar energy collector surface with cross diffusion effects, where the resulting governing equations were transformed and then solved numerically using the local nonsimilarity method and Runge-Kutta shooting quadrature. Nasrin et al. [8] performed effects of physical parameters on natural convection in a solar collector. They concluded that the highest heat transfer rate was observed for the largest value of number of wave and wave amplitude.

In this paper, we investigate numerically the aspect ratio (length/height) effect on the natural convection inside a flat plate solar collector having sinusoidal wavy absorber. The objective of this paper is to present flow and heat transfer used to harness solar energy.

2. PROBLEM FORMULATION

Figure 1 shows a schematic diagram of a solar collector. The fluid in the computation domain is air. The fluid is assumed incompressible and the flow is considered to be laminar. The solar collector is a metal box with a glass cover plate on the top surface and a dark colored sinusoidal wavy absorber plate on the bottom. The length and height of the collector are L and H respectively. The top transparent surface is continuously absorbing solar energy. The transparent cover plate has initially maintained at a constant temperature T_w, while bottom wavy wall is at temperature T_c, with $T_w > T_c$. The density of the nanofluid is approximated by the Boussinesq model. Amplitude of wave $A_m = 0.075$ and number of wave $\lambda = 3.5$ are assumed. Only steady state case is considered.

The governing equations for laminar natural convection inside a solar collector in terms of the Navier-Stokes and energy equation (dimensional form) are given as:

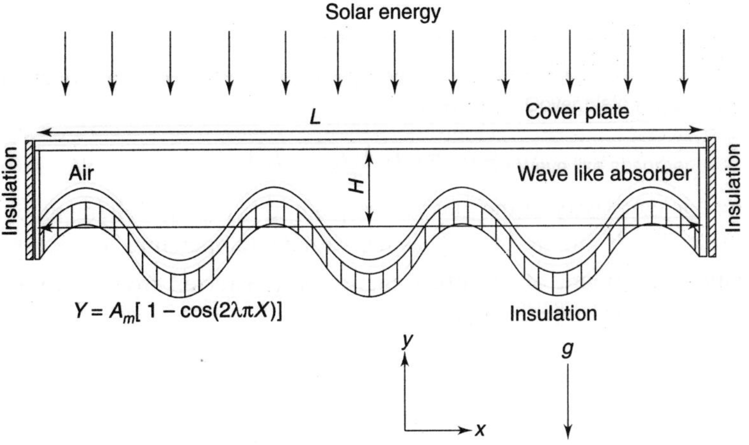

Fig. 1 Schematic diagram of the solar collector

Continuity equation:

$$\frac{\partial u}{\partial x} + \frac{\partial v}{\partial y} = 0 \tag{1}$$

x-momentum equation:

$$\rho \left(u\,\frac{\partial u}{\partial x} + v\,\frac{\partial u}{\partial y} \right) = -\frac{\partial p}{\partial x} + \mu \left(\frac{\partial^2 u}{\partial x^2} + \frac{\partial^2 u}{\partial y^2} \right) \tag{2}$$

y-momentum equation:

$$\rho \left(u\,\frac{\partial u}{\partial x} + v\,\frac{\partial u}{\partial y} \right) = -\frac{\partial p}{\partial y} + \mu \left(\frac{\partial^2 v}{\partial x^2} + \frac{\partial^2 v}{\partial y^2} \right) + g\,\rho\beta(T - T_c) \tag{3}$$

Energy equation:

$$u\,\frac{\partial T}{\partial x} + v\,\frac{\partial T}{\partial y} = \alpha \left(\frac{\partial^2 T}{\partial x^2} + \frac{\partial^2 T}{\partial y^2} \right) \tag{4}$$

Radiation heat transfer by the top glass cover surface must account for thermal radiation which can be absorbed, reflected, or transmitted. This decomposition can be expressed by,

$$q_{net} = q_{absorbed} + q_{transmitted} + q_{reflected}$$

Outside the boundary layer, the amount of energy $q_{reflected}$ is neglected.

Thus total energy of the glass cover plate becomes

$$q_{net} = q_{absorbed} + q_{transmitted}$$

Now the amount of transmitted energy is radiated from the cover plate to the bottom wavy absorber wall as:

$$q_{transmitted} = q_r = \varepsilon \ \sigma \ A(T_w^{\ 4} - T_i^4)$$

Here ε is emissivity of the glass cover plate, σ is Stefan Boltzmann constant $5.670400 \times 10^{-8} \ Js^{-1}m^{-2}K^{-4}$ and T_w is the variable temperature of the top wall.

Again, the amount of absorbed energy is transferred from cover plate to absorber by natural convection as:

$$q_{absorbed} = q_c = h_A \ (T_w - T_i)$$

So total energy gained or loosed by the cover plate is

$$q_{net} = h_A \ (T_w - T_i) + \varepsilon \ \sigma \ A(T_w^{\ 4} - T_i^4)$$

The boundary conditions are:

at all solid boundaries: $u = v = 0$

at the vertical walls: $\dfrac{\partial T}{\partial x} = 0$

at the top cover plate: $q = h_A \ (T_w - T_i) + \varepsilon \ \sigma \ A(T_w^{\ 4} - T_i^4)$

at the bottom wavy absorber: $T = T_c$

The above equations are non-dimensionalized by using the following dimensionless dependent and independent variables:

$$X = \frac{x}{L}, \ Y = \frac{y}{L}, \ U = \frac{uL}{v_f}, \ V = \frac{vL}{v_f}, \ P = \frac{pL^2}{\rho v^2}, \ \theta = \frac{T - T_c}{T_w - T_c}$$

Then the non-dimensional governing equations are

$$\frac{\partial U}{\partial X} + \frac{\partial V}{\partial Y} = 0 \tag{5}$$

$$U \frac{\partial U}{\partial X} + V \frac{\partial V}{\partial Y} = -\frac{1}{\rho} \frac{\partial P}{\partial X} + Pr \left(\frac{\partial^2 U}{\partial X^2} + \frac{\partial^2 U}{\partial Y^2} \right) \tag{6}$$

$$U \frac{\partial V}{\partial X} + V \frac{\partial V}{\partial Y} = -\frac{1}{\rho} \frac{\partial P}{\partial Y} + Pr \left(\frac{\partial^2 V}{\partial X^2} + \frac{\partial^2 V}{\partial Y^2} \right) + Ra \ Pr \ \theta \tag{7}$$

$$U \frac{\partial \theta}{\partial X} + V \frac{\partial \theta}{\partial Y} = \frac{1}{Pr} \left(\frac{\partial^2 \theta}{\partial X^2} + \frac{\partial^2 \theta}{\partial Y^2} \right) \tag{8}$$

where $Pr = \dfrac{v}{\alpha}$ is the Prandtl number, $Ra = \dfrac{g \ \beta \ L^3 \ (T_w - T_c)}{v\alpha}$ is the Rayleigh number.

The corresponding boundary conditions take the following form:

at all solid boundaries: $U = V = 0$

at the vertical walls: $\dfrac{\partial \theta}{\partial X} = 0$

at the absorber surface: $\theta = 0$

The shape of the bottom wavy absorber profile is assumed to mimic the following pattern

$$Y = A_m \left[1 - \cos\left(2\lambda\pi X\right)\right]$$

where A_m is the dimensionless amplitude of the wavy surface and λ is the number of undulations.

The solar collector aspect ratio $Ar = L/H$.

2.1 Average Nusselt number

The average Nusselt number (Nu) is expected to depend on a number of factors such as thermal conductivity, heat capacitance, viscosity, flow structure of nanofluids, volume fraction, dimensions and fractal distributions of nanoparticles.

The non-dimensional form of local convective heat transfer is

$$\overline{Nu_c} = -\frac{\partial \theta}{\partial Y}.$$

By integrating the local Nusselt number over the top heated surface, the average convective heat transfer along the heated wall of the solar collector is used by Saleh et al. [9] as

$$Nu_c = \int_0^1 \overline{Nu_c}\, dX.$$

The radiated heat transfer rate is expressed as

$$Nu_r = \int_0^1 q_r\, dX.$$

The average Nusselt number is

$$Nu = Nu_c + Nu_r$$

The mean bulk temperature and average sub domain velocity of the fluid inside the collector may be written as $\theta_{av} = \int \theta\, \dfrac{d\overline{V}}{\overline{V}}$ and $\omega_{av} = \int \omega\, \dfrac{d\overline{V}}{\overline{V}}$, where $\overline{V}$ is the volume of the collector.

3. NUMERICAL IMPLEMENTATION

The Galerkin finite element method [10, 11] is used to solve the non-dimensional governing equations along with boundary conditions for the considered problem. The equation of continuity has been used as a constraint due to mass conservation and this restriction may be used to find the pressure distribution. The finite element method is used to solve the Eqs. (6) - (8), where the pressure P is eliminated. Three points Gaussian quadrature is used to evaluate the integrals in these equations. The non-linear residual equations are solved using Newton–Raphson method. The convergence of solutions is assumed when the relative error for each variable between consecutive iterations is recorded below the convergence criterion ε such that $|\Psi^{n+1} - \Psi^{n}| \leq 10^{-4}$, where n is the number of iteration and Ψ is a function of U, V and θ.

3.1 Mesh Generation

In the finite element method, the mesh generation is the technique to subdivide a domain into a set of sub-domains, called finite elements, control volume, etc. The discrete locations are defined by the numerical grid, at which the variables are to be calculated. It is basically a discrete representation of the geometric domain on which the problem is to be solved. The computational domains with irregular geometries by a collection of finite elements make the method a valuable practical tool for the solution of boundary value problems arising in various fields of engineering. Figure 2 displays the finite element mesh of the present physical domain.

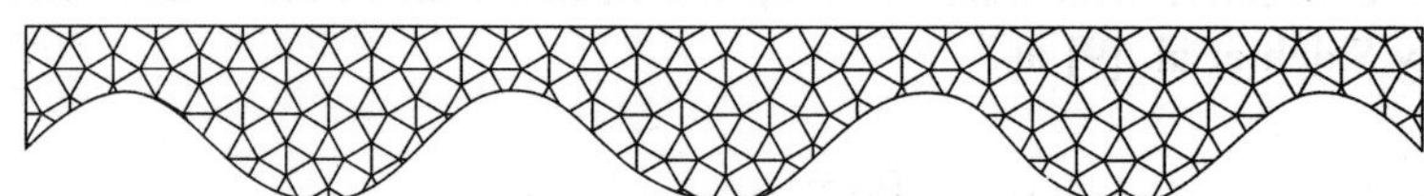

Fig. 2   Mesh generation of the collector for $Ar = 9$

3.2 Grid Independent Test

An extensive mesh testing procedure is conducted to guarantee a grid-independent solution for $Pr = 0.7$, $Ra = 10^{5}$ and $Ar = 9$ in a solar collector having undulating bottom absorber. In the present work, we examine five different non-uniform grid systems with the following number of elements within the resolution field: 2969, 5130, 6916, 9057 and 11426. The numerical scheme is carried out for highly precise key in the average convective and radiated Nusselt numbers namely Nu_c and Nu_r for the aforesaid elements to develop an understanding of the grid fineness as shown in Table 1 and Fig. 3. The scale of the average Nusselt number (convective and radiative) for 9057 elements shows a little difference

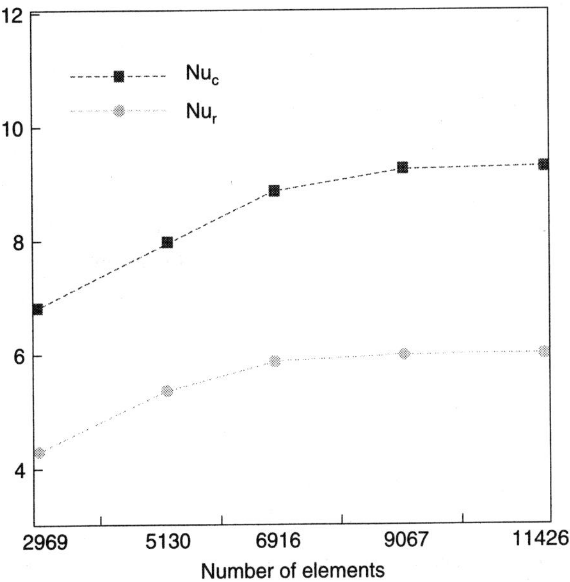

Fig. 3 Grid test for the geometry of $Ar = 9$

with the results obtained for the other elements. Hence, considering the non-uniform grid system of 9057 elements is preferred for the computation.

Table 1 Grid Sensitivity Check at $Pr = 0.7$, $Ar = 9$ and $Ra = 10^5$

Nodes (elements)	6224 (2969)	10982 (5130)	13538 (6916)	20295 (9057)	27524 (11426)
Nu_c	6.82945	7.98176	8.77701	9.088308	9.089308
Nu_r	4.32945	5.38176	5.98701	6.454909	6.455909
Time (s)	226.265	292.594	388.157	421.328	627.375

3.3 Code Validation

The present numerical solution is validated by comparing the current code results for heat transfer - temperature difference profile at $Pr = 0.73$, $Gr = 10^4$ with that of Gao et al. [12] are compared. Gao et al. [12] studied heat transfer augmentation inside a channel between the flat-plate cover and sine-wave absorber of a cross-corrugated solar air heater. Figure 4 depicts the result of numerical simulations of the effect of ΔT on the natural convection heat transfer coefficient h.

4. RESULTS AND DISCUSSION

In this section, numerical results of streamlines and isotherms for various values of collector aspect ratio (Ar) with water-alumina nanofluid in a solar collector

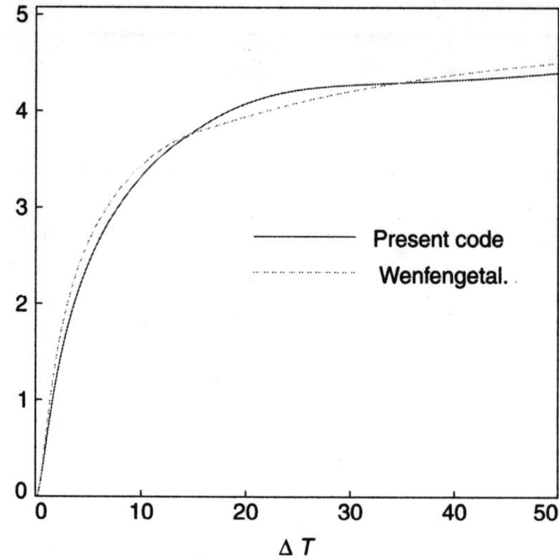

Fig. 4 Comparison between present code and Gao et al. [12] at $Pr = 0.73$ and $Ra = 10^5$

are displayed. The considered values of Ar are Ar (= 7, 9, 12 and 18) while the Prandtl number $Pr = 0.7$, $Ra = 10^5$ and the emissivity $\varepsilon = 0.9$ are kept fixed. In addition, the values of the average Nusselt number both for convection and radiation, mean bulk temperature and average sub domain velocity are shown graphically for the pertinent parameter.

The effect of solar collector aspect ratio (Ar) on the thermal is presented in Fig. 5. The red and black colored solid and dashed lines indicate water-alumina nanofluid and clear water respectively. The strength of the thermal current activities is much more activated with escalating Ar. Isotherms are almost similar to the active parts for working fluid air. Increasing Ar, the temperature lines at the middle part of the collector become horizontal whereas initially they are almost wavy pattern due to concentration of solid particles is dominated across the solar collector. With the rising values of Ar, the temperature distributions become distorted resulting in an increase in the overall heat transfer. This result can be attributed to the dominance of the radiated solar energy. This is because the ratio of length and height increases and the absorber becomes very close to the top transparent glass plate. This means that higher heat transfer rate is predicted by the fluid for closer situation of top and bottom surfaces than the lowest value of collector aspect ratio ($Ar = 7$). It is worth noting that as Ar increases, the thickness of the thermal boundary layer near the top cover plate enhances which indicates a steep temperature gradients and hence, an increase in the overall heat transfer from the transparent cover plate to the wavelike absorber.

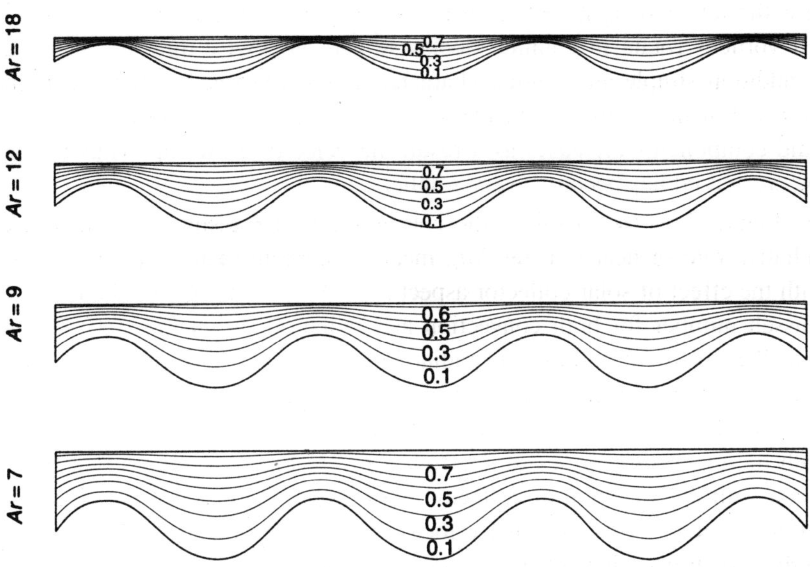

Fig. 5 Effect of *Ar* on isothermal lines

Figure 6 represents the infuence of solar collector aspect ratio (*Ar*) on the flow field. From this figure it is observed that the strength of the flow circulation rises with growing *Ar*. For the working fluid, six primary recirculation cells occupying the whole collector is found at the lowest value of the collector aspect ratio (*Ar* = 7) near the wavy zones. As well as two tiny eddies appear

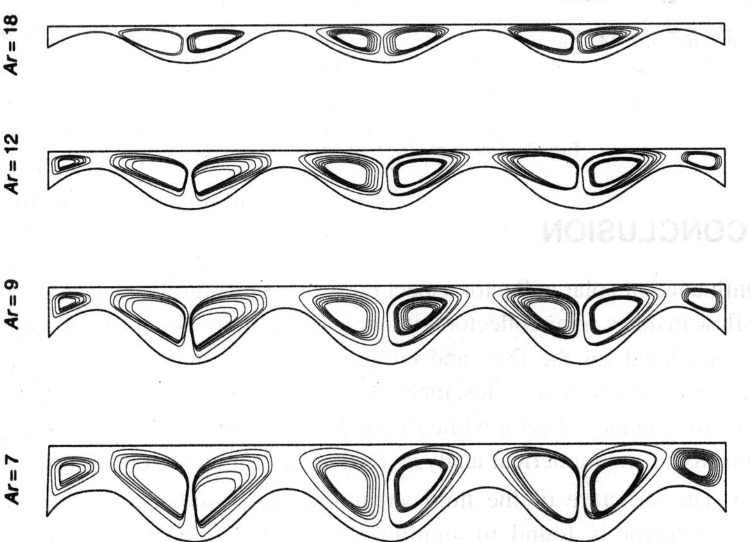

Fig. 6 Effect of collector aspect ratio *Ar* on streamlines

near the left and right vertical walls in this case. In each wave, the right and left vortices rotate in counter clockwise and clockwise direction respectively. In addition simple and regular characteristics is observed in the streamlines at $Ar = 7$. For increasing Ar, inside a tiny region the skin friction of the working fluid components increases as a result the velocity grows up. Thus the size of the created eddies becomes very smaller.

Figure 7 (i)-(iii) displays the convective heat transfer rate Nu_c as well as radiative rate of heat transfer Nu_r, mean temperature and velocity of the fluid with the effect of solar collector aspect ratio Ar. Mounting Ar enhances average Nusselt number for both convection and radiation. From Fig. 7(i), it is found that there is slight variation in radiative heat transfer whereas convective heat transfer has notable changes with different values of Ar. Rates of convective heat transfer increases by 10% respectively whereas this rate for radiation is 7% with the increasing values of Ar from 7 to 18. Fig. 7(ii) shows that θ_{av} grows down with the variation of collector aspect ratio. ω_{av} has notable changes with different values of Ar. Average velocity field inside the solar collector rises due to the mounting values of Ar.

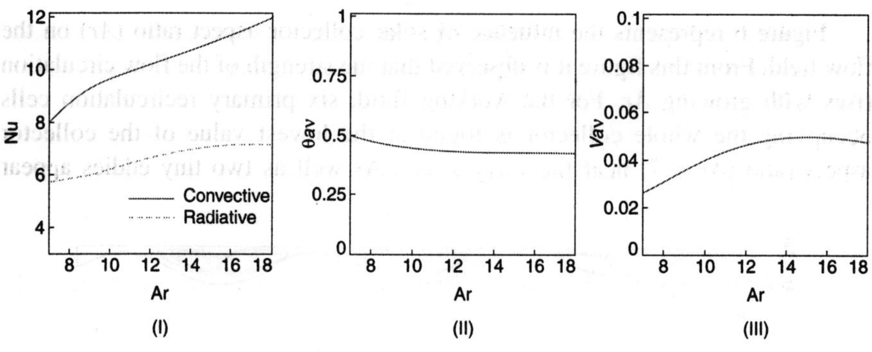

Fig. 7 Effect of *Ar* on (i) Nu (i) θ_{av} and (iii) θ_{av}

5. CONCLUSION

The influence of solar collector aspect ratio (Ar) on natural convection boundary layer flow inside a solar collector with air is accounted. Various aspect ratios have been considered for the flow and temperature fields as well as the convective and radiative heat transfer rates, mean bulk temperature of the fluids and average velocity field in the collector while Pr and Ra are fixed at 0.7 and 10^5 respectively. The results of the numerical analysis lead to the following conclusions:

- The structure of the fluid streamlines and isotherms within the solar collector is found to significantly depend upon the collector aspect ratio.

- The fluid with the highest *Ar* is established to be most effective in enhancing performance of heat transfer rate.
- Average heat transfer is obtained higher for convection than radiation.
- Mean temperature diminishes for fluid with mounting the physical parameter.
- Average velocity field grows up due to growing *Ar*.

References

1. Xiaowu, W., and Hua, B, 2005: *Energy analysis of domestic-scale solar water heaters*, Renew. Sustain Energy Rev., **9** (6), 638-645.

2. Hussain, A., 2006: *The performance of a cylindrical solar water heater*, Renew. Energy, **31** (11), 1751-1763.

3. Orel, Z.C., Gunde, M.K., and Hutchins, M.G., 2002: *Spectrally selective solar absorbers in different non-black colours*, Proceedings of WREC VII, Cologne on CD-ROM.

4. Tripanagnostopoulos, Y., Souliotis, M., and Nousia, Th., 2000: *Solar collectors with colored absorbers*, Solar Energy, **68**, 343-356.

5. Konttinen, P., Lund, PD., and Kilpi, RJ., 2003: *Mechanically manufactured selective solar absorber surfaces*, Solar Energy Mater Solar Cells, **79**(3), 273-283.

6. Varol, Y., and Oztop, H.F., 2007: *Buoyancy induced heat transfer and fluid flow inside a tilled wavy solar collector.* Building Environment, **42**, 2062-2071.

7. Bég, OA., Bakier, A., Prasad, R., and Ghosh, S.K., 2011 *Numerical modelling of non-similar mixed convection heat and species transfer along an inclined solar energy collector surface with cross diffusion effects*, World J. of Mech., **1**, 185-196.

8. Nasrin, R., Alim, M.A., and Chamkha, A.J., 2013: *Effects of physical parameters on natural convection in a solar collector filled with nanofluid*, Heat Trans.-Asian Research, **42**(1), 73-88.

9. Saleh, H., Roslan, R., and Hashim, I., 2011: *Natural convection heat transfer in a nanofluid-filled trapezoidal enclosure*, Int. J. of Heat and Mass Trans., **54**, 194-201.

10. Taylor, C., and Hood, P., 1973: *A numerical solution of the Navier-Stokes equations using finite element technique*, Computer and Fluids, **1**, 73-89.

11. Dechaumphai, P., 1999: *Finite Element Method in Engineering*, 2nd ed., Chulalongkorn University Press, Bangkok.

12. Gao, W., Lin, W., and Lu, E., 2000: *Numerical study on natural convection inside the channel between the flat-plate cover and sine-wave absorber of a cross-corrugated solar air heater*, Energy Conversion & Manag., **41**, 145-151.

Nomenclature

A Area of glass cover plate (m^2)

A_m Dimensionless amplitude of wave

Ar Aspect ratio (L/H)

C_p Specific heat at constant pressure ($J\ kg^{-1}\ K^{-1}$)

g Gravitational acceleration ($m\ s^{-2}$)

h Local heat transfer coefficient ($W\ m^{-2}\ K^{-1}$)

H Average height of the solar collector (m)

k Thermal conductivity ($W\ m^{-1}\ K^{-1}$)

L Length of the solar collector (m)

Nu Nusselt number, $Nu = hL/k_f$

Pr Prandtl number, $Pr = v/\alpha$

Ra Rayleigh number, $Ra = \dfrac{g\ \beta\ L^3\ (T_w - T_c)}{v\alpha}$

T Dimensional temperature (°K)

T_i Initial temperature of fluid (°K)

u, v Dimensional x and y components of velocity ($m\ s^{-1}$)

U, V Dimensionless velocities, $U = \dfrac{uL}{v},\ V = \dfrac{vL}{v}$

X, Y Dimensionless coordinates, $X = x/L,\ Y = y/L$

x, y Dimensional coordinates (m)

Greek Symbols

α Fluid thermal diffusivity ($m^2\ s^{-1}$)

β Thermal expansion coefficient (K^{-1})

ε Emissivity

θ Dimensionless temperature, $\theta = (T - T_c)/(T_w - T_c)$

λ Number of wave

μ Dynamic viscosity ($N\ s\ m^{-2}$)

v Kinematic viscosity ($m^2\ s^{-1}$)

ρ Density ($kg\ m^{-3}$)

σ Stefan Boltzmann constant

ω Magnitude of dimensionless velocity

Subscripts

av average
c cold
h hot
w cover plate